FIELD THEORY IN ELEMENTARY PARTICLES

Studies in the Natural Sciences

A Series from the Center for Theoretical Studies
University of Miami, Coral Gables, Florida

Orbis Scientiae: Behram Kursunoglu, *Chairman*

Recent Volumes in this Series

Volume 8 **PROGRESS IN LASERS AND LASER FUSION**
Edited by Arnold Perlmutter, Susan M. Widmayer, Uri Bernstein, Joseph Hubbard, Christian Le Monnier de Gouville, Laurence Mittag, Donald Pettengill, George Soukup, and M. Y. Wang

Volume 9 **THEORIES AND EXPERIMENTS IN HIGH-ENERGY PHYSICS**
Edited by Arnold Perlmutter, Susan M. Widmayer, Uri Bernstein, Joseph Hubbard, Christian Le Monnier de Gouville, Laurence Mittag, Donald Pettengill, George Soukup, and M. Y. Wang

Volume 10 **NEW PATHWAYS IN HIGH-ENERGY PHYSICS I**
Magnetic Charge and Other Fundamental Approaches
Edited by Arnold Perlmutter

Volume 11 **NEW PATHWAYS IN HIGH-ENERGY PHYSICS II**
New Particles—Theories and Experiments
Edited by Arnold Perlmutter

Volume 12 **DEEPER PATHWAYS IN HIGH-ENERGY PHYSICS**
Edited by Arnold Perlmutter, Linda F. Scott, Mou-Shan Chen, Joseph Hubbard, Michel Mille, and Mario Rasetti

Volume 13 **THE SIGNIFICANCE OF NONLINEARITY IN THE NATURAL SCIENCES**
Edited by Arnold Perlmutter, Linda F. Scott, Mou-Shan Chen, Joseph Hubbard, Michel Mille, and Mario Rasetti

Volume 14 **NEW FRONTIERS IN HIGH-ENERGY PHYSICS**
Edited by Arnold Perlmutter, Linda F. Scott, Osman Kadiroglu, Jerzy Nowakowski, and Frank Krausz

Volume 15 **ON THE PATH OF ALBERT EINSTEIN**
Edited by Arnold Perlmutter and Linda F. Scott

Volume 16 **HIGH-ENERGY PHYSICS IN THE EINSTEIN CENTENNIAL YEAR**
Edited by Arnold Perlmutter, Frank Krausz, and Linda F. Scott

Volume 17 **RECENT DEVELOPMENTS IN HIGH-ENERGY PHYSICS**
Edited by Behram Kursunoglu, Arnold Perlmutter, and Linda F. Scott

Volume 18 **GAUGE THEORIES, MASSIVE NEUTRINOS, AND PROTON DECAY**
Edited by Arnold Perlmutter

Volume 19 **FIELD THEORY IN ELEMENTARY PARTICLES**
Edited by Arnold Perlmutter

A Continuation Order Plan is available for this series. A continuation order will bring delivery of each new volume immediately upon publication. Volumes are billed only upon actual shipment. For further information please contact the publisher.

FIELD THEORY IN ELEMENTARY PARTICLES

Chairman

Behram Kursunoglu

Editor

Arnold Perlmutter

Center for Theoretical Studies
University of Miami
Coral Gables, Florida

PLENUM PRESS • NEW YORK AND LONDON

Library of Congress Cataloging in Publication Data

Orbis Scientiae (1982: University of Miami. Center for Theoretical Studies)
Field theory in elementary particles.

(Studies in the natural sciences; v. 19)
"Proceedings of Orbis Scientiae 1982, held January 18–21, 1982 by the Center for Theoretical Studies, University of Miami, Coral Gables, Florida"—Verso t.p.
Includes bibliographical references and index.
1. Field theory (Physics)—Congresses. 2. Particles (Nuclear physics)—Congresses.
I. Kursunoglu, Behram, 1922– . II. Perlmutter, Arnold, 1928– .
QC793.3.F507 1982 539.7'21 83-2442
ISBN 0-306-41345-0

Proceedings of Orbis Scientiae 1982, held January 18–21, 1982, by the Center for Theoretical Studies, University of Miami, Coral Gables, Florida

A Division of Plenum Publishing Corporation
233 Spring Street, New York, N.Y. 10013

Printed in the United States of America

PREFACE

We respectfully submit these proceedings of the 1982 Orbis Scientiae for your reading enjoyment. As always, the success of the conference was due to the hard work and wisdom of the moderators and dissertators. This year, in addition to the excellent overview of QCD and GUT, and the customary reports of the latest progress in theoretical and experimental particle physics, there have been discussions of new developments in astrophysics and especially of field theory and composite models.

We wish also to note here that the 1981 Orbis paper by Stephen S. Pinsky on "Death of Fractional Topological Charge" was actually co-authored by William F. Palmer of Ohio State University, whose name was inadvertently omitted from the authorship, due to a series of misunderstandings.

As in the past, this Orbis Scientiae 1982 was supported on a small scale by the Department of Energy, and this year as well by the National Science Foundation, on the same scale.

We would like to thank Mrs. Helga S. Billings for her excellent typing for the n-th time, where n is a large number. This series of proceedings is also enhanced by Linda Scott's editorial help which includes improvements in the presentation of some of the papers.

The Editors
Coral Gables, Florida
December, 1982

CONTENTS

The Present State of Gravitational Theory....................1
P. A. M. Dirac

QCD - Prospects and Problems...............................11
David J. Gross

Grand Unification: A Status Report.........................31
Howard Georgi

Majorana Neutrinos - Their Electromagnetic Properties and Neutral Current Weak Interactions..................49
Boris Kayser

Proton Decay - 1982..71
William J. Marciano

Prompt Neutrinos: Present Issues and Future Prospects......95
Gianni Conforto

Recent Results on Charmed Particle Lifetimes and Production...105
R. N. Diamond

The Weak Mixing Angle and its Relation to the Masses M_Z and M_W...................................123
Emmanuel A. Paschos

Intermediate Vector Bosons and Neutrino Cosmology..........139
Zohreh Parsa

Precessional Period and Modulation of Relativistic Radial Velocities in SS 433............................151
A. Mammano, F. Ciatti, and R. Turolla

Periodic and Secular Changes in SS 433.....................163
George W. Collins, II and Gerald H. Newsom

Superstrings.. 193
John H. Schwarz

A Geometrical Approach to Quantum Field Theory................ 213
F. R. Ore, Jr. and P. van Nieuwenhuizen

Glueballs - A Status Report.................................. 239
Daniel L. Scharre

Glueballs - Some Selected Theoretical Topics.................. 263
Carl E. Carlson

Monte Carlo Computations of the Hadronic Mass Spectrum........ 277
C. Rebbi

Learning from Mass Calculations in Quarkless QCD.............. 293
Richard C. Brower

Beyond QCD: Why and How...................................... 305
Giuliano Preparata

Large Magnetic Moment Effects at High Energies and Composite Particle Models.............................. 323
Asim O. Barut

The Next Layer of the Onion? Composite Models of Quarks and Leptons.................................... 333
O. W. Greenberg

Composite Fermions: Constraints and Tests.................... 349
Gordon L. Shaw and Dennis J. Silverman

Light Composite Fermions - An Overview........................ 365
Stuart Raby

The N-Quantum Approximation, Concrete Composite Models of Quarks and Leptons, and Problems with the Normalization of Composite Massless Bound States...................... 379
O. W. Greenberg

Monte Carlo Evaluation of Hadron Masses....................... 393
Don Weingarten

Revival of the Old String Model............................... 407
Yadin Y. Goldschmidt

QCD on a Random Lattice...................................... 419
N. H. Christ

Collective Field Theory....................................435
B. Sakita

Program..453
Participants...457
Index..461

THE PRESENT STATE OF GRAVITATIONAL THEORY

P.A.M. Dirac

Florida State University

Tallahassee, Florida 32306

The first thing to understand is that the Einstein Theory is extremely good. There are the three original tests proposed by Einstein.

(1) The motion of the perihelion of Mercury.

(2) The deflection of light passing close by the sun.

(3) The displacement of spectral lines emitted by atoms in a gravitational potential.

The success of the Einstein theory in explaining the motion of Mercury was evident right from the beginning. To check on (2) one needed a total solar eclipse, so that the sun's light is excluded and one can observe the positions of the stars nearby. Observations were made during the eclipse of 1919, which confirmed the theory with a fair amount of accuracy. Since then many further eclipse observations have been made, and the Einstein theory was always confirmed, with an accuracy as good as could be expected for the observing conditions that held at the time.

More recently, it has become possible to check on this effect with radio waves instead of light waves passing close by the sun. One does not now have to wait for a total eclipse, since the sun is not a strong emitter of radio waves. One merely has to wait for the sun to pass close in front of a star that emits radio waves and see if the star's position gets deflected. There is a complication now arising because the radio waves get deflected by the sun's corona. However, the latter effect depends on the wavelength of the radio waves, so by making observations at two different wavelengths and comparing them, one can separate the coronal effect from the Einstein effect. The Einstein effect is then again confirmed.

A similar test can be made with radar waves sent to Mars and reflected back to Earth when Mars is suitably placed so that the radar waves pass close by the sun. These observations have been made by I.I. Shapiro. The radar waves, like radio waves, get deflected by the sun's corona, and one has again to make observations at two different wavelengths to separate the coronal effect from the Einstein effect. One gets further confirmation of the Einstein theory in this way.

The test (3) of the Einstein theory can be applied to the spectra of light emitted from the sun. It is difficult to get much accuracy, because the effect is obscured by the Doppler effect coming from the motion of the gases in the sun's atmosphere. Not much is known about this motion so one cannot allow for it with great accuracy, and one gets only a rough check of the Einstein theory by this method.

However, there are other stars, the white dwarfs, for which the gravitational potential at the surface is very much greater than for the sun, so that the deflection of the spectral lines emitted from its surface is correspondingly increased and can easily be observed. If one can determine the mass of the white dwarf, through its being a member of a double star, one can check numerically on the Einstein theory, and one again gets good confirmation.

This way of checking the Einstein theory can be carried out by a purely laboratory experiment. One can get γ-rays emitted with a well-defined frequency, by the Mössbauer effect. If they are emitted from a high source and then reabsorbed lower down, one finds that the frequency has been increased corresponding to the reduction in the gravitational potential. This work has been done by R.V. Pound and G.A. Rebka.

To the three original tests of the Einstein theory a fourth has recently been added. The theory requires that when electromagnetic waves pass close by the sun, they are not only deflected by the gravitational field but are also retarded. This effect can be observed with a pulsar, where the radiation occurs in regularly spaced pulses. It also shows up with radar waves reflected from Mars when they pass close by the sun and one observes the time of the to and from journey. This has been done by Shapiro. With this test there is again a coronal effect, which can be separated from the Einstein effect by comparing the observations for two different frequencies.

With this series of tests the Einstein theory shows itself as very successful. The success must be attributed to the basic assumption that physical space is a curved space of the Riemann type. With the departure from the flat space of Euclid or Minkowski, people have often wondered whether other kinds of curved space might not be useful in physics.

The other fields of physics might require a geometrical explanation in terms of some more general kind of space than Riemann space. Einstein himself was much concerned with combining the gravitational field with the electromagnetic field, the other long-range field known in physics, and worked many years on it.

Many mathematicians have studied extensively the development of new kinds of space, generalizing Riemann space in various ways. But this work has had little success and is of no value for understanding gravitation. One cannot do better than keep to Einstein's original ideas for explaining the gravitational field.

The Einstein field equations by themselves form an incomplete theory. One needs to supplement them by boundary conditions that hold at great distances, before one has a definite scheme of equations that one can apply to specific examples. For this purpose one must make some further assumptions about the universe as a whole. One must set up a cosmological model. A cosmological model results when one tries to describe the rough distribution of matter in the universe with local irregularities smoothed out.

Einstein himself realized the need for a cosmological model and proposed one, his cylindrical model. He needed a small change in his field equations, introducing a cosmological constant into them, which he was able to do without disturbing the general symmetry properties of his theory. Einstein's model involved space being finite but unbounded, with a uniform distribution of matter in it. It did not fit in with the recession of distant matter, which was later discovered, and so it was abandoned.

Then a model, with a high degree of mathematical symmetry, was proposed by de Sitter. This model did give the recession of distant matter. But it gave zero for the average density of matter, which was obviously unacceptable, so it was also abandoned.

After that a whole series of models was proposed by various people, Friedmann, Lemaitre and others, all conforming to Einstein's field equations, some with the cosmological constant and some without. Among them there was one, proposed jointly by Einstein and de Sitter in 1932 (called the E.S. model) which I would like to call to your attention. It is the simplest model which is not obviously wrong.

With this model the metric is specified by

$$ds^2 = d\tau^2 - \tau^{\frac{4}{3}} (dx^2 + dy^2 + dz^2) \tag{1}$$

with τ as the time variable. The factor $\tau^{4/3}$ is needed in order to make the pressure zero, corresponding to the real world having the pressure, coming from electromagnetic radiation, negligibly small compared to the average mass density. The mass density with this model was worked out by Einstein and de Sitter and was found to be of the right order of magnitude. Actually, the mass density given by this model and other similar models is too large by a factor of about 20, to agree with the observed average mass density. This discrepancy is known as the problem of the missing matter. It has been much discussed by astrophysicists in recent times. They believe that the explanation must lie in the existence of invisible matter which has not yet been detected. It may be in intergalactic gas, or in black holes or other invisible masses in the centers of galaxies.

The boundary conditions that are brought into the theory by a suitable cosmological model have too small an effect to disturb the successes of the theory with regard to the four tests mentioned above. But there is a further effect which needs more discussion.

The Large Numbers Hypothesis.

Physicists tend to believe that the dimensionless numbers provided by general theory have some explanation. One such number is the ratio of the electric to the gravitational force between the proton and electron in a hydrogen atom. Its value is

$$e^2/Gm_em_p = 2.3 \times 10^{39} \tag{2}$$

It is difficult to understand how such a large number could be explained mathematically.

If one takes the age of the universe, $t = 18 \times 10^9$ years approximately, and expresses it in terms of a unit of time provided by atomic constants, say e^2/m_ec^3, one finds

$$t = 6 \times 10^{39}\, e^2/m_ec^3 \quad . \tag{3}$$

One gets about the same large number appearing as (2). One might look upon this as a remarkable coincidence. It would be more reasonable to look upon it as having some explanation which is at present unknown, but will become clear when one knows more about both atomic theory and cosmology. If this is so, the number (2) must increase as the universe gets older. We shall thus have

$$e^2/G\, m_e m_p :: t\,. \tag{4}$$

We are led to consider that quantities that are usually counted as constants may actually vary with τ, the cosmological age or epoch.

The idea that leads to this conclusion, if it is correct, should apply to all important dimensionless numbers that can be constructed from general physical theory and general cosmological considerations. If such a number is very large, it should be some power of the epoch (3) expressed in atomic units. We get in this way a general principle. I call it the Large Numbers Hypothesis (L.N.H.).

It is a principle which is somewhat vague for two reasons.

(i) It applies only to important numbers that occur in general theory.

(ii) It is uncertain just how large a number should be to count as "Large".

Subject to these uncertainties, the principle can be powerful. It asserts that there cannot be an important very large number occurring in our general description of nature that is a constant, independent of the epoch.

For example, it asserts that there cannot be a maximum size to the universe. The universe must continue to expand forever, otherwise the maximum size, expressed in atomic units, would provide a large number independent of t. It imposes severe restrictions on the model of the universe.

We may consider getting a large number from the total mass of the universe. However, this may be infinite. To get a mass that is certainly finite we take the mass of that part of the universe that is receding from us with a velocity $< \frac{1}{2}c$. Express this in proton masses to make it dimensionless and call it N. Its value is not known very accurately because of uncertainties in the amount of invisible matter, but it is somewhere of the order of 10^{78}. The L N H now allows us to infer

$$N :: t^2 \quad .$$

There is thus a continual increase in the amount of matter within that part of the universe that is receding with velocity $v < \frac{1}{2}c$. To reconcile this with conservation of mass we must have the velocity of recession continually decelerating, so that more and more galaxies are continually appearing with $v < \frac{1}{2}c$. This is what the E.S. model provides. A little calculation shows readily that the E.S. model is the only one consistent with the Large Numbers hypothesis.

According to the L N H we have equations (4) showing that G, the gravitational constant, when expressed in atomic units, is varying. According to the Einstein theory it has to be constant. How can one reconcile these two contradictory requirements? It seems that one must assume there are two metrics, one of them ds_E to be used in the Einstein theory, the other ds_A provided by atomic units. The connection between them is

$$ds_E = t\, ds_A \tag{5}$$

as follows from dimensional arguments applied to (4). The ds in equation (1) is the Einstein ds_E, and the τ is the Einstein time, connected with atomic time t by

$$\begin{aligned} d\tau &= t\, dt \ , \\ \tau &= \frac{1}{2}t^2 \ . \end{aligned} \tag{6}$$

With the metric (1) the distance of a particular galaxy (i.e. a particular x,y,z) in Einstein units is proportional to $\tau^{2/3}$. So the velocity of recession (a velocity is the same in Einstein or atomic units) is proportional to $\tau^{-1/3}$. This differs considerably from the usual picture, according to which the velocity is roughly constant. The observations are not sufficiently definite to decide between these possibilities. But the evidence from the natural microwave radiation provides good evidence to support the L N H.

The microwave radiation is a black-body radiation of temperature 2.8°K. In an expanding world black-body radiation remains black-body radiation and its temperature falls corresponding to the wave-length of each component increasing at the same rate as the distance between galaxies. With the expansion law

$$ds_E \propto \tau^{2/3} \text{ or } t^{4/3} ,$$

we have

$$ds_A \propto t^{1/3} ,$$

so

$$\lambda_A \propto t^{1/3} ,$$

and the temperature

$$T \propto \nu_A \propto \lambda_A^{-1} \propto t^{-1/3} . \tag{7}$$

Now the temperature provides a dimensionless number

$$kT/m_e c^2 , \tag{8}$$

which has the value for 2.8°K of about 2.5×10^{-13}. Its reciprocal may be considered as a large number to which the L N H applies and it should then vary proportional to $t^{-1/3}$. This agrees with (7).

To get a dimensionless number from the temperature we might

have used the mass of the proton in (8) instead of the electron. We should then get the number $kT/m_p c^2$ equal to about 5×10^{-10}. The L N H would suggest that this should vary in proportion to $t^{-1/4}$, which would not quite agree with (7). We see here an example of the inaccuracy that can arise when one uses the L N H. It just gives a rough picture and one cannot get precise equations without having a better understanding of the ratio m_p/m_e.

With the usual picture of the expansion of the universe being roughly uniform we should have T proportional to t^{-1}. This law could have been valid only after a certain decoupling time, when the radiation field became independent of the matter. The existence of such a decoupling time would provide a special epoch which would contradict the L N H.

The two metrics provide two time scales, one of which, τ, is to be used in the Einstein theory and the other, t, is the time marked out by atomic clocks and governing radioactive decay. The difference between these two times should be detectable by sufficiently accurate observations. The best chance is with observations of the moon's motion against the background of the stars. The moon's angular acceleration, produced mainly by tidal effects, has been observed accurately for some centuries. It has been recorded in ephemeris time, which is the time, τ, of the Einstein theory.

Since 1955 the moon's motion has been recorded with atomic clocks. The observations have been studied for many years by Van Flandern. His results have fluctuated over the years as his work has improved in accuracy. They have now stabilized with the figures, for the moon's angular acceleration $-\dot{n}$ in units of seconds per (century)^2,

$$\dot{n}\ (\text{ephermeris}) = 28.0 \pm 1.2$$

$$\dot{n}\ (\text{atomic}) = 22.6 \pm 1.3 \quad .$$

The difference of these two figures is evidence for the need of two time scales and supports the L N H.

One can also check on whether the theory with the two metrics is correct by observing the reflection of radar waves from the inner planets, in particular Mars. Shapiro has been working on this method. His results from Mars have confirmed the Einstein theory tests (2) and (4), but have not yet been worked out sufficiently to provide a check on the variation of G.

QCD - PROSPECTS AND PROBLEMS*

David J. Gross

Joseph Henry Laboratories, Princeton University

Princeton, New Jersey 08544

I. INTRODUCTION

If you need to be convinced of the enormous progress that we have made in understanding the strong interactions it is sufficient to read the proceedings of the Coral Gables conference of a decade ago, just prior to the emergence of QCD. I shall review here the progress that has occurred during the ensuing decade, the problems that remain and the prospects that exist for realizing the full potential of this complete theory of the strong interactions.

First a word about the status of the experimental evidence of QCD. Some people are frustrated by the lack of totally conclusive tests of QCD. This is hardly surprising, since we can only make precise predictions when $\frac{\alpha_s(Q^2)}{\pi}$ is small. Since asymptotic freedom yields only a logarithmic decrease of the coupling, to reduce it from $\approx .05$ to $Q^2 = 1000$ Gev^2 to .01 requires Q^2 of $\sim 10^{20}$ Gev^2! In fact the experimental agreement is very satisfactory and the growing sophistication in the application of QCD to short distances, "perturbative QCD" whose problems and prospects I shall discuss,

---

*Research partially supported by NSF Grant No. PHY80-19754

should provide many new tests.

The theoretical problems that must be solved by QCD are difficult and frustrating. We must explain both confinement and dynamical chiral symmetry breaking. The theory is in principle extraordinarily predictive, but for that very reason (the lack of arbitrary couplings or small parameters) it defies easy approximations. Nonetheless much progress has been made in understanding the mechanism of confinement, although this has not yet been proved. Most impressive are the recent computer calculations, using Monte Carlo methods applied to the lattice approximation to QCD, which have yielded the first reliable evidence for confinement, and the beginnings of the first calculations from first principles of hadron masses. I shall discuss some of the problems and prospects for the development of an analytic treatment of QCD. I will suggest some ways by which QCD can avoid the price of success in physics -- namely death by boredom. Finally I will draw some general lessons from QCD, and discuss their implications in extrapolating physics to higher energies.

II. DEVELOPMENTS AND PROSPECTS OF PERTURBATIVE QCD

In recent years enormous advances have been made in extending the applicability of perturbative QCD to processes hitherto believed to require nonperturbative input. These include the Drell-Yan process for small transverse momentum (of the μ pair in the COM), meson elastic form factors, perhaps baryon form factors, exclusive decay modes of J/ψ, wide angle scattering, and the small and large x regions of inclusive structure functions.

These advances are due to two developments. One is an increased ability to deal with what used to be called "wee partons", i.e. the possible invalidation of factorization of mass singularities caused by the exchange of soft gluons. It has yet to be shown that soft gluon exchange between the hadrons that collide in the Drell-Yan process cancel. However in the case of the analogous process of

$e^+e^- \to h_1+h_2+x$ Collins and Sterman have finally proved factorization.

The other development is that in many processes, in which potential contributions from soft momentum might invalidate a naive application of perturbative QCD -- by producing double logarithmic factors $(\ln Q^2)^2$, methods have been developed to sum these factors to yield exponentials of the form $\exp[-\ln Q^2 \ln\ln Q^2]$. These are called Sudakov effects, since such an exponentiation of $(\ln Q^2)^2$ first appeared in Sudakov's calculation of the behavior of the electron form factor in QED. Collins and Soper have developed methods that should allow for a complete calculation of such effects.

I would like to note two, out of many, striking new predictions.

First Collins and Soper, improving upon a leading double logarithm calculation of Parisi and Petronzio predict the behavior of the Drell-Yang cross-sections as a function of $q_\perp = \mu$ pair transverse momentum, and $Q^2 = \text{C.O.M. energy}^2$. This striking effect, the decrease and flattening of the $q_\perp^2$ distribution, should easily be seen at ISABELLE (if it is built).

Second, recent work of Mueller, building on previous work of Amati, Barsetto, Ciafaloni, Machesini, Veneziano, and Konishi, has shown that the small x-behavior of the inclusive $e^+e^- \to$ hadron+x structure functions, can be calculated. The anomalous dimension γ_J, which for $J \approx 1$ has a singular perturbative expansion and thus invalidates naive extrapolation to $x = 0$, can be evaluated to yield a square root branchpoint

$$\gamma_J = \frac{1}{4}\left[\sqrt{(J-1)^2 + \frac{8\pi C_2}{\pi}} - (J-1)\right] .$$

This produces a hadronic multiplicity growing like $\bar{n}(Q^2) \sim e^{\sqrt{\frac{2C_2}{\pi b}\ln Q^2}}$.

These are exciting developments. The application of QCD to short distance phenomena -- Perturbative QCD -- is not dead yet. It is important to continue to test the theory in this domain (after all one could disprove QCD) and to push the theoretical methods as far as possible. We can look forward in the coming years, I am sure, to more striking predictions and an increasing number of experimental tests.

III. CONFINEMENT

The central dynamical issue in QCD is that of color confinement. We have already come quite far from the original suggestion that the infrared slavery of QCD would lead to quark confinement. To be sure there exists to date no proof that QCD does indeed confine -- but no one today would be much surprised if such a proof were to soon appear. We have become comfortable with the notion of confinement. In fact young physicists, educated in the last few years, find the notion of confinement easier to swallow than our present inadequate understanding of the dynamics of the Higgs mechanism.

First we have learned of many systems which exhibit confinement or analogous behavior. These systems, to be sure, are not QCD, but they offer reassurance that confinement is a typical feature of many field theories or even classical statistical mechanical systems with an infinite number of degrees of freedom. The list of such systems is large, it includes gauge theories in 2 or 3 space-time dimensions (some of which, like 2-D QCD in the large N limit can be solved exactly), and Z_N lattice gauge theories whose confined phase can easily be pictured as due to a condensation of magnetic monopoles or a dual (electric) Meissner effect.

Second there are a host of qualitative descriptions of confinement and of hadronic structure which provide much insight. The bag model, which has proved to be an extremely successful picture of hadronic structure, assumes that the QCD vacuum cannot support colored fields and thus quarks live inside bubbles in the vacuum created at some finite energy cost per unit volume, inside of which they propagate almost freely. Thus a heavy quark-antiquark system consists of a flux tube emersed in the vacuum, with the confined colored electric field pressure balancing the vacuum pressure. Light quark hadrons pictured as spherical bags, in which the quarks are confined also due to the large mass they would possess in the ordinary vacuum. This model has not been derived from QCD, assumes both the existence of confinement and chiral symmetry breaking, and has adjustable parameters.

On the other hand it is certainly consistent with our understanding of QCD, and in fact, one can give arguments, based on semiclassical (or WKB) methods, that lead to a very similar picture. These methods indicate that the important vacuum fluctuations at intermediate scales -- instantons -- behave as permanent paramagnetic dipoles, which produce in the vacuum a small, if not vanishing dielectric constant, as well as a source of dynamical chiral symmetry breaking that produces a large quark mass. This produces the physics of the bag model and can be used to obtain rough estimates of the bag constant and the string tension. Unfortunately, like all weak coupling techniques it seems exceedingly difficult to pursue these ideas to the end and to develop a truly systematic and quantitative treatment of hadrons.

The most important evidence for confinement, and the first reliable quantitative calculations of hadronic spectrum has emerged from numerical methods (in particular Monte Carlo simulations) applied to the lattice approximation to QCD. The lattice approximation replaces the continuum action by a regularized version

$$S_w = \frac{1}{g^2(a)} \sum_{\text{plaq.}} (\mathrm{Tr}\, U_p + \mathrm{Tr}\, U_p^+) \qquad U_p = \prod_{L\varepsilon p} U_L$$

formulated on a Euclidean lattice, where the dynamical variables are unitary matrices which sit on the links of a lattice of spacing a. Expectation values of observables can be evaluated on the lattice -- in particular expectation values of Wilson loops which serve as an order parameter for confinement -- and their continuum values can be extracted by taking the continuum limit

($a \to 0$, $g^2(a) \to \frac{C}{\ell n \frac{1}{a\Lambda_L}}$). Dimensional transmutation assures us that all physical parameters can be calculated in terms of Λ_L -- the QCD scale parameter, and the notion of universality assures us that almost any lattice action will do to define the continuum theory.

Lattice gauge theories are easily solved in the unphysical limit of $g \to \infty$, whence the gauge fields are totally chaotic and random. This disorder leads automatically to a mass gap (i.e., a finite glueball mass and no massless gluon continuum) and to area behavior for Wilson loops (i.e. linear confinement). It yields, for heavy quarks, a simple intuitive picture of hadrons -- as infinitely thin flux tubes, similar to the bag or string model descriptions. The issue is then whether this behavior persists as g is decreased to zero -- or whether there is a "phase transition" to nonconfining behavior as we expect (and find) in abelian gauge theories.

Well, as you all know, there is now strong evidence due to Creutz, Rebbi, Wilson and others that there is no such phase transition. The best evidence is based on computer Monte Carlo evaluations of $\langle W_L \rangle$, on finite lattices. Typically the lattice sizes are $8^4 - 10^4 - 16\times10^3$, and calculations have been performed for SU(2) - SU(5) with a variety of different actions. The results all confirm Creutz's original calculations. One generates ensembles of U_L's with the appropriate distribution for a given $g^2(a)$, extracts $\sigma a^2\ [g^2(a)]$, and tests whether

$$\sqrt{\sigma}\, a = f(g^2) = \left(\frac{16}{11g^2}\right)^{2\ 51/121} \cdot \exp - \left(\frac{8\pi^2}{11g^2}\right) \cdot \left(\frac{\sqrt{\sigma}}{\Lambda_L}\right) .$$

These results yield for SU_3:

$$\sqrt{\sigma} = (200 \pm 60)\ \Lambda_L = 2.4\ \Lambda_{\substack{MON \\ \alpha=1}} \underset{exp}{=} 400 \text{ Mev.}$$

$$\to \Lambda_{\overline{MS}} = [60 \pm 20 \text{ (syst)} + ? \text{ (fermions)}] \text{ Mev.} \quad \pm 10 \text{ (stat)} .$$

The rapid transition from strong to weak coupling behavior, and the agreement with the predictions of AF are quite striking. It is clear, that at least as far as numerical approximations are concerned nature is kind and even with relatively small lattices and short running times good numbers can be extracted.

The magnitude of $\sqrt{\sigma}$ and Λ are quite reasonable, given the systematic errors in the calculation and the absence of fermions. $\Lambda_{\overline{MS}}$ as calculated here with no adjustable parameters is certainly consistent with that measured at short distances!

The spectrum of quarkless QCD can also be extracted from this approach in principle by measuring the rate of exponential decay of correlation functions. This is difficult, partly due to the large mass of the first excited state -- the mass gap -- the glueball mass. Since the glueball is in strong coupling a flux tube, with mass $= 4\sigma a$ a reasonable estimate of m_G is to take $a \sim \frac{1}{\sqrt{\sigma}} \rightarrow m_G \sim 4\sqrt{\sigma}$ ($\approx$ 1000 Mev). This large mass leads to a rapid decay which is hard to measure. Nonetheless for SU(2) it has been found that

$$m_G \approx (3\pm)\ \sqrt{\sigma}\quad .$$

Finally Monte Carlo calculations have been performed at finite temperature, and have confirmed that, as expected, a deconfining phase transition occurs at high temperature. The expected behavior of the theory at high temperature is that of a plasma of weakly interacting quarks and gluons, in which the potentially confining color interactions are Debye screened at large distances. Monte Carlo calculations have shown that indeed above $T_c \approx (.35\pm.1)\ \sqrt{\sigma}$, color is no longer confined -- and that the color interactions are screened at large distances. Would that we could create such temperatures in the lab and not just on a computer!

There are many other quantities that might be evaluated by these methods and should be very informative. An incomplete list includes:

(a) θ-dependence. The vacuum angle θ is a relevant and free parameter in QCD, although we usually set it (for good reasons) equal to zero. It is not easy to introduce this parameter into lattice gauge theories, since topology is destroyed by the discretization. Luscher has proposed an interesting and reasonable definition of topological charge on the lattice and it would be fascinating to explore the θ dependence of quarkless QCD. This could confirm the resolution of the U(1) problem, could settle the issue as to whether instantons play a significant role in the confinement mechanism and could determine whether QCD undergoes a phase transition near $\theta \approx \pi$.

(b) Gluon Operator Matrix Elements. Monte Carlo techniques can be used to extract the vacuum matrix elements of composite operators -- $\langle \mathrm{Tr}\, F_{\mu\nu}^2 \rangle$, $\langle \mathrm{Tr}\, D_\alpha F_{\mu\nu} D^\alpha F_{\mu\nu} \rangle$ etc. These are useful for a variety of phenomenological purposes. Here the problem is not numerical but theoretical -- i.e. how to separate the nonperturbative contributions from the infinite series of perturbative contributions that must be subtracted in going to the continuum limit.

In conclusion Monte Carlo techniques have proved to be an excellent device for convincing ourselves that QCD works as expected, and for extracting numbers from the theory. Originally it was thought by many that confinement was such a mysterious concept that it required radical new ideas and mechanisms -- Not so. It has turned out to be a rather simple phenomenon. Many of us are so convinced by all of this, that even if it turns out to be true that Fairbanks has observed fractional charge -- I, for one, would bet that this is evidence not for unconfined color -- but rather for fractionally charged leptons or hadrons composed of integer and fractionally charged quarks.

IV. FERMIONS AND CHIRAL SYMMETRY BREAKING

Fermions greatly complicate the dynamical treatment of QCD.

We not only have to deal with color confinement -- but also with dynamical chiral symmetry breaking. The basis of dynamical mechanism has been understood ever since the work of Nambu and Jano-Lasino; all approximate treatments of QCD, whether based on strong coupling expansions or semiclassical methods do produce a $\bar{\psi}\psi$ condensate by this mechanism. This then should yield all of the enormously successful phenomenology of chiral $SU_3 \times SU_3$, and, in the case of QCD, a resolution of the notorious U(1) problem.

However, quarks do make calculations in QCD much harder. In the lattice approximation there are two substantial problems. First one must develop clever ideas in order to integrate over anti-commuting variables on a computer. A variety of such methods have been recently developed to the point where it is clear that Monte Carlo calculations can be performed with a large but tolerable increase in computing time. Second it is impossible to latticize quarks without destroying chiral symmetry or multiplying the number of quarks. This is easily understood if we recall that the existence of axial anomalies in QCD is a sequence of the impossibility of achieving a chirally invariant regulation scheme -- and that the lattice is merely a particular regularization scheme. Nonetheless the methods that have been suggested to latticize fermions -- Wilson's, which recovers chiral symmetry in the continuum limit, and Susskind's, which preserves discrete chiral symmetry at the price of flavor doubling -- are quite sufficient to treat QCD. What they do not do is allow us to use similar numerical methods to explore the dynamics of other interesting theories -- such as chiral gauge theories or supersymmetric theories.

Recently the first calculation of hadronic spectrum has been performed by Weingarten, Hamber, Parisi, Marinari, Rebbi, et al. Weingarten will undoubtedly present many of these results -- so I have listed here just some of the masses calculated by Hamber and Parisi. These are based on the drastic approximation in which the fermion determinant is set =1, and careful continuum extrapolations have been shortcutted. The results, which must be accepted with some reservations, are certainly impressive. Similar results have been obtained for charmonium masses.

We can expect that these methods will be soon improved (it is only a question of time and money) to include fermion loops and that they will yield a (10%) calculation of the spectrum s-wave hadronic states, coupling constants, space like form factors, etc. I would also urge Monte-Carloists to estimate the masses of exotic hadrons -- here they can make actual predictions.

There are, however, apparent limitations to this approach, which one should try to overcome.

(1) It will be very difficult to deal with $L \neq 0$ hadrons without a large increase in lattice size.

(2) These methods are not well suited for the evaluation of real-time scattering amplitudes.

This is clearly the most promising area for major calculational advances within the next few years.

V. PROBLEMS

a. Analytical Solutions of QCD

The power and success of the combination of modern computers and the lattice approximation to QCD is undeniable. To date this is the _only_ formulation which can give _quantitative_ results -- and will undoubtedly yield, within the next few years, the first reliable calculations of the hadronic spectrum. Nevertheless I doubt whether this will ultimately satisfy most theorists, who will continue to yearn for analytic methods that will allow them to deduce the properties of hadrons with the aid of pencil and paper alone.

The search for an analytic treatment of the dynamics of QCD has occupied many of us for almost a decade. It is clearly very difficult and has been extremely frustrating. The basic problem is that QCD possesses no small parameters which could justify approximations valid for both short and large distances. Weak coupling methods, perturbation theory, are only adequate at short distances, and although the inclusion of semiclassical effects (WKB approximation or instantons) can be used to push these methods to larger distances, and do exhibit the dynamical effects that could lead to an understanding of hadronic structure, they too ultimately are insufficient.

In order to obtain a systematic, quantitative treatment of QCD one requires a formulation which is adequate both at short and large distances and thus does not rely on either strong or weak coupling methods.

The lattice approximation to QCD is only simple to treat in the unphysical limit where $a\Lambda \to \infty$ (strong coupling). Nonetheless since it is a well formulated approximation for all lattice spacings, numerical techniques can be used to extract physics in the physical $a\Lambda \to 0$ limit.

Another approach is to expand QCD in powers of $1/N$, where $N =$ number of colors $=3$. The strict $N=\infty$ theory would appear to have many properties which are approximately true (infinitely narrow resonances, an exact QZI rule, duality, etc...), and thus 3 might be effectively large. Such expansions have proved extremely useful in simpler field theories. In QCD as well the $N=\infty$ theory is much simpler, for example the hadronic spectrum can be gotten by solving a <u>classical</u> Hamiltonian system -- a major simplification. Unfortunately one is forced to work in loop space whose coordinates are Wilson loops: $\mathrm{Tr}\, e^{i\int_c A dx}$, an unfamiliar and enormous configuration space. Thus, although promising, the large N limit by itself does not appear to yield enough simplification.

There are other ideas that have been put forward that are quite promising. Much speculation exists on the relationship between QCD and various string theories. Migdal has even claimed that QCD is equivalent to a particular fermionic string theory. Unfortunately not enough is known yet of the properties of these string theories to know whether these correspondences have more than formal validity or are at all useful. Polyakov has proposed a new method for quantizing strong theories that explains why the previous light cone quantization was only valid in 26 or 10 dimensions. Finally some people, convinced that such a true and elegant theory should be exactly soluble, are attempting to apply methods developed for treating integrable field theories (mostly in 2-dimensions) to QCD.

Of course no one promised us that QCD would be susceptible to an exact analytic treatment and physics is full of situations where complicated problems have complicated solutions. However, this should not deter us from trying. Even unsuccessful attempts will produce elegant and hopefully useful physics and the ultimate payoff might be very large.

b. Hadronic Models

Finally a somewhat less ambitious program should be pursued to develop approximate models which incorporate our increased understanding of QCD, and which can be used to give simple descriptions of hadronic dynamics and structure. In particular there is a crying need for QCD based models which can deal with nonperturbative real time processes -- scattering, fragmentation, Regge behavior, etc...

c. Survival

Can QCD continue to flourish as an active area of experimental and theoretical research? Assuming, as it seems to be the case, that QCD is the correct theory of the strong interaction we may look forward to much in the next few years -- increasingly precise experimental tests, increasingly sophisticated predictions of short distance behavior, the calculation of the low energy spectrum and couplings, and a growing understanding of the dynamics of confinement and chiral symmetry breaking.

However, to stay alive as a vital field more is required. At some point we will be tempted to lose all interest in the theory unless we discover new phenomena or invent new theoretical ideas -- i.e. unless there are surprises. We can take heart from other fields of physics, where great excitement can persist even though the microscopic laws are known. Solid state physics provides such an example. Condensed matter physicists however, have the great advantages -- first the systems they deal with often are of practical interest and second the experiments are easier and cheaper. Nonetheless the continued intellectual vitality of the field requires the discovery of new phenomenon.

Where can we look for surprises in QCD? I cannot really answer this question since a real surprise is an experimental discovery, unanticipated by theorists. Theorists can suggest, nevertheless, places to look for surprises or predict qualitatively new phenomenon.

The obvious candidates are new forms of hadronic matter -- weird hadronic bound states or strange phases of quark matter.

Exotic hadrons are expected in QCD -- although we cannot yet reliably predict their properties. Here there is much room for both theorists to make predictions and for experimentalists to discover new particles using low energy accelerators.

More tantalizing is the prospect of exploring different phases of quark matter. QCD predicts that with increased temperature and/or baryon density hadronic matter will undergo a phase transition. At temperatures above ≈ 200 Mev and densities greater than 5-10 nuclear density, chiral symmetry should be restored and confinement should be lost. At high temperature and/or density we would see a plasma of weakly interacting gluons and light quarks.

Can one produce such a phase transition in the laboratory? Perhaps in high energy collisions of heavy ions! Even though it is not at all clear that this is feasible, it appears to offer the only chance to produce new phases of quarks in the lab -- and should be pursued by both experimentalists and theorists.

Are there any signs of strange quark matter? Two controversial candidates:

(1) Centauro events -- In these strange cosmic ray events 200 or so nucleons are produced with large transverse momentum and no pions!

(2) Judek effect -- 6% of fragments produced by large momentum transfer heavy ion collisions have anomalously small (~1/10) mean free paths (10× normal cross section).

As always more experimental data is required.

VI. LESSONS

In the concluding part of my talk I would like to step back and try to draw some general lessons and guidelines for future development from QCD.

a. Quantum Field Theory (QFT)

The pioneers who developed QFT in the 1930's were quick to become suspicious of its foundations and validity. The central problem was the existence of UV divergences which led many to propose that, once again as in the case of quantum mechanics, a radical revision of fundamental concepts was called for. The success of the renormalization program in QED provided a consistent perturbative scheme for EM, but gave little insight into the rich dynamical structure of quantum field theory. The absence of fundamental theories of the strong and weak interactions, in an era where both the fundamental constituents of matter (quarks) and the carriers of the forces (the gluons and massive vector mesons) were well concealed, lead to further disillusionment with QFT and thus to attempts in the 1960's to renunciate the use of a microscopic relativistic Hamiltonian and base particle theory on such general principles as unitarity, analyticity and symmetry.

Now, after a decade of the development of the gauge theories of the strong and the electroweak interactions it is safe to say that the announced death of QFT was premature. We now possess QFT theories of both the strong and electroweak interactions. Our dynamical and nonperturbative understanding of QFT has greatly matured -- we now have an expanding catalogue of dynamical mechanisms: spontaneous symmetry breaking, the Higgs mechanism, asymptotic freedom, confinement, etc., that can be used to explain the observed properties of elementary particles. The consistency and physical reality of renormalization has been confirmed by the development of renormalization group methods, the discovery of asymptotic freedom and its experimental confirmation in QCD. The path integral formulation of QM has been developed into a formidable tool for both rigorous mechanics and for the development of numerical approximations.

This is then the most important lesson that we have learned: QFT is alive and well and no revision of fundamental principles is called for.

In addition our increased dynamical understanding of QFT gives us a platform for extrapolation and speculation. Thus the notion of dynamically broken symmetries allows us to envisage new symmetries, yet unseen, which are broken at a scale beyond the present experimental range; the mechanism of confinement permits us to imagine that leptons and quarks are themselves composites of confined preons. Finally QCD itself provides a paradigm for what a complete QFT should be.

b. Renormalization and Dimensional Transmutation

One of the major advances associated with QCD is a deeper understanding of renormalization, which has yielded many benefits. First we have learned that renormalization does make sense -- it is the manifestation of the dependence of physical processes on the scale of lengths involved in the process. The existence of asymptotically free theories means not only, as is the case for all renormalizable theories, that low energy physics is independent of the very high energy structure of the theory, but that the high energy or short distance behavior of the theory is totally under control. Strong coupling problems are relegated to the domain of infrared or low energy behavior which is more accessible to both theorists or experimentalists. The ultraviolet instabilities which led Landau to believe in the demise of field theory are removed and we are in an excellent position to establish, with rigor, the mathematical existence of the theory. We now see that the UV divergences that appear in perturbation theory are physically reasonable -- in that there are no real divergences at all. The bare coupling in QCD is zero, and divergences occur only when we encounter the ratio of the coupling governing interactions at finite distances, where measurements take place, to that occurs in the ideal limit of zero distance. A finite theory would have no advantages over an AF theory -- indeed it would be boring, since in the absence of arbitrary dimensional scales it would lead to scale invariant behavior and could not contain the remarkably varied phenomena that exist in QCD as we go from small to large distances.

We have also learned from QCD the remarkable predictive power that a QFT can in principle (given enough work) possess. Due to the scale breaking effects of renormalization a QFT can in principle contain no arbitrary dimensionless parameters. Indeed to a good approximation (i.e. neglect the light quark masses and ignore the heavy ones) QCD is such a theory, in which all dimensionless constants are calculable and there are no free parameters. In this sense QCD is the most complete and predictive theory that we have.

c. Extrapolation

But physics never rests. As soon as we are satisfied that we have an explanation of the relevant phenomenon in a given energy region we raise our eyes and consider an even higher energy. Typically advances have proceeded in jumps of 10^6-6 orders of magnitude. Macroscopic physics was sufficient down to ~ micron, atomic physics to an angstrom, our present understanding goes down to 10^{-14-16} cm. Where is the next frontier? Present speculation puts it anywhere between 10^{-26} and 10^{-33} cm. This unprecedented increase of 12-18 orders of magnitude of extrapolation is due to our realization that physics is a logarithmic function of distance. There is much action at 10^{-16} cm, the scale of electroweak symmetry breaking, at 10^{-13} cm the scale of strong QCD coupling, at 10^{-8} cm ($\sim \frac{1}{\alpha m_e}$); but nothing much happens at distances smaller than 10^{-16}cm when we extrapolate present theory. When we reach 10^{-28} cm it appears reasonable that the strong and electroweak interactions are unified. The similarities between quarks and leptons, W's and gluons, the remarkable fit of the standard model within a unified SU(5) gauge theory and the joining of the weak and strong couplings at 10^{15} Gev with the resulting successful prediction of the Weinberg angle make a compelling case for unification. Of course, based on the mechanisms of confinement and symmetry breaking, we can conjure up a whole variety of new physics which could fill the "desert" from 10^3 Gev to 10^{15} Gev with technicolor, extended technicolor, supersymmetry, etc. -- all of which have been proposed to deal with the outstanding problems of the electroweak theory -- i.e. the lack of dynamical understanding of the spontaneous symmetry breaking, CP violation, the quark-lepton mass spectrum and the proliferation of ad hoc parameters. The attempts at grand unification have not yet solved these problems and in addition lead to the mass hierarchy problem: i.e. why is $\frac{\Lambda_{GUT}}{\Lambda_{QCD}}$ so big?

These problems are discussed at length in Georgi's talk. I would like to ask you to suspend for a moment all knowledge of the electroweak interactions as well as your anthropomorphic concerns and consider QCD all by itself. In many respects this is a complete theory with no apparent limitations at large or short distances. Are there any <u>clues</u> within QCD all by itself as to where and when new physics might appear?

Let me, following Einstein, write the QCD equation for the gauge field in the form:

$$D^{\mu} F^{a}_{\mu\nu} = \sum_F g \bar{\psi}_F \gamma_{\nu} \frac{\lambda^a}{2} \psi_F$$

and remark that the left-hand side contains all that is beautiful and elegant in a Yang-Mills gauge theory whereas the right-hand side seems arbitrary and ugly. Gluons are required by local gauge invariance -- quarks are unnecessary. New physics is needed to understand the right-hand-side. The existence of quarks alone, with their strange variety of flavors and masses, would surely have led us to postulate the existence of flavor dependent interactions which might explain the only dimensionless parameters ($\frac{m_a}{\Lambda}$) in QCD. But even more compelling is the question (unanswered even by GUTS) -- why do matter fields -- quark or lepton fields -- exist at all. Gauge symmetry only requires gluons and one can easily imagine QCD without quarks (in fact life would be easier). Matter is unnecessary. So who ordered quarks? The only possible answer I know to this question is based on the beautiful extension of Poincaré invariance known as supersymmetry. So far there is no experimental evidence for supersymmetry, and although many exciting proposals have been advanced, there is no compelling scenario that includes supersymmetric interactions. Nonetheless, it is my opinion that this marvelous extension of the notion of symmetry to include fermionic charges has the potential to put spin-1/2 matter sources on the left-hand side of the equation, where they belong, to explain why fermions exist at all and further reduce the number of free parameters in the theory. At some point it will surely play a role in nature.

Finally, even though due to AF one can imagine extrapolating QCD to arbitrarily small distances, we know of a scale where the theory must break down -- namely the dimensional scale associated with gravity -- the Planck scale of 10^{-33} cm. Here finally we are at a total loss, and most likely some of our basic principles will require radical modification. It is unlikely to be a coincidence that the Planck mass of 10^{19} Gev is so close to the hypothetical GUT energy scale of 10^{14-16} Gev -- and in fact it might very well be that a final unification and complete theory of particles and their interactions will require an understanding of quantum gravity. Here might lie an explanation to the perplexing question of why nature favors only gauge theories -- whereas theorists can construct other, albeit less beautiful, quantum field theories. An example of such an explanation plus true grand unification would be a Kaluza-Klein theory, in which all is geometry. Unfortunately it is difficult to imagine that we will ever be able to obtain more than a few very indirect experimental probes of this energy region.

This is quite depressing, since the history of particle physics offers little encouragement as to the ability of theorists to accept new concepts until they are forced down our throats by hard data. On the other hand, we do know a bit about physics at 10^{-33} cm -- since gravity is purely attractive and long range it gives rise to macroscopic effects which have led to a theory (general relativity) of gravity. Given what we do know about matter and its interactions and given Einstein's theory of relativity, with a few significant experimental clues and much hard work perhaps we can cross even this frontier.

GRAND UNIFICATION: A STATUS REPORT

H. Georgi*

Lyman Laboratory of Physics, Harvard University

Cambridge, Massachusetts 02138

I. INTRODUCTION

By the middle of 1974, only a few papers had been written about what we now call grand unification (GU)[1,2,3] but the main ideas were already in place. GU solved many puzzles, but two obvious puzzles remained: the flavor puzzle and the hierarchy puzzle.

The flavor puzzle existed in much the same form before unification. Why are objects with the same gauge properties, such as e and μ, repeated? After GUTs they are just combined into repeated families. What accounts for the structure of the fermion mass matrix?

The hierarchy puzzle, in a sense, also existed before unification. Why is the scale of low energy particle physics so much smaller than the Planck mass M_p.[4] After unification, we simply had another large parameter, the unification scale M_G, not too different from M_p.

*Research supported in part by the National Science Foundation under Grant No. PHY77-22864.

Now at the beginning of 1982, a few papers on grand unification are written every week. But the flavor puzzle and the hierarchy puzzle are still with us.

Today, I will look at some topics of current interest in particle theory from the vantage point of an avowed unifier, with particular emphasis on their connection with these two major puzzles.

Before embarking on this, I want to spend a little time reviewing some of the areas in which real progress has been made since 1974.

II. CP AND THE PQWWKDFS (INVISIBLE) AXION

This is a classic. The problem is the possibility of large CP violating effects in QCD. This problem, which was not even recognized in 1974, has a very elegant solution in the context of grand unified theories (GUTS).

Before instantons, we were blissfully unaware that QCD could violate CP.[5] Instead, we thought it had a different problem, the axial U(1) problem.[6]

There was great relief when 't Hooft showed that because of instantons and triangle anomalies, the axial U(1) is not a symmetry of the physical QCD theory (by which I mean the theory with θ fixed to zero).[7] But then the quark masses can violate CP because the product of the phases of all quark masses cannot be removed without the axial U(1) transformation. If this phase (called $\bar{\theta}$) is nonzero after the $SU(2) \times U(1)$ symmetry breaking has produced the quark masses, then there is strong CP violation. From the experimental bound on the neutron electric dipole moment, we know[8] $\theta < 10^{-8}$. Of course, in the simplest $SU(2) \times U(1)$ or SU(5) model, this phase is a renormalized parameter. There is no contradiction in assuming that $\bar{\theta} < 10^{-8}$; it just seems unnatural. It would be nicer if it were automatic.

Peccei and Quinn[9] were among the first to appreciate this

puzzle, and they suggested a bizarre solution in an SU(2) × U(1) model. Their theory has an approximate symmetry whose associated transformation changes the phase of the quark masses. The (Peccei-Quinn or PQ) symmetry is spontaneously broken by the Higgs VEV, v. Then $\bar{\theta}$ is actually a field, A/v, where A is the pseudoGoldstone boson associated with the broken PQ symmetry. Now the field A does not appear anywhere else in the theory. Its VEV is determined in lowest order only by the quark mass term and it adjusts itself to conserve CP.

This works. But Weinberg[10] and Wilczek[11] realized that the field A is associated with a light, semiweakly coupled neutral particle, the axion, which was looked for, but not found.[12]

The key point is that the axion field, because it is a pseudo-Goldstone boson, appears only in the combination A/v where v is the VEV which breaks the PQ symmetry. If v were much bigger, it would be more weakly coupled and might have escaped attention. Of course, the VEV would have to preserve SU(2) × U(1).

This was realized first by Kim[13] but the idea did not catch on at first because it was buried in a complicated model. Kim's model was discussed by Dicus, Kolb, Teplitz and Wagoner[14] who found that if $v < 10^9$ GeV, stars would lose too much energy to axion emission.

Dine, Fischler and Srednicki[15] first understood and stated the idea in a simple form. They and Glashow, Wise and I[16] and Nilles and Raby[17] constructed GUTs in which the PQ symmetry is broken at the GUT scale $M_G \sim 10^{15}$ GeV, so it is pretty much invisible.

One question that remains about the strong CP violation in such a theory is "Who imposed the PQ symmetry?" Hall, Wise and I[18] have shown that in a certain type of theory it need not be imposed. It is an automatic consequence of gauge invariance and renormalizability.

III. NEUTRINO MASSES AND B-L

The original SU(5) model (and also the original version of the

SO(10) model[19]) conserves baryon number minus lepton number (B-L) so that the neutrinos are exactly massless. There are other amusing possibilities.

(a) The B-L symmetry may not be a symmetry of the theory at all, as in SU(5) with a 15.

(b) The B-L may be gauged and spontaneously broken, as in the SO(10) model of Gell-Mann, Ramond and Slansky.[20] This is phenomenologically similar to a.

(c) The B-L may be a global symmetry which is spontaneously broken at a large scale (μ, M_G). Then there is a very weakly coupled Goldstone boson, called the Majoron by Chicashige, Mohapatra and Peccei,[21] who first discussed this. Except for the Majoron, which is invisible if the scale is M_G, this is phenomenlogically identical to (a) and (b).

(d) Finally, the B-L may be a global symmetry which remains unbroken down to a very small scale, less than 100 KeV. This weird possibility was first discussed by Gelmini and Roncadelli.[22] It is really different. It has another light neutral particle besides the Majoron. It has doubly and singly charged Higgs mesons with a mass ratio of $\sim \sqrt{2}$. I happen to believe that this model has nothing to do with the world. But it is wonderful fun.

In all these models, the neutrinos get potentially observable masses. All can be incorporated into GUTs.

IV. EFFECTIVE THEORIES

On of the most important effects of GUTs is that they have forced us to understand and assimilate the language of effective field theories. Ken Wilson realized how important this was all along, of course, but for the rest of us, it was important to have theories with two or more wildly different scales to underline the importance of the effective field theory (EFT) language. GUTS have served, for many of us, as a set of intellectual bar-bells. The discipline of the EFT language, which we learned in dealing with

GUTs, it also enormously useful in other field theory applications.

The use of effective field theory ideas in GUTs began with the original GQW[3] calculation of coupling constant renormalization, which we can now recognize as the first order piece in a systematic expansion incorporating the effects of physics at the unification scale on the low energy SU(3) × SU(2) × U(1) effective theory.

The idea of effective theories is to divide the momentum scale up into regions. The boundary points between regions are the important masses or symmetry breaking scales of the full theory. For example, in the simplest SU(5) model, we might have three regions $M > M_G$, $M_G > M > M_W$, and $M_W > M$. In each region, we can ignore the symmetry breaking associated with all lower mass scales and ignore heavy particles with masses of the order of any larger mass scale.

Thus for $M_G > M > M_W$ in SU(5), the appropriate theory is an SU(3) × SU(2) × U(1) theory with the heavy gauge bosons and scalars with mass of order M_G left out.

There was considerable confusion about the next order terms in the GQW calculations. Much of it was caused by me and David Politzer.[23] David and I insisted that people should use a momentum space subtraction scheme to define coupling constants in GUTs (and QCD) because the alternative MS or $\overline{\text{MS}}$ schemes, while they were more convenient, did not incorporate the decoupling of heavy degrees of freedom. This argument was right, as far as it went. But what we had missed was that by combining $\overline{\text{MS}}$ with the effective field theory idea, you get a definition of coupling constants which is both convenient and physically sensible.

Each of the effective theories is defined using the $\overline{\text{MS}}$ scheme and then the couplings are chosen so that contiguous theories describe the same physics on the boundary between each pair of regions.

This scheme was worked out in detail by Weinberg,[24] Ovrut and Schnitzer[25] and others. Hall[26] used it to calculate the next order corrections to coupling constant renormalization. Marciano incorporates the same kind of ideas in his calculations of $\sin^2\theta_W$ and the proton decay rate, which he will discuss later.[27]

The effect of the virtual exchange of heavy particles is described in this language by nonrenormalizable (NR) interactions in the effective theory. The "natural" scale for these NR interactions is the next highest scale in the theory. For example, $W^\pm$ exchange in the effective SU(3) × U(1) theory below M_W is described by four-Fermion operators and other NR interactions.

These ideas were applied to proton decay in GUTs by Weinberg[28] and Wilczek and Zee[29], who found all possible baryon number changing operators of low dimension.

The operator analysis is very nice, but one should not get carried away with it. It is the most convenient way of organizing the analysis, but nothing more.

For example, it is easy to construct unified theories in which baryon number is conserved. The same baryon number changing operators are allowed in the effective SU(3) × SU(2) × U(1) theory. But they are not produced in the unification.

The GR model[22] is a wonderful example of the utility of the effective field theory language, because the B-L symmetry breaking takes place at a small scale, less than 100 KeV. In very low energy processes, such as neutrino decay, it is appropriate to treat the B-L symmetry as spontaneously broken and the Majoron as a Goldstone boson. But for most processes, the energies involved are high enough so that it is appropriate to ignore the B-L symmetry breaking. Then the Majoron and the light Higgs particle are combined into a complex scalar field.

The effective field theory language allows us to formulate the hierarchy puzzle in a simple problem for a model with fundamental Higgs. In the effective theory below a symmetry breaking scale, we are interested in the light particles. Some of them are light for some reason, such as gauge bosons or chiral fermions. But sometimes we need other light fields, such as the Higgs doublet in the

SU(3)×SU(2)×U(1) model. The mass of each irreducible multiplet of such fields is some combination of mass parameters. Since no symmetry forbids the appearance of a large mass, we expect that some of these parameters will be proportional to the next highest scale. For each such light multiplet, we need an "unnatural" condition. A linear combination of large masses must be small. We can choose the parameters to make it so, but we do not like it very much.

V. TABLE OF UNIFIED MODELS

The rest of the talk is summarized in the following table, where I organize various models according to whether they address the hierarchy puzzle (H) and the flavor puzzle (F):

Model	H	F	+/-	Examples
SU(5), SO(10) E(6) + Higgs			+ simple, explicit models.	lots
Global sym. or larger groups	x		- not clear waht the rules are. Why particular reps?	lots
SO(18)	x		- perturbation theory breaks down.	GW-GRS-WZ
SO(14)	x		- GIM, b and τ problems.	Kim
Supersymmetry		x	?	none
Technicolor		x	? quark and lepton masses	not unified
ETC	x	x	- ΔSNC, CP violation, structure	none
Composite C	x	x		
Composite B	x			Bars
Composite A			-no evident virtues	lots
eTC	x	x	? Basic idea needs checking	GG,GM,G

VI. SU(5), SO(10), E(6) + HIGGS

It is very important to remember that the flavor puzzle and the hierarchy puzzle are only puzzles, not inconsistencies. The original SU(5) model and its simple generalizations based on SO(10) and E(6) are completely consistent. One of them may even be right. We may be fooling ourselves in demanding that our theory explain flavor and hierarchy. The world may simply be the way it is, period.

The SU(5) model is probably the most attractive of these, just because it is the simplest. The virtues of SO(10) and E(6) (if any) begin to show up only when we add extra ingredients, as we will see below.

The simplest SU(5) model is very simple indeed. It involves only 8 SU(5) multiplets three 10's and $\overline{5}$'s of left-handed (LH) fermions, and a 24 and 5 of scalars. The model is completely consistent with what we know, except perhaps for the charge-1/3 quark and charged lepton masses, where there is an apparent discrepancy in the relations $m_b = m_\tau$, $m_s = m_\mu$, $m_d = m_e$ (at M_G).[30] There are three reasonable attitudes that I know of concerning this discrepancy:

(a) The discrepancy is real and the model must be elaborated.

(b) The discrepancy is real and the model is correct as far as it goes. Small NR couplings with the scale of the Planck mass M_p fix it, as suggested by Ellis and Gaillard.[31]

(c) The discrepancy is only apparent. Perhaps we should not trust current algebra to give m_s/m_d.

One other very nice feature of the simplest SU(5) model is that there is no play in the prediction of the proton decay rate. In principle, if not yet in practice, it is completely and accurately calculable in terms of low energy parameters. This model has the great virtue that it can actually be ruled out by the present generation of proton decay experiments. Most other models have much more freedom.

Global Symmetries or Larger Groups

Many attempts have been made to solve some aspects of the flavor problem in the context of models with explicit fundamental Higgs. These models do not address the hierarchy problem at all. We still have to keep the Higgs doublet light by hand.

The common idea is that by enlarging the local gauge group, or by adding additional global symmetry one may put additional constraints on the number of families, the form of the mass matrix, or both.

This game is lots of fun. And many ingenious schemes have been invented. But there is something unsatisfying about it. The starting point always seems ad hoc, justified only by the results. What I say to myself when I want to feel better about games like this is that not all relations are possible. But to be honest, I am not really sure that that is enough.

SO(18) and SO(14)

Two games of this kind should, perhaps, be singled out for special mention, because they satisfy a very strong and appealing constraint--all the fermions are in a single irreducible representation of the gauge group. They are complete unifications of flavor.

One immediately thinks of the spinor representations of large orthogonal groups, because they are (almost) built out of repeated copies of the 16 of SO(10). The "almost" is very serious, however. There are always as many $\overline{16}$'s as 16's, and thus as many fermions with V+A weak interactions as with V-A. We must have some way to account for the absence of these "mirror fermions" or we have made the flavor problem worse rather than better.

The SO(18) idea, invented by Witten and me, independently by Gell-Mann, Ramond and Slansky, and more recently taken up by

Wilczek and Zee, is to have the mirror fermions all confined by a technicolor (TC) interaction which does not act on the normal families. In this case, the TC is based on an SO(5) subgroup. The model has three normal families and a family of SO(5) 5's which are V-A and two families of SO(5) 4's which are V+A.

The main trouble with this model is that there are so many fermions that the couplings are not asymptotically free. They get large before M_G, and it is not clear what unification means.

The SO(14) idea, invented by Kim, is more bizarre still. It is that we have already seen some of the minor fermions--the b and the τ. But the τ has V-A weak interactions you say? Yes, but suppose the LH τ is part of a $\overline{16}$ multiplet (so that the weak doublet has charges +1 and 0) but all the charges in the $\overline{16}$ are displaced down by one unit. Then, we get the right weak interactions. If we add another $\overline{16}$ with charges displaced up one unit, this bizarre scheme can all be embedded in the spinor representation of SO(14). In addition to the τ, there are lots of funny quarks, a doublet with charges 5/3 and 2/3 and another with charges -1/3 (the b) and -4/3, both with V+A weak interactions.

This model is really cute. The coupling constant renormalization is different from SU(5) and its relatives because the charges are not completely inside SU(5). But the right $\sin^2\theta_W$ and the same M_G can be obtained anyway.

The problem with this model is that the b quark is really peculiar. It has abnormal decays into τ's and/or baryons. The data from CESAR make this seem unlikely.[36]

Supersymmetry

Supersymmetry (SS) is the one larger symmetry idea that may impact on the hierarchy puzzle. In supersymmetric theories, scalar meson masses can be forbidden by the chiral symmetries of their fermion superpartners.

Just two years ago, supersymmetry was a rare disease among

particle theorists. Now it is an epidemic. I speak as one who has had the disease and been cured. The symptoms of the disease are dizziness caused by too much anticommuting, a breakdown in the ability to distinguish fermions from bosons, and most serious, a compulsion to construct models in which supersymmetry has something to do with physics.

Seriously, SS is a nice idea until you try to apply it. Then, it turns into a quicksand from which it is very difficult to escape. The problem is that to be of any use in the solution of the hierarchy puzzle, the supersymmetry must be preserved in the effective theory below M_G, down to a mass scale not much larger than M_W. This causes all kinds of problems, unless you want to give up and break the SS softly. These issues may be discussed in the SS session.[37]

TC and ETC

Technicolor was the first serious stab at the hierarchy problem. The idea was to break the SU(2) × U(1) symmetry dynamically, with no fundamental Higgs scalars at all. Since the color SU(3) interactions apparently break chiral symmetries dynamically, it seemed that all that was needed was a stronger version of color: hence, technicolor.[38,39]

It is probably possible to construct toy GUTs with a technicolor interaction that breaks SU(2) × U(1) and gives a large gauge hierarchy in a model without normal families of quarks and leptons. The technicolor mechanism works fine for giving $W^{\pm}$ and Z^0 mass. No one has bothered to do this, so far as I know, because it is not very interesting. The trouble is that technicolor, by itself, is not sufficient to give masses to the quarks and leptons, at least not masses which are large enough to be interesting. In the effective theory below M_G, the ordinary fermions are protected from getting masses by global, as well as gauge, chiral symmetries. Technicolor does not break the global symmetries.

Scalar mesons could do it, but we are trying to avoid light

fundamental scalars. We must add addition gauge interactions. Conceptually, the simplest possibility is to have "extended technicolor" (ETC) interactions which combine normal and technicolored particles in the same representation.[40,41] The ETC interaction is in turn broken at some larger scale. Then the quarks and leptons get mass from radiative corrections.

What breaks ETC? Again, we would like to avoid fundamental scalars. Perhaps it is a still stronger force, technitechnicolor or something.

Alternatively, perhaps the ETC group breaks itself. This is getting awfully close to a composite picture of quarks and leptons, because at the ETC breaking scale, they took part in the very strong interactions.

At any rate, it is very hard to build a model of either kind with any interesting structure. If flavor is to be incorporated at all, it must be incorporated in a nontrivial way. This leads to problems with flavor changing neutral currents, strong CP violation and other diseases. There are no good models.

Composite Quarks and Leptons

Here I will discuss (or at least slander) composite models from the point of view of GUTs. I have divided them into three classes according to the puzzles they address.

Most of the so called composite models on the market today, in my view, should not be called models. They are premodels, short of modeldom because they are not well-defined. Most are based on simple counting with no serious proposal for a dynamics to implement the counting.

What people interested in such questions ought to be doing is building tools, learning about the binding of chiral fermions into massless or light states. Of course, some people are doing so: I hope you will hear a lot about such tool building in the session on composites.[42] Once you have some tools, by all means try them out on a model.

The only tools we have so far are the 't Hooft[43] anomaly condition and the Preskill-Weinberg[44] persistence-of-mass condition. In applications, these conditions are not strong enough to determine the dynamics completely. But it is a start.

There are two sensible ways to try to use such tools. The less ambitious is to attack the flavor problem by constructing a model in which the quarks and leptons are bound states but in which their SU(2) × U(1) breaking masses are due to explicit couplings. Probably such models are possible. Bars[45] has constructed an interesting example in which the SU(3) × SU(2) × U(1) can be unified into SU(5) or SO(10). It may be possible to make this model realistic, but it is not clear that it can be further unified with the interactions responsible for the formation of the composites. A more ambitious idea is to have the binding mechanism (let us call it hypercolor) produce technicolor fermions as well and to have the hypercolor act as a kind of nonperturbative ETC. This is very hard to do explicitly, and it is quite likely that if such a model could be constructed, it would be subject to the usual diseases of ETC models.

Effective TC

After this review, I have a little time left to at least give an advertisement for the final entry on my chart, which is a new and different proposal for the solution to the hierarchy puzzle that turns out to address the flavor puzzle as well.

The idea is that in a theory with a nonasymptotically free TC, the Higgs boson may be both composite and fundamental. I speculate that such a theory may have a strong coupling phase in which the TC theory (which has a nontrivial UV fixed point) is equivalent to an asymptotically free (AF) eTC theory which contains the same technifermions, but also strongly coupled scalar mesons which couple to the technifermions. It is these scalars which I interpret as the Higgs mesons. They are light not because of symmetry, but because of dynamics. In the TC theory, these scalars appear as bound states of fermions. Their mass is determined by the TC theory and is thus of the order of $\Lambda_{TC} \sim 1$ TeV. But because the theory is not asymptotically free, the scalars are not soft mushy objects like the bosonic bound states of quarks and antiquarks. Instead, they behave like fundamental scalars.

This crazy idea is mostly my fault. It has its roots in the comment by Holdom[46] that if TC is not asymptotically free, the ETC scale can be larger than it is in models of AF TC. Shelly Glashow[47] and Ian McArthur[48] have collaborated with me in trying to push this scale all the way to M_G, where we can use scalar field couplings to break chiral symmetries without guilt.

Obviously, the idea needs to be checked. Probably it can be checked using Monte Carlo techniques. Meanwhile, though, the idea of the mapping from an non AF TC to an AF eTC, we can organize our ignorance of strong interactions into a few parameters. That is enough to allow for some model building.

The simplest model[49] I can build which may be realistic is based on an $SO(10)^5$ gauge group with a discrete symmetry which cyclicly permutes the SO(10) factors. Three of the SO(10)s are associated with the three normal families, another SO(10) with TC and an abnormal family containing a lepton doublet and technifermions, the final SO(10) is associated only with heavy particles. This model is discussed in reference 49.

REFERENCES

1. H. Georgi and S.L. Glashow, Phys. Rev. Lett. 32, 438 (1974).
2. J.C. Pati and A. Salam, Phys. Rev. D 10, 275 (1974). The grand unified model described in this paper is based on a peculiar version of strong interactions in which color SU(3) is broken and quarks have integral charge, but it can be adapted to the standard model. See, also, J.C. Pati and

A. Salam, Phys. Rev. D 8, 1240 (1973). No grand unified model is constructed in this paper, but some aspects of grand unification are anticipated in the introduction.

3. H. Georgi, H.R. Quinn and S. Weinberg, Phys. Rev. Lett. 33, 451 (1974).
4. See Dirac's talk and references therein.
5. S. Weinberg, Phys. Rev. Lett. 31, 494 (1973) and Phys. Rev. D 8, 4482 (1973).
6. S. Weinberg, Phys. Rev. D 11, 3583 (1975).
7. G. 't Hooft, Phys. Rev. Lett. 37, 8 (1976) and Phys. Rev. D 14, 3432 (1976).
8. V. Baluni, Phys. Rev. D 19, 2227 (1979); R. Crewther, P. Di Vecchia and G. Veneziano, Phys. Lett. 88B, 123 (1979).
9. R. Peccei and H.R. Quinn, Phys. Rev. Lett. 38, 1440 (1977).
10. S. Weinberg, Phys. Rev. Lett. 40, 223 (1978).
11. F. Wilczek, Phys. Rev. Lett. 40, 279 (1978).
12. C. Edwards, et al., SLAC-PUB-2878 (1982).
13. J. Kim, Phys. Rev. Lett. 43, 103 (1979).
14. D.A. Dicus, et al., Phys. Rev. D 18, 1829 (1978).
15. M. Dine, W. Fischler and M. Srednicki, Phys. Lett. B104, 199 (1981).
16. M.B. Wise, H. Georgi and S.L. Glashow, Phys. Rev. Lett. 47, 402 (1981).
17. P. Nilles and S. Raby, Phys. Lett. B (to be published).
18. H. Georgi, L. Hall and M.B. Wise, Nucl. Phys. B192, 409 (1981).
19. H. Georgi, Particles and Fields, 1974 (APS/DPF Williamsburg), ed. C.E. Carlson (AIP, New York, 1975). I am fairly certain that I was the first to find the SO(10) model, because I discovered it a few hours before realizing that a simpler unification could be obtained in SU(5). Glashow and I chose not to include SO(10) as a footnote in the SU(5) paper (ref. 1). It was discovered independently by

H. Fritzsch and P. Minkowski, Ann. of Phys. 93, 193 (1975), and I suspect also by J. Pati and A. Salam (private communication).

20. M. Gell-Mann, P. Raymond and R. Slansky, in Supergravity, ed. by P. van Nieuwenhuizen and D. Freedman (North-Holland, 1979).
21. Y. Chikashige, R.N. Mohapatra, and R.D. Peccei, Phys. Lett. 98B, 265 (1981).
22. G.B. Gelmini and M. Roncadelli, Phys. Lett. 99B, 411 (1981). See also, H. Georgi, S.L. Glashow and S. Nussinov, Nucl. Phys. B193, 297 (1981).
23. H. Georgi and H.D. Politzer, Phys. Rev. D 14, 1829 (1976).
24. S. Weinberg, Phys. Lett. 91B, 51 (1980).
25. B. Ovrut and H. Schnitzer, Nucl. Phys. B179, 381 (1981).
26. L. Hall, Nucl. Phys. B178, 75 (1981).
27. See Marciano's talk and S. Dawson, J.S. Hagelin and L. Hall, Phys. Rev. D 23, 2666 (1981).
28. S. Weinberg, Phys. Rev. Lett. 43, 1566 (1979).
29. F. Wilczek and A. Zee, ibid., 43, 1571 (1979).
30. A. Buras, J. Ellis, M.K. Gaillard, and D.V. Nanopoulos, Nucl. Phys. B135, 66 (1978).
31. J. Ellis and M.K. Gaillard, Phys. Lett. 88B, 315 (1979).
32. H. Georgi and E. Witten, unpublished.
33. M. Gell-Mann, P. Ramond and R. Slansky, reference 20.
34. F. Wilczek and A. Zee, See lectures by A. Zee in Proceedings of the Fourth Kyoto Summer Institute on Grand Unified Theories, ed. M. Konuma and T. Maskawa (World Sci. Pub., Singapore, 1982).
35. J. Kim, Phys. Rev. Lett. 45, 1916 (1980).
36. Report to Bonn Conference.
37. See the talk by S. Dimopoulos.

38. L. Susskind, Phys. Rev. D 20, 2619 (1979). See also, the review by E. Farhi and L. Susskind, Physics Reports 74, 277 (1981).

39. S. Weinberg, Phys. Rev. D 13, 974 (1976) and D 19, 1277 (1979).

40. S. Dimopoulos and L. Susskind, Nucl. Phys. B155, 237 (1979).

41. E. Eichten and K. Lane, Phys. Lett. 90B, 125 (1980).

42. See the talks by Eichten and Raby.

43. G. 't Hooft, Lectures at the Cargese Summer Institute (1979).

44. J. Preskill and S. Weinberg, Phys. Rev. D 24, 1059 (1981).

45. I. Bars, Phys. Lett. 106B, 105 (1981).

46. B. Holdom, Phys. Rev. D 24, 1441 (1981).

47. H. Georgi and S.L. Glashow, Phys. Rev. Lett. 47, 1511 (1981).

48. H. Georgi and I.N. McArthur, HUTP-81/A054.

49. H. Georgi, HUTP-81/A057.

MAJORANA NEUTRINOS - THEIR ELECTROMAGNETIC PROPERTIES AND NEUTRAL CURRENT WEAK INTERACTIONS

Boris Kayser

National Science Foundation

Washington, D.C. 20550

In grand unified theories, the possibility that neutrinos are massive Majorana particles is a rather attractive option. Thus, it becomes interesting to try to build up a picture of the kind of animal a Majorana neutrino is. With that aim in mind, I would like to talk about two aspects of Majorana neutrinos: their electromagnetic properties, and their neutral current weak interactions. Before I do that, I would like to very carefully define what I will mean by a Majorana neutrino, and say a little bit about Majorana-Dirac confusion.

We will be talking throughout about neutrinos that have mass. Suppose there exists a massive neutrino ν_-^D with negative helicity, as considered at the extreme left of Fig. 1(a): As Fig. 1(a) indicates, invariance under CPT then implies that there must also exist the CPT mirror-image of this particle, a positive-helicity antineutrino. In addition, if the negative-helicity neutrino has a mass, then it travels slower than light, so an observer can overtake it. If he does, then in his frame the neutrino is going the other way, but still spinning the same way, so it has been converted by a Lorentz transformation into a positive-helicity particle. Now, this positive-helicity object may or may not be the same as the

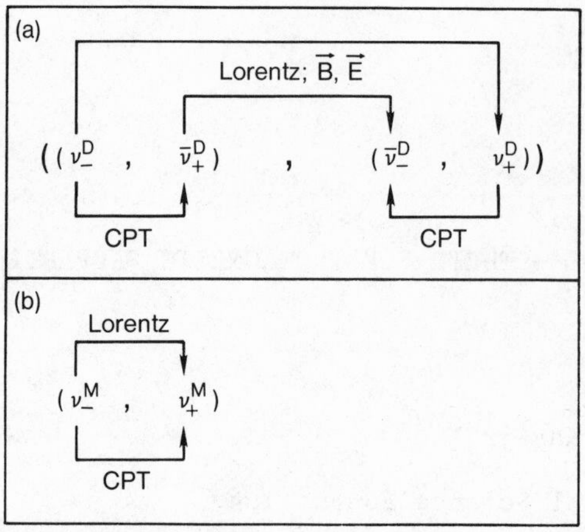

Fig. 1 a) The four distinct states of Dirac neutrino ν^D.
b) The two distinct states of a Majorana neutrino ν^M.

CPT mirror-image of the original particle. Let us suppose first that, as imagined in Fig. 1(a), it is <u>not</u> the same. Then it has its own CPT mirror-image, and there are four states with a common mass in all. This foursome is called a Dirac neutrino ν^D. In general, a Dirac neutrino will have a magnetic dipole moment, and perhaps even an electric dipole moment. Thus, the negative-helicity Dirac neutrino $\nu_-{}^D$ and its CPT-conjugate, $\bar{\nu}_+{}^D$, can be converted, respectively, into their opposite helicity partners $\nu_+{}^D$ and $\bar{\nu}_-{}^D$, not just through a Lorentz transformation, but, alternatively, through the action of an external $\vec{B}$ or $\vec{E}$ field.

The second possibility, pictured in Fig. 1(b), is that when you reverse the momentum of the original negative-helicity particle by a Lorentz transformation, the positive-helicity object that you obtain <u>is</u> identical to the CPT mirror-image of the original state. In that case there are only two states with a common mass, and this twosome is called a Majorana neutrino ν^M. In the rest frame, CPT applied to either of the two spin states of such a neutrino simply reverses its spin (due to the time reversal). A subsequent 180° rotation then brings the neutrino back to its original state. It is in this sense that a Majorana neutrino is its own antiparticle.

Now, so long as the mass is not zero, then regardless of how small it is, the pair of Dirac particles $(\nu_-^D, \bar{\nu}_+^D)$ is accompanied by the pair $(\bar{\nu}_-^D, \nu_+^D)$, whereas the pair of Majorana particles (ν_-^M, ν_+^M) is not accompanied by any other states. Thus, there is a distinction between Dirac and Majorana neutrinos no matter how small the mass is. Nevertheless, there is a state of affairs which I would like to summarize by asserting the validity of a

<u>Practical Dirac-Majorana Confusion Theorem:</u>

Suppose the mass of some neutrino is very small compared to all other mass and energy scales in the problem, and that all weak currents are left-handed. Suppose further that all experiments on this neutrino are done with one of two incoming states--a state "ν_-" of negative helicity, or its CPT--conjugate, "$\bar{\nu}_+$." (All neutrino experiments have, of course, been done this way.) Then it is practically impossible to tell experimentally whether "ν_-" and "$\bar{\nu}_+$" are really $(\nu_-^D, \bar{\nu}_+^D)$, two of the four states of a Dirac neutrino, or (ν_-^M, ν_+^M), the two states of a Majorana neutrino.

We will see illustrations of this theorem doing its evil work, first in the case of electromagnetic interactions, and then in the case of the neutral weak ones.

(When the neutrino mass m is identically zero, there is no distinction at all between $(\nu_-^D, \bar{\nu}_+^D)$ and (ν_-^M, ν_+^M). When $m = 0$, one can no longer flip the helicity of a neutrino by going to another Lorentz frame. Nor can one any longer flip it by putting the neutrino in an external magnetic or electric field, as we shall see. Consequently, the pair $(\nu_-^D, \bar{\nu}_+^D)$ gets disconnected from the pair $(\bar{\nu}_-^D, \nu_+^D)$, and the latter pair need not even exist. One just has two states, and whether one refers to them as a Dirac neutrino and its antiparticle, or as the two spin states of a Majorana neutrino, is arbitrary.)

Let us now turn to the electromagnetic properties of Majorana neutrinos,[1] beginning with the static properties. It has been argued on various grounds that a Majorana neutrino cannot have a magnetic or electric dipole moment. It seems not to have been noticed, however, that this absence of dipole moments already follows just from CPT-invariance in a trivial way. The argument is as follows: Suppose we have a Majorana neutrino at rest in external, static, uniform $\vec{E}$ and $\vec{B}$ fields. If it had a magnetic dipole moment μ_{Mag} and an electric dipole moment μ_{El}, then its interaction energy in these external fields would obviously be $-\mu_{Mag}\langle\vec{s}\cdot\vec{B}\rangle - \mu_{El}\langle\vec{s}\cdot\vec{E}\rangle$, where $\vec{s}$ is its spin operator. Now, under CPT, $\vec{B}$ and $\vec{E}$ fields just go into themselves. However, we just agreed that the effect of CPT on a Majorana neutrino is to reverse its spin. Consequently, the dipole interaction energy changes sign. Thus, if CPT invariance holds, μ_{Mag} and μ_{El} must vanish.

Notice that this argument does not use the property that the Majorana neutrino is an eigenstate of C. I made a point of not using that property because it is not obvious that a Majorana neutrino <u>is</u> indeed an eigenstate of C, even though, as you know, in free field theory that is how a Majorana neutrino is defined. What I am questioning is whether a <u>physical</u>, <u>dressed</u> Majorana neutrino can maintain the property of being C selfconjugate, which it had in free field theory, once the C-violating weak interactions are turned on. Let me explain why I question that.

Consider the Majorana neutrino of negative helicity in Fig. 2. Since it is a Majorana particle, it can go virtually not only into a charged lepton ℓ^- and a W^+, but also into a charged antilepton ℓ^+ and a W^-. Now, the states ℓ^-W^+ and ℓ^+W^- are the C mirror-images of each other. However, if all weak currents are left-handed, then the vertices in the ℓ^+W^- diagram are highly suppressed by helicity, compared to those in the ℓ^-W^+ diagram. Consequently, these two C mirror-image states occur with very unequal amplitudes in the neutrino wave function, so it is not obvious that this neutrino can be C selfconjugate.

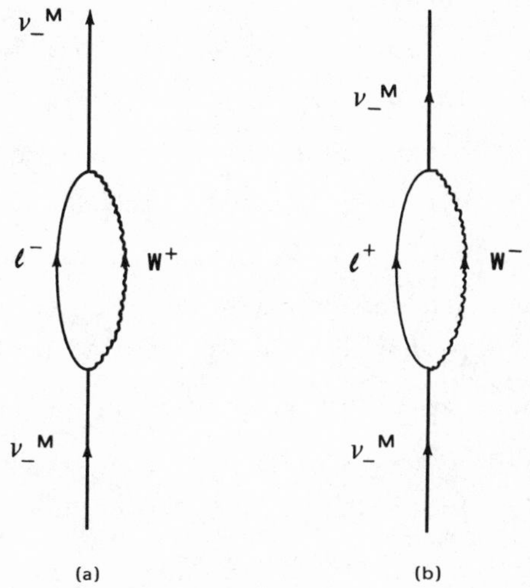

Fig. 2 Virtual components in the state of a Majorana neutrino ν_-^M. The symbol ℓ denotes a charged lepton, and W the charged weak boson.

For many purposes, this possible complexity of the C properties of a Majorana neutrino does not make any difference. However, we shall assume that the only way a photon couples to a neutrino is by coupling to a charged particle in an intermediate state such as those in Fig. 2. Consequently, if these intermediate states vitiate the C selfconjugacy of a Majorana neutrino, one can hardly disregard that fact when trying to understand its electromagnetic properties. Therefore, I will not assume C selfconjugacy in anything that follows.

Let us now derive the most general expression for $\langle\nu^M|J_\mu^{EM}|\nu^M\rangle$, the matrix element of the electromagnetic current of a Majorana neutrino. To do that, instead of looking at the s-channel process $\gamma + \nu \to \nu$ which this matrix element describes, let us look at the related t-channel process, $\gamma \to \nu + \bar{\nu}$. We do this because in the t-channel, it will be very easy to see the consequence of having neutrinos of Majorana type. Note that the crossing properties of local field theory imply that the "$\bar{\nu}$" in $\gamma \to \nu + \bar{\nu}$ is the CPT (rather than C) conjugate of the incoming "ν" in $\gamma + \nu \to \nu$. Thus, in the Majorana case, the t-channel process has two identical particles in the final state.

Now, in $\gamma(q) \to \nu + \bar{\nu}$, involving an off-shell photon of momentum q, how many different independent amplitudes are there? First, it must be recognized that the conserved electromagnetic current will only produce $\nu\bar{\nu}$ pairs with J = 1, never with J = 0, despite the fact that the photon is off-shell. To see this, note that if J_μ^{EM}, acting on the vacuum, produces some state $|f_{J=0}(q)\rangle$ with J = 0 and momentum q, then $\langle f_{J=0}(q)|J_\mu^{EM}|0\rangle = aq_\mu$, where a is a form factor. Current conservation then requires that $q_\mu \langle f_{J=0}(q)|J_\mu^{EM}|0\rangle = aq^2 = 0$, so that a must vanish. Now, in the Dirac case, the t-channel process is $\gamma \to \nu^D + \bar{\nu}^D$. In the nonrelativistic limit, there are four J = 1 states that the $\nu^D\bar{\nu}^D$ pair could be in: $^3D_1$, $^3P_1$, $^3S_1$ and $^1P_1$. Thus, a Dirac neutrino has <u>four</u> electromagnetic form factors. In the Majorana case, the t-channel process is $\gamma \to \nu^M + \nu^M$. Since the two final particles are identical, the final state must be antisymmetric. Of the four states just mentioned, only one, $^3P_1$, meets this requirement. Consequently, a Majorana neutrino only has <u>one</u> electromagnetic form factor.

Returning to the s-channel, one can show with standard techniques that for a Dirac neutrino, the electromagnetic matrix element may be written in the form

$$\langle \nu^D(p_f,s_f)|J_\mu^{EM}|\nu^D(p_i,s_i)\rangle = i\bar{u}_f[F_D(q^2)\gamma_\mu$$
$$+ G_D(q^2)(q^2\gamma_\mu - 2miq_\mu)\gamma_5 + M_D(q^2)\sigma_{\mu\nu}q_\nu + E_D(q^2)i\sigma_{\mu\nu}q_\nu\gamma_5]u_i \quad . \tag{1}$$

Here p_i and s_i (p_f and s_f) are the initial (final) neutrino momentum and spin projection, $u_i(u_f)$ is the initial (final) Dirac spinor, and $q = p_f - p_i$. The particular choice of Dirac form factors embodied by Eq. (1) includes an electric charge distribution-type form factor F_D, an axial vector-type form factor G_D (multiplying a covariant constructed to guaranteee current conservation), and magnetic and electric dipole form factors M_D and E_D. Before one imposes the requirement that $\gamma \to \nu^M + \nu^M$ yield an antisymmetric state, the electromagnetic matrix element $\langle \nu^M | J_\mu^{EM} | \nu^M \rangle$ for a Majorana neutrino can also be written in the form of Eq. (1). That form, after all, is just the most general expression for the matrix element of a conserved operator with one Lorentz index. Arguing similarly for the t-channel Majorana matrix element, we find that, for outgoing momenta and spins p_1, s_1 and p_2, s_2,

$$\langle \nu^M(p_1,s_1)\nu^M(p_2,s_2)|J_\mu^{EM}|0\rangle = i\bar{u}_2[F_M\gamma_\mu + G_M(q^2\gamma_\mu - 2miq_\mu)\gamma_5 + M_M\sigma_{\mu\nu}q_\nu + E_M i\sigma_{\mu\nu}q_\nu\gamma_5]v_1 \quad . \tag{2}$$

Here F_M, G_M, M_M, and E_M are Majorana form factors, and now $q=p_1+p_2$. Also, the u spinor of Eq. (1) has been replaced by a v spinor, as appropriate for a t-channel matrix element.

Now, we saw that Fermi statistics has the consequence that actually only one linear combination of the four terms in Eq. (2) survives; namely, the one that corresponds to production of the $^3P_1$ state in the nonrelativistic limit. To see which linear combination this is, let us take the nonrelativistic limits of the various terms. More specifically, let us look at the nonrelativistic limit of $\eta_\mu \langle \nu^M \nu^M | J_\nu^{EM} | 0 \rangle$, which represents the amplitude for an off-shell photon of momentum q and polarization η to decay into two Majorana neutrinos. We work in the photon rest frame, so that $\vec{p}_2 = -\vec{p}_1 \equiv \vec{p}$, and choose a gauge in which $q_\mu \eta_\mu = 0$. Suppose S is the amplitude for the two-neutrino singlet spin state, and the m'th component of $\vec{T}$ is the amplitude for the triplet spin state with $S_z = m$. In terms of S and $\vec{T}$, the nonrelativistic limits of the four terms in Eq. (2) are as given in Table I. Now, if you are familiar with the Cartesian tensor representation of angular momentum states, you will immediately recognize that $\vec{\eta} \cdot \vec{T} \times \hat{p}$ is the amplitude corresponding to production of $^3P_1$. This is an amplitude for a P-wave because it

Table I

Nonrelativistic limits of terms in the t-channel electromagnetic matrix element. The final column lists the states produced by each term.

Term	Nonrelativistic Limit	States Produced
γ_μ	$(1 + \frac{p^2}{12m^2})\vec{\eta}\cdot\vec{T} - \frac{p^2}{6m^2}(3\vec{\eta}\cdot p\ \vec{T}\cdot p - \vec{\eta}\cdot\vec{T})$	$^3S_1$, $^3D_1$
$(q^2\gamma_\mu - 2miq_\mu)\gamma_5$	$\vec{\eta}\cdot\vec{T}\times\hat{p}$	$^3P_1$
$\sigma_{\mu\nu}q_\nu$	$(1 - \frac{\vec{p}^2}{12m^2})\vec{\eta}\cdot\vec{T} + \frac{\vec{p}^2}{6m^2}(3\vec{\eta}\cdot\hat{p}\ \vec{T}\cdot\hat{p} - \vec{\eta}\cdot\vec{T})$	$^3S_1$, $^3D_1$
$\sigma_{\mu\nu}q_\nu\gamma_5$	$\vec{\eta}\cdot\hat{p}\ S$	$^1P_1$

involves one power of the momentum, it is clearly a triplet amplitude, and the final state quantities $\vec{T}$ and $\hat{p}$ are combined into a vector (J = 1) product. From Table I, we see that $\vec{\eta}\cdot\vec{T}\times\hat{p}$ only occurs in the nonrelativistic limit of the axial vector-type term, so it is only this term that survives in the Majorana case. Notice that the Table makes the symmetry properties of the various terms explicit. The vector $\hat{p}$ is antisymmetric under $1\leftrightarrow 2$, while $\vec{T}$ and S are, respectively, symmetric and antisymmetric, so $\vec{T}\times\hat{p}$ is antisymmetric, while all the other terms in the Table are symmetric. Of course, we need not have gone to the nonrelativistic limit to find the symmetry properties of the terms in Eq. (2). We could have looked directly at that expression and shown that the $(q^2\gamma_\mu - 2miq_\mu)\gamma_5$ term is odd under $1\longleftrightarrow 2$, and all the other terms are even.

We conclude that the most general expression for the matrix element of the electromagnetic current of a Majorana neutrino is

$$<\nu^{M}(p_f,s_f)|J_\mu^{EM}|\nu^{M}(p_i,s_i)> = i\bar{u}_f G_M(q^2)(q^2\gamma_\mu - 2miq_\mu)\gamma_5 u_i \; . \qquad (3)$$

This expression involves only one form factor, of essentially axial-vector character.

That Eq. (3) is correct can be confirmed in various ways. I verified it essentially by brute force; that is, by constructing the most general effective electromagnetic current J_μ^{eff} that can be built out of free Majorana fields χ and their derivatives up to second order. (One has to go to that order before getting anything.) The result is

$$J_\mu^{eff} = - \Box^2(\bar{\chi}\gamma_\mu\gamma\ \chi) + 2m\partial_\mu(\bar{\chi}\gamma_5\chi) \; . \qquad (4)$$

Obviously, the matrix element of this J_μ^{eff} between free Majorana neutrino states is the right-hand side of Eq. (3).

A still different derivation of the matrix element $<\nu^M|J^{EM}|\nu^M>$, with a result in agreement with Eq. (3), has been given by Nieves.[2] Also, an effective photon-Majorana neutrino interaction Hamiltonian, expressed in a two-component formalism and involving one form factor, has been written down by Schechter and Valle.[3] Now, our t-channel analysis shows that there can only be one form factor. Thus, their effective Hamiltonian must be equivalent to our J_μ^{eff}, and Schechter has now demonstrated that it is indeed equivalent.[4] Finally, by the time of preparation of the written version of this talk, two additional derivations of the matrix element $<\nu^M|J^{EM}|\nu^M>$ have appeared, one by Shrock,[5] and one in a revised version of the preprint by Nieves.[2] The latter derivation, and a similar one shortly to appear by McKellar,[6] obtain the whole electromagnetic matrix element through a less trivial application of the approach we adopted for the dipole moments; namely, the imposition of CPT-invariance.

A close look at the effective current of Eq. (4) leads us to an interesting digression on the parity properties of Majorana neutrinos. While it is not obvious by any means, it can be shown that under parity a free Majorana field transforms in just the same way as does a Dirac field; that is,

$$\chi(\vec{x},t) \xrightarrow{P} \eta_P \gamma_4 \chi(-\vec{x},t) \quad , \tag{5}$$

where η_P is a phase factor. Thus, one can immediately read off the parity properties of J_μ^{eff}, Eq. (4), by pretending that the fields χ in it are Dirac fields. Since an axial vector and a pseudoscalar have positive and negative parity, respectively, $\vec{J}^{eff}$ is obviously a positive parity operator. This means that when acting on the vacuum, it will produce a state of positive parity. Ah, but in the t-channel analysis we discovered which state that is. It is the state which in the nonrelativistic limit is $^3P_1$. Consequently, we learn that

$$P|\nu^M\nu^M;\ ^3P_1> = +|\nu^M\nu^M;\ ^3P_1> \quad . \tag{6}$$

That is, the parity of a pair of identical Majorana neutrinos in a P-wave is <u>even</u>! Now, that might surprise you, but it is true. It is a reflection of the fact that the intrinsic parity of one Majorana neutrino at rest is ±i,[7]

$$P|\nu^M(\vec{p}=0,s)> = \pm i|\nu^M(\vec{p}=0,s)> \quad , \tag{7}$$

so that the intrinsic parity of two identical Majorana neutrinos is $(\pm i)^2$. Note that the phase factor ±i in Eq. (7), and the consequent even parity of the two-neutrino P-wave, is not a convention, but something with measureable consequences. For example, if you have a parity-conserving world in which the decay $\psi(2^+) \to \nu^M\nu^M$ of a boson with $J^P = 2^+$ occurs, the final state neutrinos will have odd L, not even L, and this will affect the outgoing angular distribution. Now, why is the phase factor in Eq. (7) ±i? One way to see why is to notice that the parity transformation law obeyed by the field χ,

Eq. (5), induces a related law for the charge-conjugate field $\chi^c \equiv C\bar{\chi}^T$, where $C = \gamma_4\gamma_2$ is the charge conjugation matrix. Insertion of Eq. (5) into the definition of χ^c yields

$$\chi^c(\vec{x},t) \xrightarrow{P} -\eta_P^* \gamma_4 \chi^c(-\vec{x},t) \quad . \tag{8}$$

Of course, the point is that for a Majorana neutrino the free field χ and its charge-conjugate are the same. Consequently, the results of operating with parity on these two objects must be equal. Comparing Eq. (8) to Eq. (5), we see that this requires that $\eta_P = \pm i$. If one expands χ in terms of momentum states, this implies the intrinsic parity of $\pm i$ in Eq. (7).

Inspecting our general result for the electromagnetic matrix element, Eq. (3), we see that in the Majorana case not only do the electric and magnetic dipole moments vanish, but the entire magnetic and electric dipole form factors vanish as well. So does the "electric charge distribution" form factor, the coefficient of γ_μ.

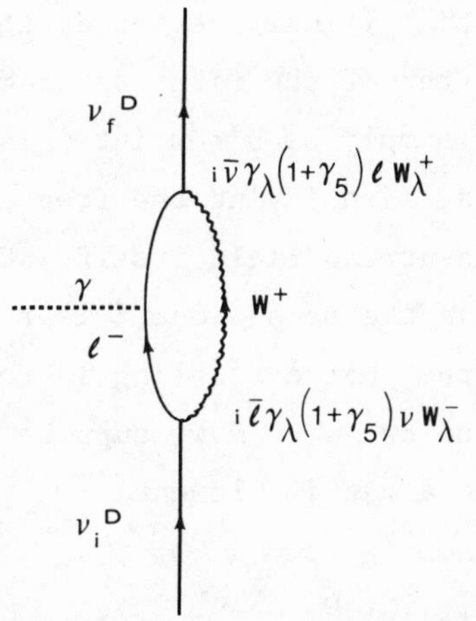

Fig. 3 One-loop diagram for the electromagnetic interaction of a Dirac neutrino ν^D with initial momentum and spin projection values i and final ones f. The term in the weak Lagrangian which is active at each vertex is written next to it.

Now, we all know that for Dirac neutrinos, form factors, and in particular the magnetic dipole moment, have been calculated in $SU(2)_L \times U(1)$ in terms of loop diagrams such as that in Fig. 3.[8] (Any such diagram is, of course, to be understood to be accompanied by another in which the photon couples to the W rather than to the ℓ.) The various form factors of the Dirac neutrino, and in particular its dipole moment, do not vanish when calculated from these loop diagrams. Therefore, one may well ask how the diagram in Fig. 3 is smart enough to know that when the external particle is replaced by a Majorana neutrino, all of the form factors forbidden to such a neutrino must now vanish. The answer to this question is that, as we noted when discussing the C properties of a Majorana neutrino, if the external particle is of Majorana character, then besides the diagrams which one has in the Dirac case, there are additional diagrams in which the incoming neutrino produces an (antileptonic) ℓ^+W^- virtual state, rather than a (leptonic) ℓ^-W^+ one. This is shown in Fig. 4, in which the term in the Lagrangian which is active at each weak vertex is indicated next to it. In the second diagram in Fig. 4, special to the Majorana neutrino, the term acting at the initial (final) vertex is the one which acted at the final (initial) vertex of the first diagram. In the second diagram, this term has simply been rewritten in terms of charge-conjugate lepton fields, identifying the free charge-conjugate neutrino field as the neutrino field itself. Comparing the two diagrams, we see that in the second one ℓ is replaced by ℓ^c and W^+ by W^-, so the sign of the photon coupling is reversed, and, in addition, γ_5 is replaced by $-\gamma_5$. Now, suppose the first diagram gives in the Dirac case a matrix element

$$\langle \nu_f^D | J_\mu^{EM} | \nu_i^D \rangle = \bar{u}_f \Gamma_\mu (\gamma_5) u_i \quad , \tag{9}$$

with $\Gamma_\mu(\gamma_5)$ an expression of the form indicated by Eq. (1), involving γ_5. Then the sum of the two diagrams will give in the

Majorana case

$$\langle \nu_f^M | J_\mu^{EM} | \nu_i^M \rangle = \bar{u}_f[\Gamma_\mu(\gamma_5) - \Gamma_\mu(-\gamma_5)]u_i \quad . \tag{10}$$

Consequently, in the Majorana case, terms that do not involve a γ_5 do not survive. In particular, then, the magnetic dipole coupling, $\sigma_{\mu\nu}q_\nu$, does not survive, nor does the "electric charge distribution" coupling γ_μ. In addition, the electric dipole term, $\sigma_{\mu\nu}q_\nu\gamma_5$, does not survive because in this model it was not present in the first diagram to begin with. It is CP violating, and there is no CP violation in these one-loop diagrams. We see that the loop diagrams do conform in a simple way to the general result that certain form factors must vanish for a ν^M.

At the beginning of this talk, it was said that in the massless limit, the difference between Majorana and Dirac behavior becomes invisible. Let us see, now, how this happens with respect to the electromagnetic interactions. It is only supposed to occur when all

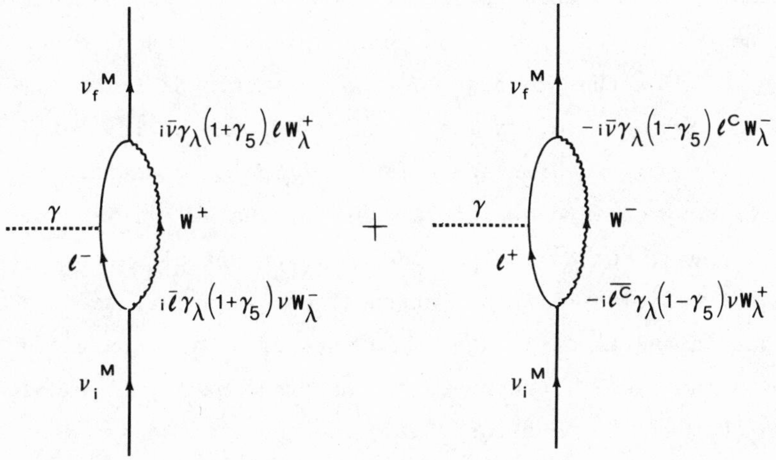

Fig. 4 One-loop diagrams for the electromagnetic interaction of a Majorana neutrino ν^M. The symbol ℓ^c denotes the charge-conjugate of the field ℓ.

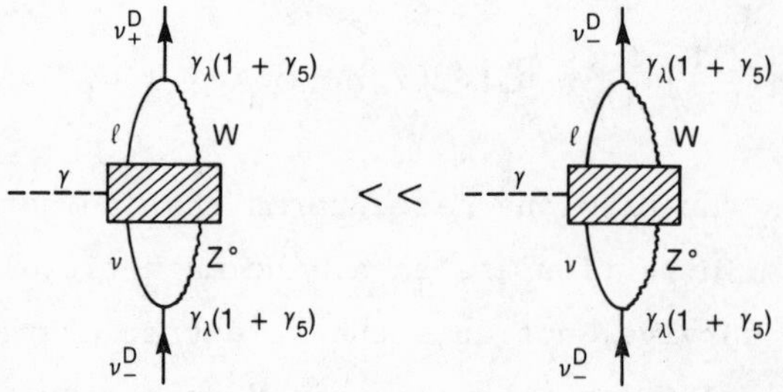

Fig. 5 Electromagnetic interactions of a highly relativistic Dirac neutrino, when all weak currents are left-handed, as indicated at the initial and final vertices. The symbol Z^o denotes the neutral weak boson, and the shaded area an arbitrary structure.

weak currents are left-handed, so we are assuming that this is the case. Let us first compare the electromagnetic interactions of the negative-helicity particles ν_-^D and ν_-^M. The interaction of a Dirac neutrino with a photon has the general structure illustrated by the diagrams in Fig. 5. The first thing that happens, when there are no right-handed currents, is that the neutrino encounters some left-handed current, either neutral or charged. Similarly, the last thing that happens is an encounter with some left-handed current, either neutral or charged. The photon plugs into some black box in the middle of the process. Now, suppose the neutrino mass m is very small compared to its momentum $|\vec{p}|$. Then, as indicated in Fig. 5, the amplitude for the helicity-flipping transition $\nu_-^D \to \nu_+^D$ is highly suppressed compared to that for the helicity-preserving one, $\nu_-^D \to \nu_-^D$, due to the handedness of the final vertex. As $m \to 0$, the helicity-flipping interactions disappear, so in particular the magnetic and electric dipole moments of a Dirac particle go to zero with the mass. Turning to the Majorana neutrino, we see from Eq. (3) that in the massless limit, the coupling to a photon becomes purely axial vector. This means, of course, that it becomes helicity-preserving in this limit. Furthermore, not only do the electromagnetic transitions of both ν_-^D and ν_-^M become helicity-preserving when m is small, but the surviving helicity-preserving amplitudes in the two cases become equal:

$$\langle \nu_-^{\;M} | J_\mu^{EM} | \nu_-^{M} \rangle \xrightarrow[m \to 0]{} \langle \nu_-^{D} | J_\mu^{EM} | \nu_-^{D} \rangle \quad . \qquad (11)$$

That is very easy to see at the one-loop level. As Fig. 4 illustrates, if there are only left-handed currents, then the currents which act in the extra diagram peculiar to the Majorana neutrino are effectively right-handed. Therefore, if the incoming and outgoing Majorana neutrinos are left-handed, this extra diagram is suppressed compared to the first one in Fig. 4, which is common to Dirac and Majorana neutrinos. Eq. (11) then follows.

In a similar way, one can show that m → 0 the electromagnetic interactions of the positive-helicity particles $\bar{\nu}_+^{D}$ and ν_+^{M} become helicity-preserving and equal. Thus, we learn that electromagnetic interactions conform to the "Practical Dirac-Majorana Confusion Theorem." When m is small, the electromagnetic interactions of a Majorana pair, (ν_-^{M}, ν_+^{M}), and those of the pair $(\nu_-^{D}, \bar{\nu}_+^{D})$ which is part of a Dirac neutrino $((\nu_-^{D}, \bar{\nu}_+^{D}), (\bar{\nu}_-^{D}, \nu_+^{D}))$, become indistinguishable.

It is amusing to ask what happens to this situation if there are both left- and right-handed weak currents. From Eq. (3), we see that the electromagnetic interactions of a Majorana neutrino still become helicity-preserving in the massless limit. However,

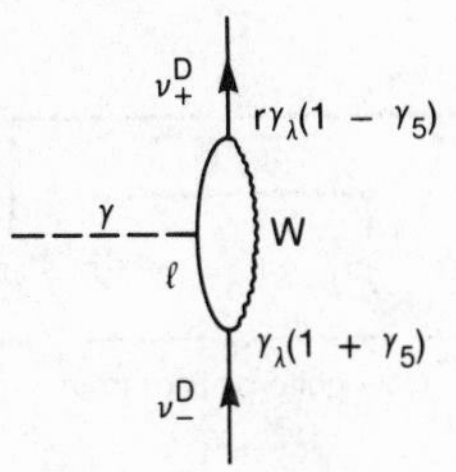

Fig. 6 A helicity-flipping transition involving the action of a right-handed current of strength r.

for a Dirac neutrino, one can now have diagrams such as that in Fig. 6, in which there is a right-handed current with some strength r at the final vertex. This diagram can continue to flip helicity, turning ν_-^D into ν_+^D with non-zero amplitude, even when $m \to 0$. The magnetic and electric dipole moments of ν^D need no longer vanish when $m = 0$, nor need the dipole form factors at $q^2 = 0$.[8] Even when $m = 0$, the particles ν_-^D and $\bar{\nu}_+^D$ can be converted into their helicity partners ν_+^D and $\bar{\nu}_-^D$ in an external $\vec{B}$ or $\vec{E}$ field (see Fig. 7). On the other hand, the Majorana neutrino has only two states, not four, as does ν^D, and the helicity of these states will not flip in $\vec{B}$ or $\vec{E}$ fields, so Majorana and Dirac neutrinos are now quite distinct, even when the mass vanishes. One can also show that when right-handed currents are present, the helicity-conserving electromagnetic amplitudes of Majorana and Dirac particles do <u>not</u> become equal in the massless limit:

$$\langle \nu_-^M | J_\mu^{EM} | \nu_-^M \rangle \not\xrightarrow[m \to 0]{} \langle \nu_-^D | J_\mu^{EM} | \nu_-^D \rangle \quad . \tag{12}$$

Thus, when there are right-handed currents, there is a big difference between a Majorana particle and a Dirac particle, even when the mass goes to zero. However, if all weak currents are left-handed, then, in spite of the fact that a Dirac neutrino has four form factors and a Majorana neutrino has only one, and a Dirac neutrino can have dipole moments while a Majorana neutrino cannot, one cannot observe these differences when the neutrino mass approaches zero.

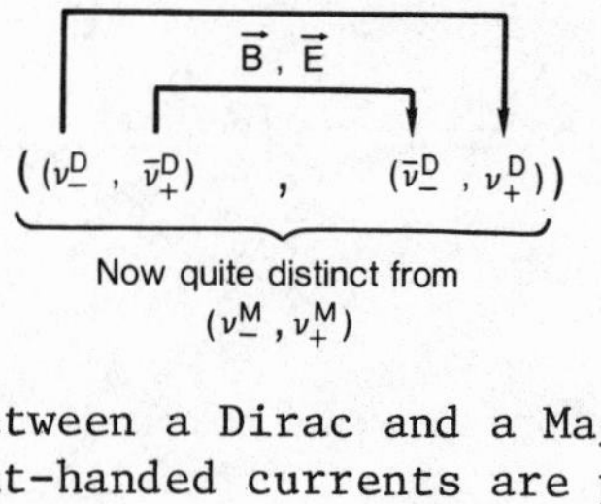

Fig. 7 Distinction between a Dirac and a Majorana neutrino when $m = 0$ but right-handed currents are present.

I would like to turn now to the neutral current interactions of Majorana and Dirac neutrinos, and discuss an analysis by Robert Shrock and myself.[9] Consider a neutral-weak process $\nu + A \to \nu + B$ involving some target A, and some system of outgoing particles B. In $SU(2)_L \times U(1)$, this process is described by a Hamiltonian

$$\mathcal{H} \sim [\bar{\nu}\gamma_\mu(1 + \gamma_5)\nu]M_\mu(A,B) \tag{13}$$

coupling a left-handed neutrino current to some other current M_μ pertaining to A and B. If the neutrino is a Majorana particle, then $\bar{\nu}\gamma_\mu\nu$, the vector part of the neutrino neutral current, vanishes. Roughly speaking, this is because $\bar{\nu}\gamma_4\nu$ is the density of neutrinos minus the density of antineutrinos, which is zero when neutrinos and antineutrinos are identical. Secondly,

$$\langle\nu|\bar{\nu}\gamma_\mu\gamma\ \nu|\nu\rangle\Big|_{\nu=\nu^M} = 2\langle\nu|\bar{\nu}\gamma_\mu\gamma\ \nu|\nu\rangle\Big|_{\nu=\nu^D} . \tag{14}$$

The reason that the matrix element of the axial vector part of the neutrino neutral current is twice as big in the Majorana case as in the Dirac one is this: In the Dirac case only the field ν can annihilate the incoming particle, and only the field $\bar{\nu}$ can create the outgoing one. However, in the Majorana case, roughly speaking, there is no difference between ν and $\bar{\nu}$, so each of these fields can play both the role of a creator and that of an annihilator, and there are two terms in the matrix element instead of one. These terms are equal, which explains the factor of two.

If there is such a big difference between Majorana and Dirac neutral weak interactions, why is it that analyses of neutral current data have not already told us whether the neutrinos are Majorana or Dirac? The answer, of course, is that nasty "Practical Dirac-Majorana Confusion Theorem" again. Let us see how it works here. For a highly relativistic left-handed neutrino ν_- with Dirac spinor u_-, we have $\gamma_5 u_- \overset{\sim}{=} u_-$. Thus,

$$<\nu_-|\bar{\nu}\gamma_\mu(1+\gamma_5)\nu|\nu_-> = \bar{u}_-\gamma_\mu(1+\gamma_5)u_- \overset{\sim}{=} 2\bar{u}_-\gamma_\mu u_-, \quad \nu = \nu^D, \tag{15.D}$$

and, in view of Eq. (14),

$$<\nu_-|\bar{\nu}\gamma_\mu(1+\gamma_5)\nu|\nu_-> = 2\bar{u}_-\gamma_\mu\gamma_5 u_- \overset{\sim}{=} 2\bar{u}_-\gamma_\mu u_-, \quad \nu = \nu^M \quad . \tag{15.M}$$

Note that the matrix element of the neutrino neutral current in the Dirac case is indistinguishable from its counterpart in the Majorana case. This is why conventional neutral current experiments cannot discriminate between the two cases.

In an effort to get around the "Practical Confusion Theorem," we might consider going to the nonrelativistic limit. Let us look, in particular, at the process $\nu + \mathrm{Nuc}(0) \rightarrow \nu + \mathrm{Nuc}(0)$, where Nuc(0) stands for a spinless nucleus. In this case (see Eq. (13))

$$M_\mu \sim (p^i_{\mathrm{Nuc}} + p^f_{\mathrm{Nuc}})_\mu \quad , \tag{16}$$

where $p^{i(f)}_{\mathrm{Nuc}}$ is the incoming (outgoing) nuclear momentum. Thus, in the nonrelativistic limit, only the fourth component of M_μ survives. By contrast, in this same limit, $\bar{\nu}\gamma_\mu\gamma_5\nu$, which is the entire neutrino neutral current in the Majorana case, has a vanishing fourth component. (Recall that $\bar{\nu}\gamma_\mu\gamma_5\nu \rightarrow (\Phi^+\vec{\sigma}\Phi, 0)$ in this limit.) Hence,

$$\frac{d\sigma}{d\Omega}(\nu^M + \mathrm{Nuc}(0)) \xrightarrow[\text{Non. Rel.}]{} 0 \ . \tag{17}$$

By comparison, one finds that

$$\frac{d\sigma}{d\Omega}(\nu^D + \mathrm{Nuc}(0)) \xrightarrow[\text{Non. Rel.}]{} \frac{G_F^2\, M^2_{\nu,\mathrm{red}}}{8\pi^2} \ , \tag{18}$$

where $M_{\nu,\mathrm{red}}$ is the reduced mass of the neutrino. Thus, the ν^M and ν^D cross sections are quite different in principle. Similarly, in the elastic scattering of nonrelativistic neutrinos from protons, the Majorana cross section is about four times the Dirac cross section. The difficulty with the nonrelativistic approach is, of course, that nobody has yet detected a nonrelativistic neutrino, much less done an experiment with one.

There is another possible way to evade the theorem, and that is by avoiding the condition in it that one works only with two incoming states, a left-handed one and its CPT-conjugate. Let us, instead, try to get our hands on right-handed relativistic neutrinos for use as incoming particles. That is, let us try to find particles which are right-handed and either Majorana, so that there is no distinction between neutrino and antineutrino, or, if they are Dirac particles, then they are in particular neutrinos as a result of the way they were made. To see how such particles may perhaps be obtained, consider the π^+ decay in Fig. 8. In π decay, the neutrino is normally very light compared to the positron, as in Fig. 8(a). Thus, if the weak currents are left-handed, the neutrino takes on its preferred negative helicity, and then the spinlessness of the pion forces the positron to be left-handed as well. Now, suppose that there is in fact a neutrino which is heavy compared to the positron, and which is produced in π decays (or in K_{e2} decays) some small (as yet undetected) fraction of the time. When such a neutrino is emitted, the positron is the most relativistic

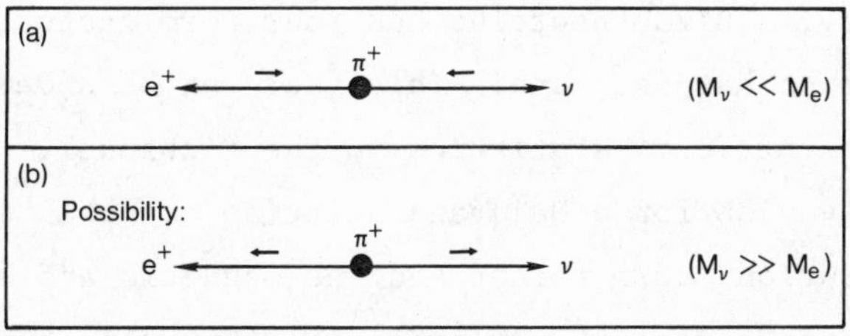

Fig. 8 a) Normal π^+ decay. The arrows above the particle lines indicate helicity.
b) A π^+ decay involving a neutrino much heavier than a positron.

particle in the decay. Then, as illustrated in Fig. 8(b), it is the positron which takes on its favored helicity, which is positive, and the neutrino is forced to have positive helicity as well. This positive-helicity neutrino is either a Majorana particle, or a Dirac particle, in which case it is a neutrino, not an antineutrino, since it was born together with a positron. If it is in fact a Dirac neutrino ν_+^D, then when it tries to interact with some target via the left-handed neutral current in Eq. (13), the handedness of the current and the helicity of the neutrino will result in a drastically reduced cross section. In fact, given the degree of positive polarization that this neutrino from π or K decay can have,[10] the suppression of the cross section could be by four orders of magnitude! By contrast, if this positive-helicity neutrino is actually a Majorana particle ν_+^M, then it is just what we normally call a right-handed antineutrino, and when interacting with matter it will do what a right-handed antineutrino always does. That is, it will interact with full weak strength. (Another way to see this is to note that when the neutrino is Majorana, the vector part of the neutrino neutral current in Eq. (13) is not present, so the projection factor $(1 + \gamma_5)$ which favors left-handed neutrinos is absent.) Thus, the interaction cross sections are dramatically different in the Majorana and Dirac cases, and one could try to look for this distinction in deep inelastic scattering experiments. Of course, for this scheme to work at all somebody has first to find a neutrino which weighs more than an electron.[10]

In summary, a Dirac neutrino has four form factors, but a Majorana neutrino has only one. This fact can be understood very easily in terms of Fermi statistics in the t-channel. The form factors which vanish for a Majorana neutrino are the "electric charge distribution" form factor and the magnetic and electric dipole form factors, and we saw explicitly how this happens in one-loop diagrams. Unfortunately, in both neutral current and electromagnetic interactions, Dirac and Majorana behavior is very similar when the neutrinos are relativistic. However, Majorana and Dirac neutrinos are really very different objects, and they behave in very different ways when, like the neutrinos all around us from the big bang, they are nonrelativistic.

REFERENCES

1. Our discussion of the electromagnetic properties, and that of Majorana-Dirac confusion, are based on B. Kayser, SLAC preprint SLAC-PUB-2879, to be published in Phys. Rev. D.
2. J. Nieves, University of Puerto Rico preprint.
3. J. Schechter and J. Valle, Phys. Rev. D 24, 1883 (1981).
4. J. Schechter, private communication.
5. R. Shrock, Stony Brook preprint ITP-SB-82-2.
6. We thank B. McKellar for giving us a draft of his paper.
7. E. Majorana, Nuovo Cimento 14, 171 (1937), and G. Racah, *ibid.*, p. 322. For a nice discussion, see P. Carruthers, *Spin and Isospin in Particle Physics* (Gordon and Breach, New York, 1971), p. 132.
8. See B. Lee and R. Shrock, Phys. Rev. D 16, 1444 (1977).
9. B. Kayser and R. Shrock, Phys. Letters 112B, 137 (1982).
10. R. Shrock, Phys. Rev. D 24, 1232 (1981).

PROTON DECAY - 1982

William J. Marciano

Brookhaven National Laboratory

Upton, New York 11973

ABSTRACT

Employing the current world average $\Lambda_{\overline{MS}} = 0.160$ GeV as input, the minimal Georgi-Glashow SU(5) model predicts $\sin^2\hat{\theta}_W(m_W) = 0.214$, $m_b/m_\tau \simeq 2.8$ and $\tau_p \simeq (0.4 \sim 12) \times 10^{29}$ yr. The first two predictions are in excellent agreement with experiment; but the implied proton lifetime is already somewhat below the present experimental bound. In this status report, uncertainties in τ_p are described and effects of appendages to the SU(5) model (such as new fermion generations, scalars, supersymmetry etc.) are examined.

I. INTRODUCTION

I presented my first theoretical overview of proton decay at the 1980 Orbis Scientiae Meeting.[1] The main theme of my talk was that several detailed renormalization group analyses of the Georgi-Glashow[2] SU(5) model had been completed and we needed only to wait for experimental verification (or negation) of its predictions[3,4] regarding $\sin^2\hat{\theta}_W(m_W)$ and τ_p, the proton lifetime. Since then, the experimental value of $\sin^2\hat{\theta}_W(m_W)$ has been lowered and rendered rather precise, primarily by the inclusion of electroweak radiative corrections.[5] It is now in remarkably good agreement with the SU(5) prediction (see Sections II and III).

In the case of the proton lifetime, using $\Lambda_{\overline{MS}} = 0.5$ GeV as input (the prevailing 1980 value of the QCD mass scale suggested by scaling violations in deep-inelastic scattering experiments), the minimal SU(5) model predicted[1] $\tau_p \approx 10^{31} \sim 10^{32}$ yr. This lifetime was only about one or two orders of magnitude larger than the 1980 experimental bound[6] $\tau_p^{exp} > 10^{29} \sim 10^{30}$ yr. So, with several new dedicated proton decay experiments just beginning, we anxiously anticipated an exciting discovery.[7] Two of those experiments have now completed initial runs and published their findings. The Homestake gold mine collaboration has reported a bound[8]

$$\tau_p \gtrsim 3 \times 10^{31} \text{ yr} \quad (\mu^+ \text{ in final state}) \qquad (1.1a)$$

for proton decays with a final state muon. In some models where proton decay is mediated by scalar mesons, one expects μ^+ rather than e^+ in the decay products and (1.1a) is then a stringent constraint. However, in the standard minimal SU(5) model roughly 10% of all proton decays lead to a μ^+. In such a theory (1.1a) becomes

$$\tau_p \gtrsim 3 \times 10^{30} \text{ yr} . \qquad (1.1b)$$

I should remark that this collaboration did observe a few proton decay candidates; but since those events were consistent with anticipated backgrounds, they interpret their results in terms of a bound. A second new experiment which was quickly assembled and began running in a remarkably short time at the Kolar gold field has published 3 proton decay candidate events.[9] Excluding one because of its proximity to the apparatus edge, they find that their two remaining events correspond to $\tau_p \simeq 8 \times 10^{30}$ yr. Several subsequent months of running have recently yielded a fourth very clean new candidate.[10] It is, however, consistent with the anticipated background due to neutrinos. A very conservative interpretation of these findings is that they roughly correspond to the bound

$$\tau_p \gtrsim 6 \times 10^{30} \text{ yr} . \tag{1.2}$$

The fact that both experiments find candidate proton decay events at a rate consistent with the 1980 SU(5) prediction should be encouraging news for advocates of Grand Unified Theories (GUTS). However, an amusing development occurred during the last two years which has had a dramatic effect on the SU(5) prediction for τ_p. Most experiments now favor a lower value for $\Lambda_{\overline{MS}}$. Indeed, the current world average[11] is about 0.16 GeV. (see Section II). Since the proton lifetime scales approximately as $\Lambda_{\overline{MS}}^4$, this reduction in $\Lambda_{\overline{MS}}$ lowers the minimal SU(5) model's prediction to $\tau_p \simeq (0.4 \sim 12) \times 10^{29}$ yr. Does this mean that the SU(5) model is now ruled out? No, because there are still other sources of uncertainty in the SU(5) prediction (see Section IV). However, it is clear that the confrontation between experiment and SU(5) theory has already reached an exciting stage. If the minimal SU(5) model is correct and $\Lambda_{\overline{MS}} \simeq 0.16$ GeV, then proton decay probably has been observed or will certainly be seen during the coming year when the Brookhaven-Irvine-Michigan experiment gets underway.[7] That experiment is capable of pushing the bound on τ_p up to $10^{32} \sim 10^{33}$ yr. and can readily detect a more rapid rate of decay. They should thereby determine the fate of SU(5).

Given the above motivation for closely scrutinizing the SU(5) model's predictions, I have organized my talk as follows: In Section II I present updated experimental values and uncertainties for α, $\sin^2\hat{\theta}_W(m_W)$ and $\Lambda_{\overline{MS}}$, the fundamental $SU(3)_C \times SU(2)_L \times U(1)$ coupling parameters. Then in Section III the SU(5) renormalization group analysis initiated by Georgi, Quinn and Weinberg[3] is reviewed. Using $\Lambda_{\overline{MS}}$ and α as input, predictions for $\sin^2\hat{\theta}_W(m_W)$ and τ_p (as well as other related parameters) are given. The minimal SU(5) model's prediction for m_b/m_τ is also illustrated. In Section IV uncertainties in τ_p are discussed and effects due to intermediate mass scalars are described. Supersymmetry implications for SU(5) predictions are outlined in Section V. Finally, in Section VI some conclusions are presented.

II. $SU(3)_C \times SU(2)_L \times U(1)$ COUPLINGS

The standard model of strong and electroweak interactions is based on the gauge group $SU(3)_C \times SU(2)_L \times U(1)$ (i.e. QCD × the Weinberg-Salam model). Associated with this model are three independent fundamental gauge couplings g_3, g_2, and g_1. They are generally reparametrized and discussed in terms of $\Lambda_{\overline{MS}}$, $\alpha = e^2/4\pi$, and $\sin^2\theta_W = e^2/g_2^2$. It is interesting that the experimental values of all three couplings have changed since 1980, in part due to the calculation of higher order radiative corrections. Because of their importance in obtaining SU(5) predictions and testing that model, I present an updated phenomenological profile of these fundamental couplings.

i) <u>The Fine Structure Constant α</u>

In 1980, the Josephson effect provided the best purely experimentally measurement of α. Now the quantized Hall effect is beginning to become competitive. The average value of α obtained from those two types of experiments is[12]

$$\alpha = 1/137.035965(12) \quad \text{(Josephson \& Hall effects).} \tag{2.1}$$

Recently, Kinoshita and Lindquist[13] presented a numerical calculation of the $O(\alpha^4)$ corrections to $(g_e - 2)/2$, the anomalous magnetic moment of the electron. Using their calculation and the very precise experimental value obtained by the University of Washington group,[14] Kinoshita and Lindquist found[13]

$$\alpha = 1/137.035993(10) \quad ((g_e - 2)/2) \quad . \tag{2.2}$$

The slight disagreement between (2.1) and (2.2) is probably a spurious effect that will eventually go away. Alternatively, it might be a faint signal of new weak interaction phenomena or perhaps a first indication of lepton structure, very speculative but exciting possibilities.

What is of interest to us (for use in the SU(5) model) is the short distance coupling $\hat{\alpha}(m_W)$, the QED running coupling defined by $\overline{\text{MS}}$ (modified minimal subtraction) using dimensional regularization with the 't Hooft unit of mass $\mu = m_W$, the W boson mass.* A renormalization group analysis gives[4,15,16]

$$\hat{\alpha}^{-1}(m_W) = \alpha^{-1} - \frac{2}{3\pi}\sum_f Q_f^{\;2}\ \ln(m_W/m_f) + \frac{1}{6\pi} + \dots , \qquad (2.3)$$

where the sum is over all fermions with mass $m_f < m_W$ and electric charge Q_f. The ... are meant to indicate higher order terms which, although not illustrated, are numerically included. For three generations of fermions with $m_t \simeq 20$ GeV and $m_W = 83$ GeV one finds from (2.1) or (2.2) and a dispersion relation analysis of $e^+e^- \rightarrow$ hadrons (to determine the "effective" light quark masses)[15,16]

$$\hat{\alpha}^{-1}(m_W) = 127.54 \pm 0.30 \quad , \qquad (2.4)$$

where the uncertainty comes primarily from the light quark sector.

ii) $\underline{\sin^2\hat{\theta}_W(m_W)}$

In the standard $SU(2)_L \times U(1)$ electroweak model $\sin^2\theta_W^0 \equiv e_0^2/g_{2_0}^2 = g_{1_0}^2/(\frac{5}{3}g_{1_0}^2 + g_{2_0}^2)$ is an infinite bare parameter. Defining the

*We choose to compare all couplings at $\mu = m_W$, which is a particularly convenient scale since in the standard theory it is the dividing point between $SU(3)_C \times SU(2)_L \times U(1)$ and the effective $SU(3)_C \times U(1)_{em}$ theory.

renormalized weak mixing angle $\sin^2\theta_W(m_W)$ by $\overline{MS}$ with $\mu = m_W$, and computing the electroweak radiative corrections, one can extract $\sin^2\hat{\theta}_W(m_W)$ from experiment. At present the most precise determination of $\sin^2\hat{\theta}_W(m_W)$ comes from a measurement of $R_\nu \equiv \sigma(\nu_\mu+N\to\nu_\mu+X)/\sigma(\nu_\mu + N \to \mu^- + X)$. After including radiative corrections one finds (averaging existing data)[5,15,17,18]

$$\sin^2\hat{\theta}_W(m_W) = 0.215 \pm 0.014 + \frac{3\alpha}{16\pi\sin^2\theta_W(m_W)}\frac{m_t^2}{m_W^2} \quad , \tag{2.5}$$

where m_t is the t quark's mass.

So, assuming that $m_t^2/m_W^2 << 1$, (2.5) becomes

$$\sin^2\hat{\theta}_W(m_W) = 0.215 \pm 0.014 \quad . \tag{2.6}$$

Note that if $m_t \gtrsim m_W$ (or if there is a large mass difference in some other as yet undiscovered weak isodoublet) $\sin^2\hat{\theta}_W(m_W)$ may be somewhat larger. From a combination of R_ν and $R_{\bar{\nu}}$ data leaving the ρ parameter free and thus allowing for such effects, one finds[18,19]

$$\rho = 1.010 \pm 0.020 \quad , \tag{2.7a}$$

$$\sin^2\hat{\theta}_W(m_W) = 0.236 \pm 0.030 \quad , \tag{2.7b}$$

which yields the (not very tight) constraint $m_t \lesssim 400$ GeV.

Employing the result in (2.6), one obtains the rather precise mass predictions for the $W^\pm$ and Z^0 bosons[4,5,15,16,20]

$$m_W = \frac{38.5 \text{ GeV}}{\sin^2\hat{\theta}_W(m_W)} = 83.0 \pm 2.4 \text{ GeV} \quad , \tag{2.8a}$$

$$m_Z = \frac{77.1 \text{ GeV}}{\sin^2\hat{\theta}_W(m_W)} = 93.8 \pm 2.0 \text{ GeV} \quad . \tag{2.8b}$$

When m_W or m_Z is precisely measured, the formulas in (2.8) (which include radiative corrections) will provide the best available determination of $\sin^2\hat{\theta}_W(m_W)$. Indeed, it should then be known to within 1% accuracy (as compared with the present-day 6% uncertainty).

iii) $\Lambda_{\overline{MS}}$ - The QCD Mass Scale

The QCD coupling $\hat{\alpha}_3(\mu) = \hat{g}_3^2(\mu)/4\pi$ is also defined by $\overline{MS}$. It is traditionally parametrized in terms of the mass scale $\Lambda_{\overline{MS}}$, such that for $m_c < \mu < m_b$ (i.e., an effective 4 flavor regime)[21]

$$\hat{\alpha}_3(\mu) = \frac{12\pi}{25\ \ln(\mu^2/\Lambda^2_{\overline{MS}})}\left[1 - \frac{462}{625}\ \frac{\ln\ \ln(\mu^2/\Lambda^2_{\overline{MS}})}{\ln(\mu^2/\Lambda^2_{\overline{MS}})}\right] . \tag{2.9}$$

Using this formula, scaling violations in deep-inelastic scattering now tend to give $\Lambda_{\overline{MS}} \simeq 0.16$ GeV.[11] In addition, Lepage and Mackenzie[22] have computed the QCD corrections to Upsilon decay. Their analysis when compared with experiment indicates $\Lambda_{\overline{MS}} \simeq 0.100$ GeV. (I personally believe that the Upsilon decay analysis presently represents the best perturbative determination of $\Lambda_{\overline{MS}}$.) At the level of nonperturbative strong coupling analysis, lattice calculations[23] of the hadronic spectrum also indicate $\Lambda_{\overline{MS}} \simeq 0.1$ GeV. Clearly, $\Lambda_{\overline{MS}}$ has been lowered considerably since 1980 when 0.5 GeV was the accepted value. In this talk I will employ the world average quoted by A. Buras[11]

$$\Lambda_{\overline{MS}} = 0.160\ ^{+0.100}_{-0.080}\ \text{GeV} , \tag{2.10}$$

which for $m_t = 20$ GeV and $m_W \simeq 83$ GeV translates into

$$\hat{\alpha}_3(m_W) = 0.1088\ ^{+0.0087}_{-0.0110} . \tag{2.11}$$

However, I emphasize that the smaller values ($\Lambda_{\overline{MS}} \simeq 0.100$ GeV

$\rightarrow \hat{\alpha}_3(m_W) \simeq 0.1015$) are beginning to be favored.

To summarize the results of this section, one finds in the standard model with $m_t = 20$ GeV and $m_W = 83$ GeV

$$\begin{aligned} \hat{\alpha}^{-1}(m_W) &= 127.54 \pm 0.30 \quad , \\ \sin^2\hat{\theta}_W(m_W) &= 0.215 \pm 0.014 \quad , \\ \hat{\alpha}_3(m_W) &= 0.1088 \begin{array}{l} + 0.0087 \\ - 0.0110 \end{array} \quad , \end{aligned} \tag{2.12}$$

or using $\hat{\alpha}_2(m_W) = \hat{\alpha}(m_W)/\sin^2\hat{\theta}_W(m_W)$ and $\hat{\alpha}_1(m_W) = 3\hat{\alpha}(m_W)/5\cos^2\hat{\theta}_W(m_W)$, the central values in Eq. (2.12) correspond to

$$\begin{aligned} \hat{\alpha}_2(m_W) &= 0.0365 \quad , \\ \hat{\alpha}_1(m_W) &= 0.0166 \quad . \end{aligned} \tag{2.13}$$

III. SU(5) PREDICTIONS

In Grand Unified Theories (GUTS) such as the Georgi-Glashow SU(5) model,[2] the three bare couplings are equal, $g_{3_0} = g_{2_0} = g_{1_0}$ (which implies $\sin^2\theta_W^0 = 3/8$). However, the effective low energy couplings $\hat{\alpha}_i(m_W) = \hat{g}_i^2(m_W)/4\pi$, $i = 1,2,3$ differ by large finite radiative corrections. Such effects can be readily computed using renormalization group techniques. This approach was initiated in the pioneering work of Georgi, Quinn, and Weinberg[3] and subsequently refined and extended by others.[1,4,15,16,24] Assuming no new masses between m_W and m_S, the superheavy unification mass scale (i.e., all particles added to the standard model in order to complete the SU(5) multiplets are taken to have mass m_S), one finds for n_g fermion generations and N_H light Higgs isodoublets (with masses $\lesssim m_W$) the following relationships[15,26]

$$\frac{\hat{\alpha}(m_W)}{\hat{\alpha}_3(m_W)} = \left[1 - \frac{66+N_H}{6}\frac{\hat{\alpha}(m_W)}{\pi}\ln\frac{m_S}{m_W} + \frac{\hat{\alpha}(m_W)}{2\pi} + \frac{\hat{\alpha}(m_W)}{4\pi}\left\{\frac{272-\frac{176}{3}n_g}{11-\frac{4}{3}n_g}\ln\frac{\hat{\alpha}_3(m_S)}{\hat{\alpha}_3(m_W)}\right.\right.$$

$$\left.\left. + \frac{-\frac{136}{3}+\frac{40}{3}n_g+\frac{11}{3}N_H}{\frac{22}{3}-\frac{4}{3}n_g-\frac{1}{6}N_H}\ln\frac{\hat{\alpha}_2(m_S)}{\hat{\alpha}_2(m_W)} + \frac{\frac{4}{3}n_g+\frac{3}{5}N_H}{\frac{4}{3}n_g-\frac{1}{10}N_H}\ln\frac{\hat{\alpha}_1(m_S)}{\hat{\alpha}_1(m_W)}\right\}\right], \quad (3.1)$$

$$\sin^2\hat{\theta}_W(m_W) = \frac{3}{8}\left[1 - \frac{110-N_H}{18}\frac{\hat{\alpha}(m_W)}{\pi}\ln\frac{m_S}{m_W} + \frac{5\hat{\alpha}(m_W)}{18\pi} + \frac{\hat{\alpha}(m_W)}{4\pi}\left\{\frac{-\frac{16}{9}n_g}{11-\frac{4}{3}n_g}\ln\frac{\hat{\alpha}_3(m_S)}{\hat{\alpha}_3(m_W)}\right.\right.$$

$$\left.\left. + \frac{\frac{680}{9}-\frac{236}{9}n_g-\frac{19N_H}{9}}{\frac{22}{3}-\frac{4}{3}n_g-\frac{1}{6}N_H}\ln\frac{\hat{\alpha}_2(m_S)}{\hat{\alpha}_2(m_W)} + \frac{\frac{16}{9}n_g-\frac{1}{5}N_H}{\frac{4}{3}n_g-\frac{1}{10}N_H}\ln\frac{\hat{\alpha}_1(m_S)}{\hat{\alpha}_1(m_W)}\right\}\right]. \quad (3.2)$$

Using the $\hat{\alpha}_i(m_W)$ in Eqs. (2.12) and (2.13) as input, one can self-consistently solve the renormalization group Eqs. for $\hat{\alpha}_i(m_S)$ (which give $\hat{\alpha}_i(m_S) = 0.0242$ for $n_g = 3$) in conjunction with either (3.1) or (3.2) and thereby determine m_S/m_W. Presently, one first employs (3.1) to find m_S/m_W and then "predicts" $\sin^2\hat{\theta}_W(m_W)$ from (3.2). However, when $\sin^2\hat{\theta}_W(m_W)$ is eventually measured to within 1%, we will reverse this procedure and predict $\Lambda_{\overline{MS}}$.

After m_S is determined, the SU(5) prediction for the proton lifetime, τ_p, can be estimated, since m_S is the physical mass of the $X^{\pm 4/3}$ and $Y^{\pm 1/3}$ gauge bosons which mediate proton decay. Including radiative enhancements,[27] one finds

$$\tau_p \simeq 2C \times 10^{-29}\,(m_S/\text{GeV})^4\ \text{yr}\ , \quad (3.3)$$

where the constant C depends on $\Lambda_{\overline{MS}}$, n_g and most importantly the particular proton decay matrix element calculation one chooses to believe. C ranges from a low of 1 found by Berezinsky, Ioffe and Kogan[28] to a high of 30 obtained independently by Donoghue[29] and

Golowich[30] (for $\Lambda_{\overline{MS}} = 0.16$ GeV and $n_g = 3$). These extremes illustrate the present degree of uncertainty in the calculation of τ_p for a given m_S. (See Section IV for further discussion.)

From the formulas in Eqs. (2.8), (3.1), (3.2) and (3.3) (with the range of C mentioned above), one finds for $n_g = 3$ and $N_H = 1$ the minimal SU(5) model's predictions illustrated in Table I. (By minimal I mean only Higgs' $\underline{5}$ and $\underline{24}$ plets are included.)[15]

The agreement between the SU(5) prediction for $\sin^2\hat{\theta}_W(m_W)$ in Table I (when $\Lambda_{\overline{MS}} = 0.16$ GeV) and the experimental value $\sin^2\hat{\theta}_W(m_W)^{exp} = 0.215 \pm 0.014$ (see Eq. (2.6)) is extremely impressive. However, the corresponding value of τ_p is below the experimental bound in Eq. (1.2). How serious is this discrepancy? That question will be addressed in Section IV.

I conclude this section by updating another of the SU(5) model's successes, its prediction for m_b/m_τ. As pointed out by Georgi and Glashow,[2] if only a $\underline{24}$ and $\underline{5}$ of Higgs are used to break the SU(5) gauge symmetry (the so-called minimal model), then the quark-lepton bare mass relations $m_d^0 = m_e^0$, $m_s^0 = m_\mu^0$, $m_b^0 = m_\tau^0$ naturally follow. These lowest order relationships are strongly renormalized by higher order self-energy corrections; a feature first studied by

Table I: Minimal SU(5) predictions for a given $\Lambda_{\overline{MS}}$

$\Lambda_{\overline{MS}}$ (GeV)	m_S (GeV)	$\sin^2\hat{\theta}_W(m_W)$	m_W (GeV)	m_Z (GeV)	τ_p (yr)
0.10	1.3×10^{14}	0.2164	82.8	93.6	$(1 \sim 30) \times 10^{28}$
0.16	2.1×10^{14}	0.2136	83.3	94.1	$(4 \sim 120) \times 10^{28}$
0.20	2.7×10^{14}	0.2124	83.5	94.3	$(1 \sim 30) \times 10^{29}$
0.40	5.5×10^{14}	0.2084	84.3	94.9	$(2 \sim 60) \times 10^{30}$
0.50	6.9×10^{14}	0.2070	84.6	95.1	$(5 \sim 150) \times 10^{30}$

Chanowitz, Ellis and Gaillard[31] and further analyzed by Buras, Ellis, Gaillard and Nanopoulos.[27] Their renormalization group analysis yields[27]

$$m_b/m_\tau \simeq 2.8 \tag{3.4}$$

for the minimal SU(5) model with 3 generations of fermions and $\Lambda_{\overline{MS}} = 0.16$ GeV. This is to be compared with the experimental value

$$m_b/m_\tau \simeq 2.6 \sim 2.9 \quad (\text{Exp}) \quad . \tag{3.5}$$

The agreement is very good ($\Lambda_{\overline{MS}} = 0.1$ GeV $\rightarrow m_b/m_\tau \simeq 2.6$). Appending additional generations tends to increase this SU(5) prediction. For a given $\Lambda_{\overline{MS}}$, each new generation of fermions (with masses $\simeq m_W$) increases the m_b/m_τ prediction by about 7%. (However, $n_g = 4$ and $\Lambda_{\overline{MS}} = 0.1$ GeV implies $m_b/m_\tau \simeq 2.8$ just as in (3.4); so one cannot strongly argue that 3 generations if favored by such an analysis.) Unfortunately, the minimal SU(5) model also predicts $m_s/m_d \simeq m_\mu/m_e = 207$ whereas current algebra estimates seem to indicate $m_s/m_d \simeq 22$. The solution to this dilemma may require the introduction of additional Higgs multiplets[2] such as a $\underline{45}$. The effect of additional Higgs scalars on SU(5) predictions will be further examined in Section IV.

IV. UNCERTAINTIES IN τ_p

The current world average $\Lambda_{\overline{MS}} = 0.160$ GeV implies a proton lifetime of about $\tau_p \approx (0.4 \sim 12) \times 10^{29}$ yr. in the minimal SU(5) model (i.e., 3 generations of fermions and $\underline{5}$ and $\underline{24}$ Higgs multiplets). This prediction is already somewhat below the present experimental bound $\tau_p^{exp} \gtrsim 6 \times 10^{30}$ yr; so it is worth re-examining the uncertainties in τ_p. That will be the topic of this section.

i) Proton Decay Matrix Elements: To determine τ_p in the SU(5) model requires several ingredients. First the effective baryon number violating Hamiltonian is calculated. It is proportional to $\hat{\alpha}_i(m_S)/m_S^2$ and includes enhancement factors due to radiative corrections.*[27] Then the proton decay maxtrix elements of the effective Hamiltonian must be evaluated and transition rates calculated. Unfortunately, there is some uncertainty regarding the wavefunction overlap of quarks inside the nucleus and opposite views regarding the importance of the three quark fusion mechanism. To illustrate the differences that exist, I have given in Table II the minimal SU(5) model predictions obtained by several distinct groups.

The bag model calculations by Donoghue and Golowich give the longest lifetimes. Their estimates are actually upper bounds since they only include two particle final states and neglect 3 quark fusion contributions. In contrast, Berezinsky et al., claim that the three quark fusion mechanism dominates and obtain from it the

Table II. Comparison of different groups' predictions for τ_p using $\Lambda_{\overline{MS}} = 0.16$ GeV $\rightarrow m_S = 2.1 \times 10^{14}$ GeV as input.

Group	τ_p(yr)
Berezinsky-Ioffe-Kogan[28]	4×10^{28}
Goldman-Ross[4]	6×10^{28}
Jarlskog-Yndurain[32]	9×10^{28}
Ellis-Gaillard-Nanopoulos-Rudaz[33]	14×10^{28}
Donoghue[29]	120×10^{28}
Golowich[30]	123×10^{28}

*The enhancement factors increase the proton decay rate by a factor of 12 for $\Lambda_{\overline{MS}} = 0.4$ GeV; however, they monotonically decrease to about 7.5 for $\Lambda_{\overline{MS}} = 0.1$ GeV. Such changes have been incorporated in the τ_p predictions of Table I.

shortest lifetime prediction illustrated in our table. The truth probably lies somewhere between these extremes. The calculations of τ_p illustrated in Table II vary by about a factor of 30. Hopefully, further theoretical analysis can reduce this uncertainty in the future.

ii) $\Lambda_{\overline{MS}}$ and $\hat{\alpha}(m_W)$: Although $\Lambda_{\overline{MS}} = 0.16$ GeV is favored, there is still some degree of uncertainty in the parameter. Using $\Lambda_{\overline{MS}} = 0.16\,^{+0.10}_{-0.08}$ GeV $\rightarrow \hat{\alpha}_3(m_W) = 0.1088\,^{+0.0087}_{-0.0110}$ and $\hat{\alpha}^{-1}(m_W) = 127.54 \pm 0.30$, one finds

$$m_S = (2.1\,^{+1.7}_{-1.2}) \times 10^{14}\ \text{GeV}\ . \tag{4.1}$$

This range in the allowed values of m_S becomes magnified in τ_p, since it is proportional to m_S^4. So the uncertainty in $\Lambda_{\overline{MS}}$ and $\hat{\alpha}(m_W)$ implies about a factor of 10 uncertainty in τ_p. Taking the geometric mean of the predictions in Table II and accepting the range for m_S in (4.1), one finds

$$\tau_p = 2 \times 10^{29\pm 2}\ \text{yr}\quad , \tag{4.2}$$

which can be consistent with experiment. However, the + 2 uncertainty in (4.2) is rather generous; so I feel that a bound of $\tau_p \gtrsim 10^{32}$ yr would rule out the minimal SU(5) model.

iii) m_t: In my analysis, I took $m_t = 20$ GeV; what if the actual value is much larger? It turns out that in a leading log approximation the SU(5) predictions are independent of m_t.[15] The next to leading log dependence is very weak. Hence, this uncertainty is insignificant. I must, however, reemphasize that a very large m_t changes the phenomenological determination of $\sin^2\hat{\theta}_W(m_W)$. (See Eq. (2.5).) The requirement that the SU(5) prediction for $\sin^2\hat{\theta}_W(m_W)$ be compatible with experiment gives the constraint $m_t \lesssim 240$GeV.

iv) N_H Higgs Doublets: If there are $N_H > 1$ light Higgs isodoublets such that the physical scalars coming from these multiplets have mass $\simeq m_W$ (all other scalars still have mass m_S), then the proton lifetime decreases by a factor of[15]

$$10^{-0.76(N_H-1)} \quad . \tag{4.3}$$

Hence, one cannot merely add new light Higgs doublets (say as parts of $\underline{5}$-plets) to the minimal SU(5) model, without worsening the conflict with τ_p^{exp}.

v) Generation Mixing: By introducing $\underline{45}$ plets (in addition to the $\underline{5}$), one relaxes constraints on the fermion mixings. A very contrived choice of mixing angles could then increase the proton lifetime τ_p.[25] (i.e., the proton might prefer to decay into a τ^+ rather than an e^+; but is kinematically unable to do so.) This scenario is too unnatural and contrived to be considered a viable possibility; so I will not consider it further.

vi) A 4th Generation: Is there a fourth generation of fermions? It has been suggested that to explain the observed baryon asymmetry of the universe within the framework of SU(5) may require an additional generation of heavy fermions.[34] What would their effect be on the low energy predictions of SU(5)? We already mentioned in Section III, that a fourth generation of fermions (with charged fermion masses $\gtrsim m_W$) increases m_b/m_τ by $\simeq 7\%$ or less; not a very significant effect. In the case of $\sin^2\hat{\theta}_W(m_W)$ and τ_p the modifications are even smaller.[15,35] A fourth generation reduces $\sin^2\hat{\theta}_W(m_W)$ by 0.0004 or less. It increases τ_p by 30% or less. Similar changes occur when a fifth generation is added (eg. For $n_g = 5$, the predictions for τ_p in Table I increase by a factor $\lesssim 2$.) Obviously, the SU(5) predictions are fairly insensitive to the addition of one or two new generations of fermions.

vii) Nuclear Physics Effects: All proton decay experiments involve nuclei rather than free nucleons. What are the nuclear physics effects on τ_p? This is a rather complicated subject. Sparrow[36] has estimated an effective suppression of about 50% in the rate for $p^+ \to e^+\pi^0$ due to real π^0 absorption. On the other hand, Dover, Goldhaber, Trueman and Chau[37] find some decay rate enhancements due to virtual π^0 absorption. One expects nuclear effects to introduce about a factor of 2 uncertainty[38] in τ_p, although some controversy regarding this matter still exists. A detailed study of nuclear physics effects will become a high priority once proton decay is truly established.

viii) Higgs Scalars: The minimal SU(5) model contains a real 24 and a complex 5 of Higgs scalars. To simplify the renormalization group analysis and get definite predictions we generally assume that all residual physical scalars have mass m_S except for the usual neutral member of the isodoublet component of the 5 which has mass $\approx m_W$. What if we relax this constraint? Or what happens if new scalar multiplets are added? I will discuss a few possibilities.

The simplest modification of the minimal SU(5) analysis is to allow the color triplet charge ± 1/3 components of the 5 to have arbitrary mass $m_H \neq m_S$. Since these scalars can mediate proton decay, they can not be too light. The bound in Eq. (1.2) implies (roughly) $m_H \gtrsim 10^{10}$ GeV. Furthermore, the Coleman-Weinberg[39] mass generation mechanism suggests $m_H/m_S \gtrsim 0.1$; otherwise fine-tuning of parameters is required. For m_H arbitrary, one finds[40] that τ_p is increased by about a factor of $(m_S/m_H)^{\frac{4}{67}}$. So, $m_S/m_H = 10$ leads to a 15% increase in τ_p. The extreme possibility $m_S/m_H \simeq 10^4$ implies a 70% increase in τ_p. The corresponding change in the weak mixing angle prediction

$$\Delta\sin^2\hat{\theta}_W(m_W) = \frac{11\hat{\alpha}(m_W)}{201\,\pi}\ln(m_H/m_S) \qquad (4.4)$$

is not very significant.

Another possibility is to add a complex 45 of Higgs scalars to the minimal model. Indeed, I already mentioned that such a 45 may be required to have the low mass fermion spectrum come out right. The 45 decomposes under $SU(3)_C \times SU(2)_L \times U(1)$ as follows:

$$\underline{45} = (1,2\ -1) + (3,1,2/3) + (3,3,2/3) + (\bar{3},1,-8/3) + (\bar{3},2,7/3) + (\bar{6},1,2/3) + (8,2,-1) \quad , \tag{4.5}$$

where the last number denotes the U(1) hypercharge with $Y = 2Q - 2T_3$. In principle, each of the 7 components in (4.5) could have different independent masses. Labeling their masses by m_i, $i = 1,2..7$ (corresponding to the order in (4.5)), I find that the 45 modifies the prediction for τ_p by a factor of (approximately)

$$\left(\frac{m_1 m_3^3 m_4^4 m_5^7}{m_2 m_6^6 m_7^8}\right)^{\frac{4}{67}} \quad , \tag{4.6}$$

for which $\sin^2\hat{\theta}_W(m_W)$ changes by

$$\Delta\sin^2\hat{\theta}_W(m_W) \simeq -\frac{\hat{\alpha}(m_W)}{804\,\pi} \ln \frac{m_1^{44}\, m_3^{534}}{m_2^{44}\, m_4^{159}\, m_5^{94}\, m_6^{197}\, m_7^{84}} \quad . \tag{4.7}$$

For all m_i degenerate, there is no change in the predictions for τ_p and $\sin^2\hat{\theta}_W(m_W)$ (in leading log approx.). However, for arbitrary m_i, the changes may be significant. What guides do we have regarding the values of m_i? One expects $m_2, m_3, m_4, m_5 \gtrsim 10^{10}$ GeV since they can mediate proton decay. In addition, the Coleman-Weinberg[39] effect suggests that no two masses should differ by more than a factor of $\simeq 10$. Using this constraint and maximizing (4.6), one finds that τ_p can increase by a factor of 8 while $\Delta\sin^2\hat{\theta}_W(m_W) = -0.002$. These are not terribly large changes. However, without the Coleman-Weinberg constraint, the modifications can be much larger. For example, if $m_6 \simeq m_W$ while all other $m_i = m_S$ (a case considered by Ibanez[41]), then τ_p increases by $\simeq 3 \times 10^4$ while $\Delta \sin^2\hat{\theta}_W(m_W) \simeq -0.017$. The increase in τ_p is significant; but so is the reduction in $\sin^2\hat{\theta}_W(m_W)$. So the weak mixing angle can provide a very useful constraint on mass hierarchies in the Higgs sector.

It has been suggested[42] that there may also be 10, 15, 50, 75 etc. multiplets of Higgs scalars. If all multiplets are approximately degenerate, the SU(5) predictions are only slightly modified. However, for totally arbitrary masses, the predictability is lost. As a working constraint, I will accept the Coleman-Weinberg induced mass effect to imply that scalar mass ratios should be $\lesssim 10$. Then one does not expect much flexibility in $\sin^2\hat{\theta}_W(m_W)$ and anticipates (at most) a factor of 10 uncertainty in τ_p.

V. SUPERSYMMETRY

A possible way of increasing the proton lifetime is to impose a supersymmetry constraint on the theory.[43] One assumes that every boson (fermion) of the standard model has a supersymmetric fermion (boson) partner which has not yet been observed. In such supersymmetric extensions of grand unified theories, coupling constant renormalizations change; hence the unification mass m_S extracted using low energy experimental data as input may be significantly altered. Since τ_p is proportional to m_S^4, it is very sensitive to such changes. Indeed, a rough estimate by Dimopoulos, Raby and Wilczek[44] found (neglecting Higgs multiplets) that $\tau_p \simeq 10^{45}$ yrs while $\sin^2\hat{\theta}_W(m_W)$ was essentially unchanged in supersymmetric extensions of SU(5). However, it was later noted that m_S and hence τ_p exhibits a very strong dependence on N_H the number of relatively light Higgs isodoublets in the model.[45] Since in realistic supersymmetric theories (if such things exist) $N_H = 2,4...$ an even number due to the anomaly cancellation requirement for their fermionic partners, τ_p generally turns out to be much smaller than the 10^{45} yr estimate and $\sin^2\hat{\theta}_W(m_W)$ is somewhat larger than the ordinary SU(5) prediction.

Table III. Predictions of supersymmetric grand unified theories. N_H is the number of light Higgs isodoublets.

	$\Lambda_{\overline{MS}}$ (GeV)	$\sin^2\hat{\theta}_W(m_W)$	m_s (GeV)	τ_p (yr)
$N_H = 2$	0.1	0.239	4.8×10^{15}	$(4 \sim 120) \times 10^{33}$
	0.2	0.235	1.1×10^{16}	$(9 \sim 270) \times 10^{34}$
	0.4	0.232	2.4×10^{16}	$(1.6 \sim 48) \times 10^{36}$
$N_H = 4$	0.1	0.260	2.6×10^{14}	$(4 \sim 120) \times 10^{28}$
	0.2	0.258	5.5×10^{14}	$(6 \sim 180) \times 10^{29}$
	0.4	0.255	1.2×10^{15}	$(1 \sim 30) \times 10^{31}$

Assuming supersymmetry breaking occurs at a mass scale of about m_W, then a detailed renormalization group analysis yields the predictions in Table III[19,45]. Of course, in specific supersymmetric models, τ_p may be substantially smaller than the estimates in Table II if superheavy Higgs induced proton decay amplitudes are significant. In any case, some values of τ_p in Table III are within experimental reach (at least for $N_H = 4$).*

The clear distinction between the SU(5) model and its supersymmetric extensions lies in their predictions for $\sin^2\hat{\theta}_W(m_W)$. The values of $\sin^2\hat{\theta}_W(m_W)$ in Table III are significantly larger than their ordinary SU(5) counterparts. Are such supersymmetric models therefore already ruled out by the experimental constraint $\sin^2\hat{\theta}_W(m_W) = 0.215 \pm 0.014$ given in Eq. (2.6)? The answer (as pointed out by G. Senjanovic and myself[19]) is yes; unless $\rho > 1$. A 1% increase in ρ manifests itself as a 4% increase in the value

*For $N_H > 6$, the prediction for τ_p is well below the experimental bound and hence not illustrated.

of $\sin^2\hat{\theta}_W(m_W)$ extracted from R_ν data.[5,46] So, the supersymmetric predictions for $\sin^2\hat{\theta}_W(m_W)$ can be reconciled with experiment if one introduces effects that cause deviations $\delta\rho \simeq 2 \sim 3\%$. One possibility is to have $m_t \gtrsim 240$ GeV (see Eq. (2.5)); there are, of course, many others. A more precise determination of ρ^{exp} and new independent measurements of $\sin^2\hat{\theta}_W(m_W)$ are clearly called for if one is to test the idea of supersymmetric grand unification.*

VI. CONCLUSIONS

The minimal SU(5) model with 3 generations of fermions, one light Higgs doublet and only two boson mass scales m_W and m_S predicts $\sin^2\hat{\theta}_W(m_W) = 0.214$ and $\tau_p \simeq (0.4 \sim 12) \times 10^{29}$ yr for $\Lambda_{\overline{MS}} = 0.16$ GeV. The weak mixing angle prediction is in excellent agreement with the experimental average $\sin^2\hat{\theta}_W(m_W)^{exp} = 0.215 \pm 0.014$; however, the implied proton lifetime is somewhat below the present bound $\tau_p^{exp} \gtrsim 6 \times 10^{30}$ yr. Taking a conservative view of the uncertainties involved in obtaining τ_p and allowing for the range of input values $\Lambda_{\overline{MS}} = 0.16^{+0.10}_{-0.08}$ GeV and $\hat{\alpha}^{-1}(m_W) = 127.54 \pm 0.30$, I found (using a geometric mean formula)

$$\tau_p \simeq 2 \times 10^{29\pm2} \text{ yr} \quad . \tag{6.1}$$

Additional uncertainties due to generation mixing, the value of m_t, new fermion generations, nuclear physics effects etc. appear to be small. Therefore, it seems that the minimal SU(5) model will be well tested by the coming proton decay experiments which will thoroughly explore the $\tau_p \simeq 10^{30} \sim 10^{32}$ regime.

---

*Of course, if supersymmetry breaking occurs at a mass scale of order m_S none of the ordinary SU(5) predictions are changed. i.e., My discussion only applies to theories with supersymmetry breakdown at relatively low energies.

Enlarging the Higgs sector of the SU(5) model with 45-plets or other more exotic representations diminishes the models predictability. For unconstrained physical scalar masses, the values of τ_p and $\sin^2\hat{\theta}_W(m_W)$ can become essentially arbitrary. However, using a hierarchy condition (i.e., the Coleman-Weinberg effect) to restrict all new physical scalar mass ratios to be $\lesssim 10$, leads to about another factor of 10 uncertainty in τ_p and a rather negligible shift in $\sin^2\hat{\theta}_W(m_W)$. This additional order of magnitude uncertainty when combined with Eq. (6.1) demonstrates the importance of pushing experiments into the more difficult region $\tau_p \simeq 10^{32} \sim 10^{33}$ yr.

Supersymmetry extensions of grand unified theories are not yet on a firm theoretical footing. In any case, rather general renormalization group analysis indicate that their predictions for τ_p tend to exhibit a very sensitive dependence on the light Higgs scalar content. So, there is no definite prediction regarding τ_p. The best way of testing these models is to carry out new precise measurements of $\sin^2\hat{\theta}_W(m_W)$ which is predicted to be larger than in ordinary SU(5). In addition, one expects that such models should have $\rho > 1$ if they are to be reconciled with deep-inelastic neutrino scattering data. A very precise determination of ρ^{exp} (say to within 1%) should be a high priority.

Finally, there are many alternative theories (such as SO(10)) which are bigger than the SU(5) model and can thus easily accommodate several intermediate mass scales between m_W and m_S. Such theories generally contain many free parameters and are, therefore, incapable of making precise predictions. They will, of course, become much more attractive if the coming generation of experiments turn out to be inconsistent with ordinary SU(5).

ACKNOWLEDGEMENT

The SU(5) calculations described in this talk were obtained in collaboration with A. Sirlin while the supersymmetry analysis was carried out with G. Senjanovic. I also wish to thank C. Dover, M. Goldhaber and P. Langacker for informative discussions and comments on the subject of proton decay.

This research was supported by the U.S. Department of Energy under Contract No. DE-AC02-76CH00016.

REFERENCES

1. W. Marciano, in "Orbis Scientiae, Recent Developments in High Energy Physics", Jan. 1980 edited by A. Perlmutter and L. Scott (Plenum, New York) p. 121.
2. H. Georgi and S. Glashow, Phys. Rev. Lett. 32, 438 (1974).
3. H. Georgi, H. Quinn and S. Weinberg, Phys. Rev. Lett. 33, 451 (1974).
4. W. Marciano, Phys. Rev. D20, 274 (1979); T. Goldman and D. Ross, Phys. Lett. 84B, 208 (1979); Nucl. Phys. B171, 273 (1980).
5. A. Sirlin and W. Marciano, Nucl. Phys. B189, 442 (1981); C. Llewellyn Smith and J. Wheater, Phys. Lett. 105B, 486 (1981).
6. J. Learned, F. Reines and A. Soni, Phys. Rev. Lett. 43, 907 (1979).
7. For a review of ongoing and proposed proton decay experiments see: "Proceedings of the Second Workshop on Grand Unification" April 1981 edited by J. Leveille, L. Sulak and D. Unger (Birkhäuser, Boston).
8. M.L. Cherry et al., Phys. Rev. Lett. 23, 1507 (1981).
9. M. Krishnaswamy et al., Phys. Lett. 106B, 339 (1981).
10. Private communication from P. Langacker.
11. A. Buras, Proceedings of the 1981 Lepton Photon Symposium, Aug. 1981, FNAL preprint 81169-THY.
12. D.C. Tsui et al., Phys. Rev. Lett. 48, 3 (1982).
13. T. Kinoshita and W. Lindquist, Phys. Rev. Lett. 47, 1573 (1981).
14. R.S. Van Dyck, Jr., P.B. Schwinberg and H. Dehmelt, Bull. Am. Phys. Soc. 24, 758 (1979).

15. A detailed description of the SU(5) calculations is given in: W. Marciano and A. Sirlin, "Proceedings of the Second Workshop on Grand Unification", see Ref. 7, p. 151.

16. W. Marciano and A. Sirlin, Phys. Rev. Lett. 46, 163 (1981).

17. W. Marciano and A. Sirlin, Phys. Rev. D22, 2695 (1980).

18. J. Kim, P. Langacker, M. Levine and H. Williams, Rev. Mod. Phys. 53, 211 (1980); I. Liede and M. Roos, Nucl. Phys. B167, 397 (1980).

19. W. Marciano and G. Senjanovic, "Predictions of Supersymmetric Grand Unified Theories" BNL preprint 30398, 1981.

20. A. Sirlin, Phys. Rev. D22, 971 (1980).

21. Cf. A. Buras, Rev. Mod. Phys. 52, 199 (1980).

22. P. Mackenzie and G.P. Lepage, Phys. Rev. Lett. 47, 1244 (1981).

23. H. Hamber and G. Parisi, Phys. Rev. Lett. 47, 1792 (1981).

24. S. Weinberg, Phys. Lett. 91B, 51 (1980); P. Binetruy and T. Schucker, Nucl. Phys. B178, 293, 307 (1981); L. Hall, Nucl. Phys. B178, 75 (1981); C. Llewellyn Smith, G. Ross and J. Wheater, Nucl. Phys. B177, 263 (1981); I. Antoniadis, C. Bouchiat and J. Iliopoulos, Phys. Lett. 97B, 367 (1980). Other references can be found in Ref. 25.

25. P. Langacker, Phys. Rep. 72, 185 (1981).

26. The results in Eqs. (3.1) and (3.2) have been updated to include the two loop contribution of Higgs scalars that was calculated by D.R.T. Jones, Phys. Rev. D25, 581 (1982).

27. A. Buras, J. Ellis, M.K. Gaillard and D. Nanopoulos, Nucl. Phys. B135, 66 (1978).

28. V. Berezinsky, I. Ioffe and Ya. Kogan, Phys. Lett. 105B, 33 (1981).

29. J. Donoghue, Phys. Lett. 72B, 53 (1977).

30. E. Golowich, Phys. Rev. D22, 1148 (1980).

31. M. Chanowitz, J. Ellis and M.K. Gaillard, Nucl. Phys. B128, 506 (1977).

32. C. Jarlskog and F. Yndurain, Nucl. Phys. B149, 29 (1979).

33. J. Ellis, M.K. Gaillard, D. Nanopoulos and S. Rudaz, Nucl. Phys. B176, 61 (1980).

34. G. Segre and M. Turner, Phys. Lett. 99B, 399 (1981).

35. M. Fischler and C. Hill, Nucl. Phys. B193, 53 (1981).

36. D. Sparrow, Phys. Rev. Lett. 44, 625 (1980).

37. C. Dover, M. Goldhaber, T.L. Trueman and L.-L. Chau, Phys. Rev. D24, 2886 (1981).

38. C. Dover, private communication.

39. S. Coleman and E. Weinberg, Phys. Rev. D7, 1888 (1973).

40. G. Cook, K. Mahanthappa and M. Sher, Phys. Lett. 91B, 369 (1981).

41. L. Ibanez, Nucl. Phys. B181, 105 (1981).

42. See Ref. 25 for a review.

43. For a review of supersymmetry see P. Fayet and S. Ferrara, Phys. Rep. 32C, 249 (1977).

44. S. Dimopoulos, S. Raby and F. Wilczek, Phys. Rev. D24, 1681 (1981).

45. S. Dimopoulos and H. Georgi, Nucl. Phys. B193, 150 (1981); L. Ibanez and G. Ross, Phys. Lett. 105B, 439 (1981); M. Einhorn and D.R.T. Jones, Univ. of Mich. preprint UMHE 81-55, (1981).

46. W. Marciano, in Particle and Fields - 1979, edited by B. Margolis and D.G. Stairs, AIP Conference Proceedings No. 59 (American Institute of Physics, N.Y., 1980) p. 373.

PROMPT NEUTRINOS: PRESENT ISSUES AND FUTURE PROSPECTS

Gianni Conforto

Istituto Nazionale di Fisica Nucleare

Firenze, Italy

I. INTRODUCTION

Prompt neutrinos are those originating from the semileptonic decays of particles whose decay lengths are much shorter than their interaction lengths in matter.

They are produced by the interactions of a high energy proton beam from an accelerator in a target, the "dump", many interaction lengths thick. Because of their long decay lengths, copiously produced long-lived particles such as pions and kaons are mainly reabsorbed in the dump, thus largely reducing the yield of ordinary neutrinos originating from their decays. This allows the detectors placed downstream of the dump to be sensitive to the much weaker sources of prompt neutrinos.

The first evidence for prompt neutrinos was obtained at CERN in two series of experiments performed in 1977 and 1979.[1-7] Figure 1 shows the lay-out of the 1977 arrangement. A 2m copper dump was used in place of the beryllium target for ordinary neutrino operation.

In neutrino detectors it is not always possible to obtain a positive identification of electron-neutrino charge current events.

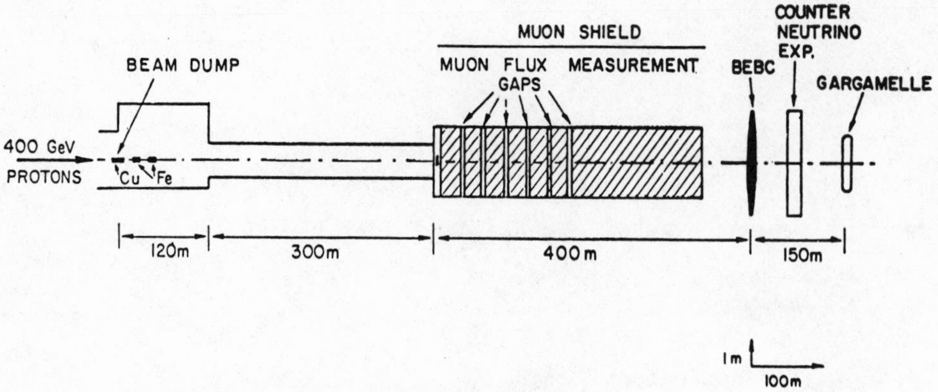

Fig. 1 Layout of the 1977 CERN prompt neutrino experiments.

In this case electron-neutrino interactions are observed as the "excess" of muonless events remaining after subtraction of those due to muon-neutrino neutral current interactions.

Prompt neutrino rates are obtained either by subtraction of the calculated nonprompt contributions or by extrapolation to infinite density of the rates observed at various dump densities.

II. PRESENT ISSUES

Prompt neutrinos are currently interpreted in terms of the production and subsequent semileptonic decays of charmed particles.

Specifically, D-mesons are taken to be centrally produced according to the invariant differential cross section

$$\frac{d^3\sigma}{dp^\sigma} \propto (1 - x_F)^n \, e^{-bp_\perp} \quad ,$$

where $x_F \equiv p_{||}^{cm}/p_{max}^{cm}$ and, typically, n=3 and b=2. Furthermore, it is assumed that charged and neutral D's are produced in p-N interactions with the same abundance as in e^+e^- collisions and consequently that the semileptonic branching ratios measured at

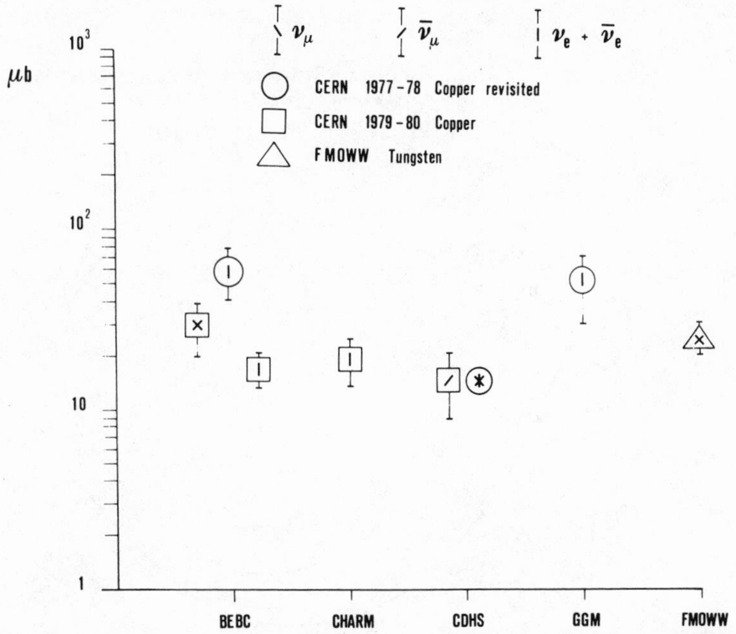

Fig. 2 Summary of the total charm production cross sections obtained in prompt neutrino experiments. Different symbols are used to represent results obtained from μ^-, μ^+ and $e^{\pm}$ events or combinations thereof. Circles indicate values derived[8] from the data of references 1-3, squares those of references 4-7 and the triangle that of reference 9. The nuclear cross section is always assumed to be proportional to A.

electron machines apply also to beam dump experiments.

This picture correctly describes the gross features of prompt neutrino production. There are however some still outstanding problems:

1) The values for the total charm production cross section derived from the CERN beam dump data (see figure 2) are somewhat larger than the upper limits for the same quantity derived[10] from 300 GeV emulsion experiments data for a charm lifetime of $2\text{-}5 \times 10^{-13}$ s (see figure 3). This could be a consequence of the assumption made for the longitudinal form of the invariant cross

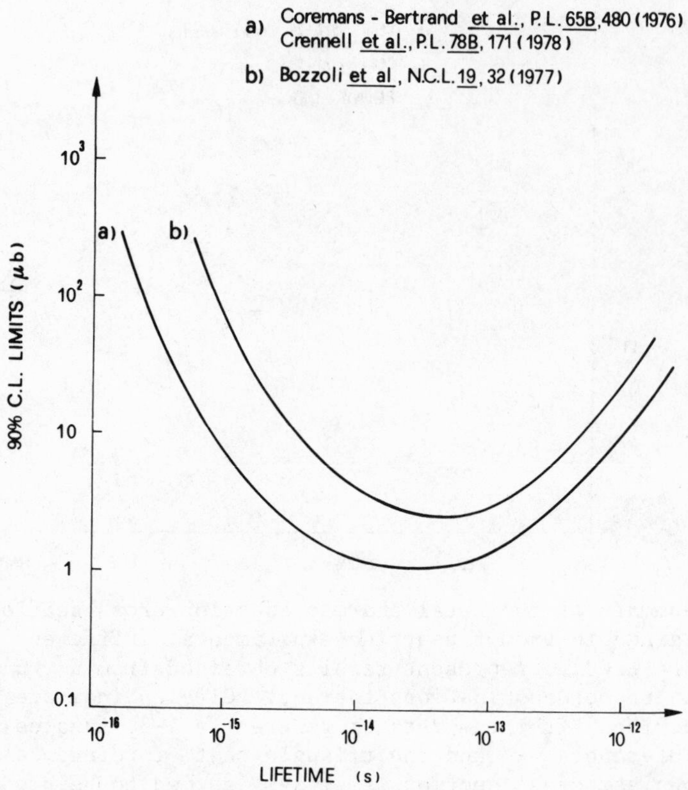

Fig. 3 90% confidence level limits to the total charm production cross section at 300 GeV as a function of the charmed particles lifetime.[10]

section and of the large dependence of the experimental value on this choice due to the very small angular acceptance of the experiments.

2) There are some questions as to whether the measured value of the prompt $\bar{\nu}_\mu$ to ν_μ flux ratio is consistent with unity as required by the central production hypothesis. The CERN ν_μ charged current results (see figure 4) are in good agreement with each other while some inconsistency seems to exists for the $\bar{\nu}_\mu$ data.

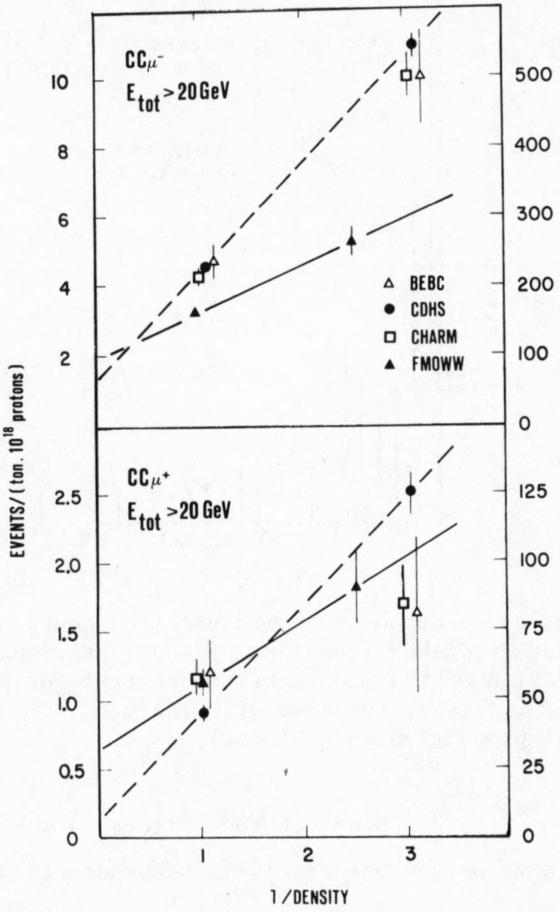

Fig. 4 The μ^- and μ^+ event rates as measured for various dump densities in the CERN experiments (BEBC, CDHS and CHARM, scale on the left) and in that of reference 9 (scale on the right).

The CDHS prompt $\bar{\nu}_\mu$ signal obtained by extrapolation to infinite density (dashed line in figure 4) is consistent with zero.

3) The energy spectrum of the prompt electron-neutrino events is not in good agreement with the prediction of the $D\bar{D}$ central production model for energies below 20 GeV (see figure 5).

4) The value of the prompt $\nu_e + \bar{\nu}_e$ to $\nu_\mu + \bar{\nu}_\mu$ flux ratio is

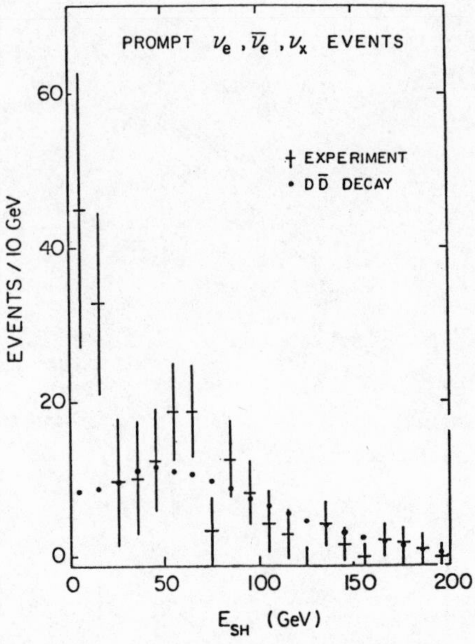

Fig. 5 The energy spectrum of prompt electron-neutrino interactions obtained by CHARM[5] from that of muonless events after subtraction of the muon-neutrino neutral current contribution (0-μ excess) compared with the prediction of the $D\bar{D}$ central production model (dots).

consistently found to be about 0.6 (see figure 6) while unity is required by electron-muon universality independently of any dynamical effect.

In recent years an experiment specifically designed to study prompt neutrinos, has been set up at Fermilab by a Firenze, Michigan, Ohio, Washington, Wisconsin (FMOWW) collaboration. The detector, an approximately 150t lead-scintillator calorimeter followed by a muon spectrometer (see figure 7), is about only 60m from the tungsten dump. As a result, the angular acceptance extends up to about 40 mrad.

Preliminary results, based on an analysis of 1-μ events for energies greater than 20 GeV, are now beginning to appear.[9]

Assuming central $D\bar{D}$ production and linear A dependence, the

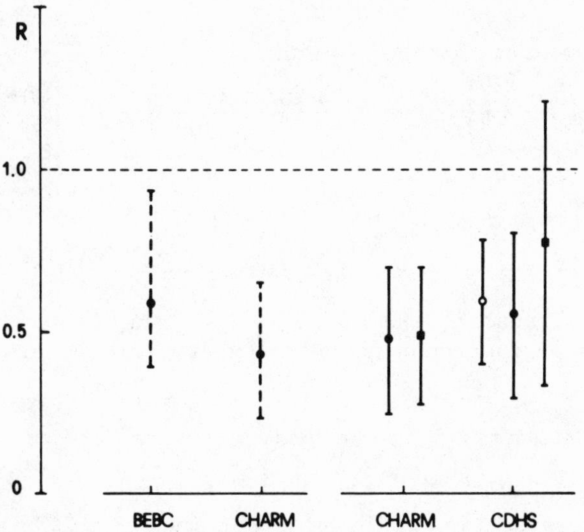

Fig. 6 The prompt $\nu_e + \bar{\nu}_e$ to $\nu_\mu + \bar{\nu}_\mu$ flux ratio as measured in the CERN experiments. Dashed-line error bars indicate results based on events in which electrons were positively identified, full-line error bars those obtained from an electron signal derived by subtraction from muonless events. Circles and squares indicated the different procedures used to determine the prompt electron and/or muon signal. Among the black points only one entry per experiment is statistically independent. The white point represents the value obtained in the 1977 CDHS experiment.

total charm production cross section is determined to be 25 ± 5 μb. This value is compared in figure 2 with those from the CERN experiments.

The μ^- and μ^+ event rates measured for the two dump densities are shown in figure 4 together with the results of the 1979 CERN experiments. The $\bar{\nu}_\mu$ prompt signal turns out to be 29 ± 9 events $/10^{16}$ protons on target, more than three standard deviations away

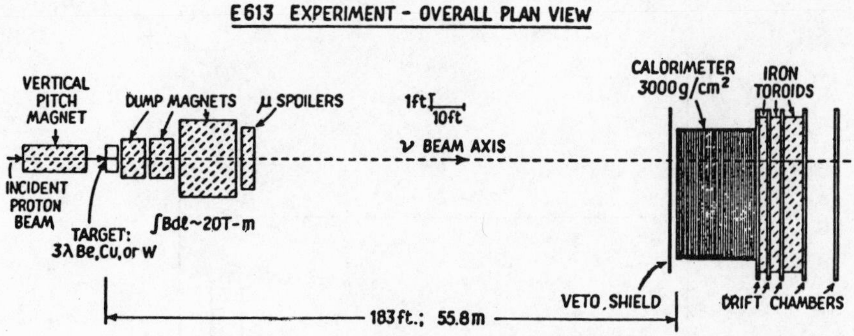

Fig. 7 The layout of the Firenze, Michigan, Ohio, Washington, Wisconsin prompt neutrino experiment at Fermilab.

from zero. The prompt $\bar{\nu}_\mu$ to ν_μ flux ratio is determined to be 0.80 ± 0.19. The agreement of the new value of the total charm production cross section with those obtained at CERN in copper and with a much smaller angular acceptance and the consistency of the prompt $\bar{\nu}_\mu$ to ν_μ flux ratio with unity are well in line with the hypotheses that charm is mainly centrally produced and that the total cross section depends linearly on A. Results from this experiment on the prompt $\nu_e + \bar{\nu}_e$ to $\nu_\mu + \bar{\nu}_\mu$ flux ratio are eagerly awaited.

III. FUTURE PROSPECTS

A second data run of the FMOWW experiment at Fermilab and another series of CERN experiments with the dump placed about 400m closer to the detector at the end of the decay tunnel (see figure 1) will take place in the first half of 1982. Thus, a wealth of new information, including perhaps the first detection of the tau-neutrino from F-meson decays,[11] will be available soon.

Work will start soon at Fermilab on the construction of the prompt neutrino beam which will become operational at Tevatron energies in a few years. The schematic layout is shown in figure 8. The full energy, full intensity proton beam from the

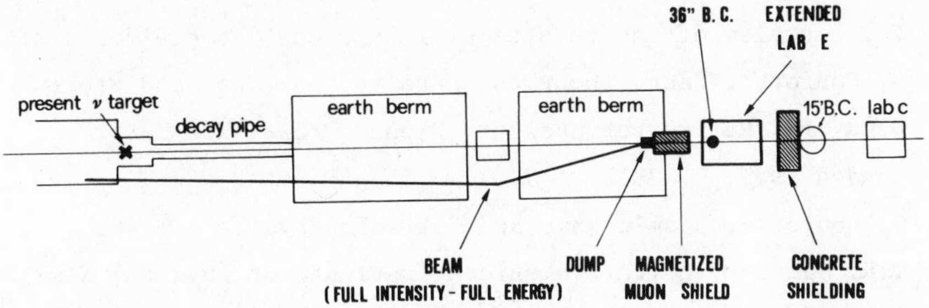

Fig. 8 The schematic layout of the prompt neutrino beam at the Fermilab Tevatron. Thick lines represent new additions to the existing neutrino area.

Tevatron is transported, by means of conventional magnets, to a variable density beryllium, copper or tungsten dump located about 200m upstream of the 15' bubble chamber. The beam can be steered in the range 0-40mrad. The magnetized muon shield downstream of the dump, similar in concept to that first used by FMOWW (see figure 7), is designed to reduce the muon flux to the low levels required for bubble chamber operation.

Two experiments have already been approved to run in this beam. One[12] uses a 36" freon high resolution bubble chamber supplemented by an external particle identifier, the other[13] the 15' bubble chamber filled with heavy neon. By detecting hundreds of tau-neutrino interactions and visible tau decays and thousands of electron-neutrino induced events, they will be able to explore extremely interesting new areas of neutrino physics.

REFERENCES

1. P. Alibran et al, Phys. Lett. 74B, 134 (1978).
2. T. Hansl et al, Phys. Lett. 74B, 139 (1978).
3. P.C. Bosetti et al, Phys. Lett. 74B, 143 (1978).
4. P. Fritze et al, Phys. Lett. 96B, 427 (1980).
5. M. Jonker et al, Phys. Lett. 96B, 435 (1980).

6. CDHS Collaboration, to be published.

7. For a review of the results of references 1-6 see for instance: G. Conforto, Gauge Theories, Massive Neutrinos and Proton Decay, A. Perlmutter, Editor, Plenum Press, New York, London 1981, p. 281.

8. G. Conforto, Cosmic Rays and Particle Physics - 1978, T.K. Gaisser, Editor, American Institute of Physics, New York 1979, p. 221.

9. B. Ball et al, Proceedings of the European Physical Society Study Conference on the Search for Charm, Beauty and Truth at High Energies, Erice, Italy, 15-22 Nov. 1981.

10. G. Conforto, Proceedings of the European Physical Society Study Conference on the Search for Charm, Beauty and Truth at High Energies, Erice, Italy, 15-22, Nov., 1981.

11. G. Myatt et al, CERN SPSC/P143.

12. I.A. Pless et al, Fermilab Experiment 636.

13. C. Baltay et al, Fermilab Experiment 646.

RECENT RESULTS ON CHARMED PARTICLE LIFETIMES AND PRODUCTION

R.N. Diamond

Florida State University

Tallahassee, Florida 32306

ABSTRACT

Recent results on the lifetimes of charmed particles observed in emulsion and high resolution bubble chambers are discussed and compared with theoretical expectations. While experiment BC73 at SLAC observes charged-to-neutral lifetime ratio much closer to unity than do the other experiments, all experiments are consistent with $\tau_{D^{\pm}} / \tau_{D^{o}} = 2.5$. Pair rather than associated production seems to be the dominant charmed particle production mechanism. Photon-gluon fusion models are able to predict the charm photoproduction cross section.

In 1979, at the time of the previous Lepton Photon Conference, there were a number of interesting questions concerning charmed particle production:

1. What were the charmed particle production cross sections?
2. What were the lifetimes of the charmed particles?
3. What were the production mechanisms involved in charmed particle production? In particular, were charmed particles produced primarily as pairs ($D\bar{D}$) or via associated productions $D\Lambda_c$?

4. What were the decay modes of the charmed particles?

In 1982 these questions are still interesting. I will touch on the first three questions. The decay modes of charmed particles are discussed in the excellent review article by Trilling.[1] The decay modes which will be considered are primarily the hadronic multiprong decay modes:

$$D^0(\bar{D}^0) \rightarrow K^-\pi^+ \quad (K^+\pi^-)$$
$$(n\pi^0)$$
$$K^\pm\pi^\mp\pi^+\pi^-$$
$$D^\pm \rightarrow K^\mp\pi^\pm\pi^\pm \quad (\pi^+\pi^- n\pi^0)$$
$$\Lambda_c^+ \rightarrow \Lambda\pi^+\pi^+\pi^-$$
$$K^0\ p\ \pi^+\pi^- \quad .$$

These decays are described by the socalled spectator model shown in fig. 1a in which the charmed quark decays to a strange quark via W boson radiation. This diagram, when compared with the diagram for muon decay shown in fig. 1b, yields a decay rate given by $\Gamma c/\Gamma\mu \sim 5(M_c/M_\mu)^5$. The second factor is a phase space factor while the first factor arises from the three quark colors and two leptons which the W boson produces. Using a quark mass $M_c = 1.8$ GeV leads to a charm lifetime $\tau_c \sim 2\times10^{-13}$ sec. Since the spectator quark flavor is immaterial for the lifetime calculation based on fig. 1a, the first order result is that $\tau_{D^\pm}=\tau_{D^0}$.

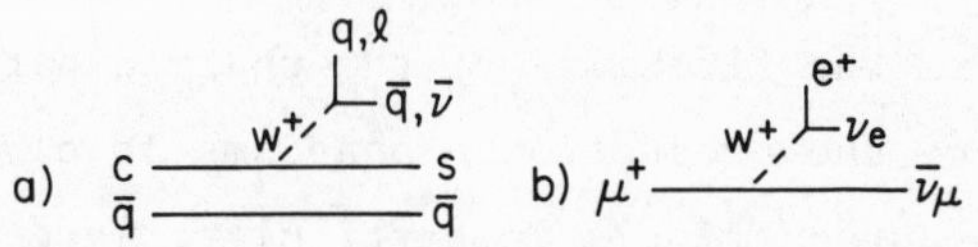

Fig. 1 Spectator diagrams describing charmed particle decays.

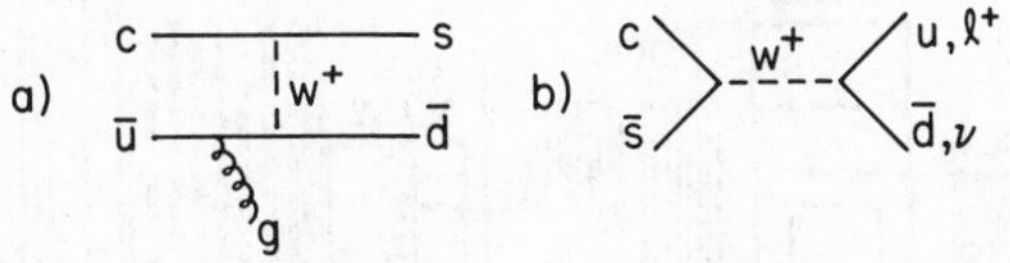

Fig. 2 Additional diagrams contributing to (a) D^o decays (b) $F^\pm$ decays.

Early experimental results on charmed particle lifetimes indicated that the charged D lifetime might be as much as an order of magnitude greater than the neutral D lifetime. This necessitated theoretical consideration of other diagrams. The W exchange diagram in fig. 2a which requires gluon emission in order to overcome helicity suppression is only allowed for the D^o and would imply that $\tau_{D^o}<\tau_{D^\pm}$. Similarly the annihilation diagram in fig. 2b, which is only allowed for $F^\pm$ decay, would imply an $F^\pm$ lifetime less than the $D^\pm$ lifetime, and perhaps comparable with that of the D^o.

Experiments which directly measure charmed particle lifetimes have involved two major techniques: nuclear emulsions and high resolution bubble chambers. The vertex detectors are usually followed by a downstream spectrometer which identifies the events and provides the kinematic information for event reconstruction.

WA17 is an emulsion experiment[2] at CERN using a wide band neutrino beam. The emulsion stack is followed by BEBC which is equipped with an external muon identifier. From a 205K picture exposure, 935 predictions of neutrino interactions were made and 169 charged current events were located in the emulsion. Three D^o decays and 2D^+ decays with lifetimes of $053^{+.57}_{-.25}\times10^{-13}$ sec and $2.5^{+2.2}_{-1.1}\times10^{-13}$ sec respectively were observed yielding a lifetime ratio $\tau_{D^+}/\tau_{D^o} = 4.7^{+12}_{-3.5}$. WA58 , an emulsion experiment[3] also at CERN, uses a tagged photon beam with energy ranging from 20 to 80 GeV and the Omega Prime spectrometer. In 1980 the experiment reported 3 D^o decays with a mean lifetime of 0.5×10^{-13} sec. Now, with six

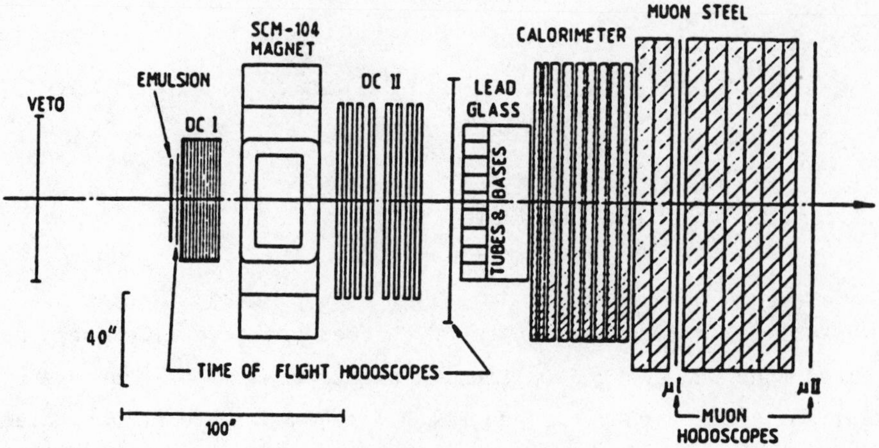

Fig. 3 Spectrometer system for experiment E531.

double decay events found from the Omega Prime predictions, of which two are not yet fully analyzed, the reported lifetime ratio is $\tau_{D^\pm}/\tau_{D^0} = 2.7^{+2.4}_{-1.8}$ ($\tau_{D^\pm} = 3.56\pm1.26\times10^{-13}$ sec and $\tau_{D^0}=1.34^{+1.2}_{-0.4}\times10^{-13}$ sec). In both these experiments, because of ambiguities in reconstructing the decays, the quoted errors should probably be taken with a grain of salt.

The emulsion experiment with the largest statistics is experiment E531, a U.S., Japanese, Korean, and Canadian collaboration operating in the Fermilab wide band neutrino beam. The detector is 23ℓ. of emulsion followed by the downstream spectrometer system shown in fig. 3 which consists of drift chambers before and after the spectrometer magnet, time of flight counter for particle identification, lead glass for photon or electron detection, hadron calorimeter, and muon filter. The emulsion stack has two accurately positioned fiducial sheets of emulsion which are changed often enough so that the density of background tracks is low. Tracks in the drift chambers, as shown in fig. 4, are extrapolated into the emulsion to within 300 μm and measurements in the fiducial sheets

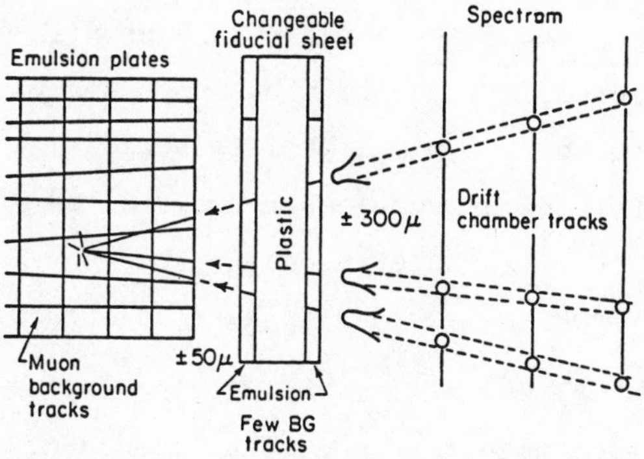

Fig. 4 Detail of the E531 scan back system for finding events in the emulsion.

to within 50μm of the primary vertex.

Early results from E531 indicates[4] a charged D lifetime $\tau_{D^\pm}=10.3^{+10.5}_{-4.1}\times10^{-13}$ sec based on 5 decays and neutral D lifetime $\tau_{D^o}=1.0^{+0.53}_{-0.31}\times10^{-13}$ sec based on 7 decays. The charged decays are located by following the charged tracks from the primary vertex, but the neutral decays are found in a volume scan of radius .3mm and length 1mm downstream. The suspicion was that longer lived D^o's were systematically missed, and indeed recent improvements making use of the drift chambers to locate D^o's has resulted in finding 11 more decays. The "fitted" D^o decays balance transverse momentum but generally have one or more particles not identified by time-of-flight, etc., and most of the 3-constraint fits have a π^o in the lead glass which could be from the primary vertex. Eighteen $D^o(\bar{D}^o)$ decays including three semileptonic decays have a mean lifetime of $3.29^{+.99}_{-.77}\times10^{-13}$ sec. With the same 5 charged decays, the lifetime ratio is now $\tau_{D^\pm}/\tau_{D^o} = 3.1^{+5.0}_{-1.7}$. One wonders if even longer lived neutral particles might be found. E531 also reports the lifetimes $\tau_{F^\pm} = 1.48^{+1.77}_{-0.81}\times10^{-13}$ sec and $\tau\Lambda_c^+ = 1.39^{+.76}_{-.44}\times10^{-13}$ sec.

Two bubble chamber experiments from CERN have reported on charmed particle lifetimes. NA16 , using LEBC[5], followed by the

European Hybrid Spectrometer, was exposed to a beam of 360 GeV π^- and protons. The chamber is operated at 33Hz with 40μm diameter bubbles. Fifty percent of the film has been analyzed, yielding 10-12 events/μb for both π^-p and pp interactions. Seven events have been found with both decay vertices reconstructed and seven with one vertex reconstructed. The eleven charged D's have a lifetime of $\tau_{D^\pm}=9.3^{+4.6}_{-2.3}\times10^{-13}$ sec and seven D^0's have a lifetime of $\tau_{D^0}=3.0^{+1.8}_{-0.8}\times10^{-13}$ sec which gives a lifetime ratio $\tau_{D^\pm}/\tau_{D^0} = 3.1^{+3.1}_{-1.7}$. It is interesting to note that one long lived neutral particle which is consistent with $D^0 \to K^-\pi^+\pi^0$, having a lifetime of 31×10^{-13} sec, has been omitted from the calculation of τ_{D^0}. Including it would raise τ_{D^0} to ~ 7.6×10^{-13} sec, but I have heard recently[5] that this decay may in fact be a poorly measured K^0. NA18, A Bern-Munich collaboration,uses a 6.5cm diameter bubble chamber, BIBC, filled with heavy liquid. The 20μm diameter bubbles hold out the possibility of high resolution, but there is much background due to the heavy liquid. Their lifetimes, based on only a few decays are $\tau_{D^\pm}$ ~ 19.8×10^{-13} sec and τ_{D^0} ~ 12.3×10^{-13} sec.

The experiment with which I have been associated is BC73, a large collaboration involving physicists from U.S., Japanese, British, and Israeli institutions. The experiment uses the SLAC Hybrid Facility (SHF) and the SLAC 40 in. hydrogen bubble chamber, which is equipped with a high resolution camera and exposed to a backscattered laser beam. The beam, shown in fig. 5, is obtained by backward scattering a photon beam at an angle of 2 mrad from the primary 30 GeV electron beam to produce a nearly monoenergetic 20 GeV photon beam (FWHM ~ 2 GeV) which passes through a quadrant detector for positioning, a pair spectrometer for monitoring the beam flux and energy, and into the bubble chambers. The SHF shown in fig. 6 consists of downstream PWC's, two gas Cerenkov counters, and lead glass array,[7] all the components of which have a deadened region along the beamline so that e^+e^- conversion pairs which fan out in the vertical beam plane do not swamp the detectors. The

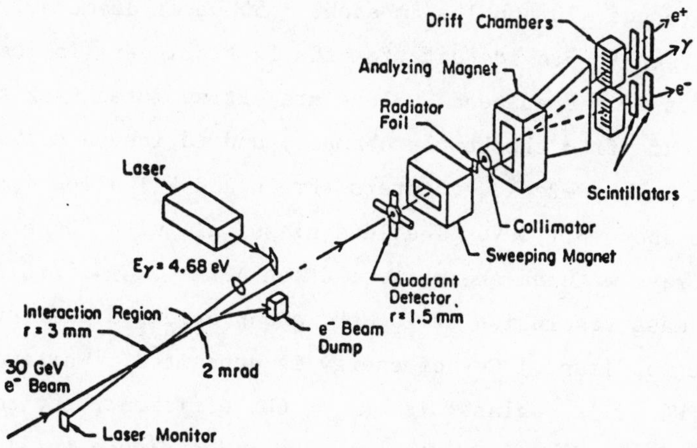

Fig. 5 The SLAC BC73 backscattered laser beam producing monochromatic plane polarized 20 GeV γ rays.

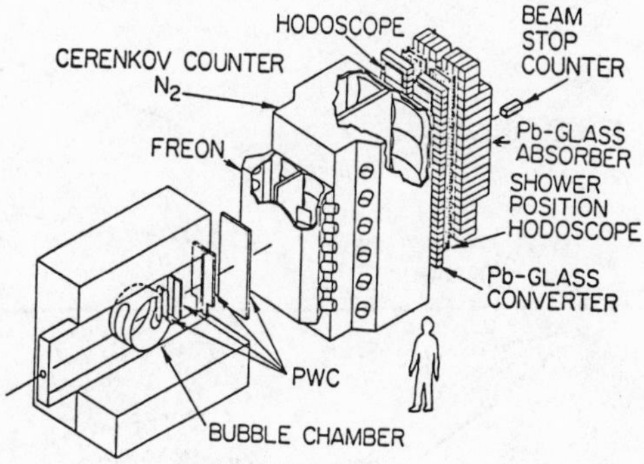

Fig. 6 The SLAC 40 in. bubble chamber hybrid facility (SHF) used for BC73.

photon beam of ~25 γ/pulse is plane polarized, which holds out the possibility of our being able to measure the asymmetry in the charmed particle production plane, a QCD effect predicted by Duke and Owens.[8]

The 40 in. chamber is run at 10Hz at a temperature of 29K

(hot) to produce ~70 bubbles/cm each, ~ 55 μm in diameter. The PWC's are used to provide data for the fast trigger (in less than 150μsec hits in the nonbend γ plane are extrapolated back to see if a "track" came from the bubble chamber) and to increase the momentum resolution. The Cerenkov counters (Freon and N_2) allow K,π separation above 3.1 GeV. The lead glass wall with 204 blocks detects γ rays with an energy resolution $\Delta E/E=(.84+4.8/\sqrt{E})\%$ and π^o's with mass resolution of 10 MeV, and provides a fast trigger signal when at least 1 GeV of energy is deposited. While each trigger (PWC or lead glass) is only ~ 60% efficient, the combined efficiency for hadronic events is ~ 90% and yields a total hadronic cross section of 100 μb. Monte Carlo estimates for various charm channels indicate an efficiency of $80\pm10\%$. At this time 1.2M pictures with 205K hadronic events have been scanned for charmed particles. By March, 1982 we shall have an additional 1.2M pictures to scan for charm.

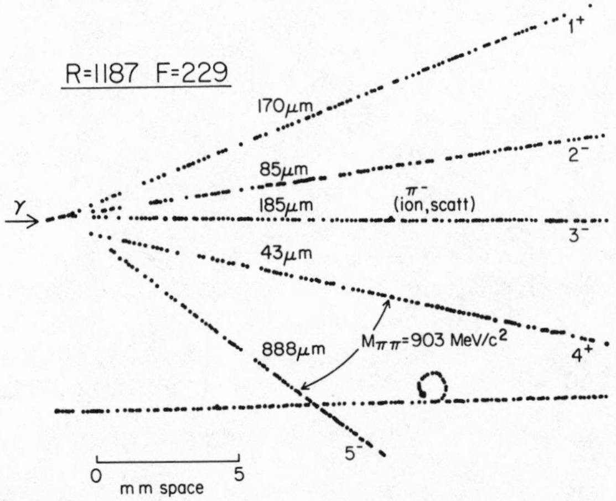

Fig. 7 An event in BC73 with two charmed particle decays: a three-prong decay of a charged particle and a two prong neutral decay. Neither decay is fully reconstructed because of missing neutrals. Each track is labelled with its impact distance from the primary vertex. The neutral decay distance is 1.8mm ($\tau=23\times10^{-13}$ sec) and the charged decay distance is 0.9mm ($\tau=6\times10^{-13}$ sec).

Figure 7 is an example of the charmed particle events for which we are scanning. Two decay vertices are clearly seen and neither decay is consistent with being strange particle decays. However, typical of charmed decays, missing neutrals are required before the charmed particles can be reconstructed. Each track in fig. 7 is labelled with an impact distance which is its distance of closest approach to the primary vertex.

The basic scanning philosophy is to look for clear decay vertices and tracks which have an impact distance greater than two bubble diameters. This figure was arrived at via Monte Carlo simulation of bubble chamber film. Serious charmed particle candidates were measured on the conventional 70mm film (300μm resolution) and hybridized with PWC information in order to obtain the kinematic quantities. They were also measured on the 35mm high resolution optics film (50μm resolution) for impact parameter studies. Twenty-nine events have been found with good evidence of charm, corresponding to a cross section $\sigma_c = N_c \sigma_h / \varepsilon N_h = 40^{+40}_{-20}$ nb. Eleven of these events have two decays detected and 18 events have one decay detected.

In order to clean up the data, we apply three cuts as depicted in fig. 8, in which a charged track has a 3-prong decay. For each decay track we calculate the impact distance from the primary vertex and require: that the track with the greatest impact distance, d^{max}, have $d^{mas} > 110\mu m$, that the track with the second greatest impact parameter, d^2, have $d^2 > 40\mu m$, and that the decay length ℓ be greater than 500μm. The first cut assures reasonable detection efficiency, the second cut assures that the decay is a multiprong decay, and the third cut removes most of the ambiguities between charged and neutral decays. There are 21 events which pass these cuts, 14 with one well-defined vertex, and 7 with two well-defined vertices, and of these decays 9 charged, 11 neutral, and 3 ambiguous decays are used in the lifetime study.

Assuming an exponential distribution in decay times, the mean

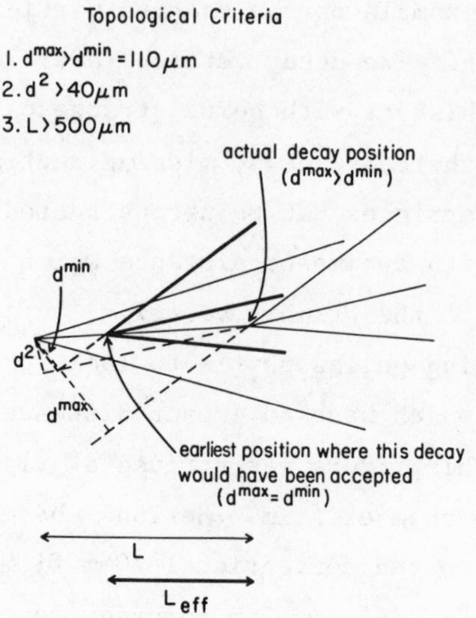

Fig. 8 Illustration of topological criteria for selecting events for charmed particle lifetime studies. The darkened tracks are the projection of the three prong decay to the earliest position at which it would have passed our selection criteria.

decay time $<\tau>$ given by the equation

$$<\tau> = \frac{\ell}{N} \cdot \sum_{i=1}^{N} [(1/p)\ m/c]_i$$

is the best estimator of the lifetime. To obtain $<\tau>$ one must know the momentum, mass, and decay length on an event by event basis. While our experiment is not sensitive to very short decays our mean lifetime will not be biased if instead of the actual decay length, ℓ, we use an effective length, ℓ_{eff}. As shown in fig. 8, ℓ_{eff} is the distance to the observed decay from the closest point at which that decay could have occurred and satisfied our selection criteria. The charmed particle momentum is usually not fully determined but bounds can often be set and an average of P_{max} and P_{min} can be used.

The charmed particle lifetime can be estimated in other ways. The maximum impact distance d^{max} is almost model independent, provided the decay is isotropic and the charm mass is much greater than the decay masses. Monte Carlo programs have been used to generate events according to a variety of models with the mean charm lifetime as a parameter. Comparisons of the quantities $<\tau^{eff}>$, $<d^{max}>$, $<d^{max}>$, $<\ell>$, and $<\ell^{eff}>$, with the data yield the lifetime, shown in Table I. This approach has the advantage that all decays, not just the fully constrained decays, can be used. It is interesting to note that while the lifetimes, obtained in this manner vary by as much as 30% depending on the variable being compared, the lifetime ratio $\tau_{D^\pm}/\tau_{D^0}$ does not vary nearly so much. The results obtained are

Table I

Charmed Particle Lifetimes
Determined from Various
Monte Carlo Models

Variable	$\tau_{D^\pm}(\times 10^{-13}\text{sec})$	$\tau_{D^0}(\times 10^{-13}\text{sec})$	$\tau_{D^\pm}/\tau_{D^0}$
$<\tau_{eff}>$	6.8	6.4	1.1
$<d^{max}>$	9.6	7.4	1.3
$<\ell>$	8.4	6.8	1.2
$<\ell_{eff}>$	8.1	6.7	1.2
Constrained decays	$8.2^{+6.5}_{-2.5}$	$9.4^{+12.3}_{-4.0}$	0.9

$$\tau_{D^o} = 6.7^{+3.0}_{-2.5}\times10^{-13} \text{ sec} \quad ,$$

$$\tau_{D^\pm} = 8.2^{+4.5}_{-3.0}\times10^{-13} \text{ sec} \quad ,$$

and
$$\tau_{D^\pm}/\tau_{D^o} = 1.2^{+1.0}_{-3.0} \quad .$$

The lifetimes obtained from a smaller sample of fully constrained decays is also shown in Table I. Figure 9 shows the charmed particle lifetimes obtained using the effective length ℓ_{eff} and best estimate of the momentum. The lifetime curves obtained from the Monte Carlo comparison are also shown. A graphical comparison of charmed particle lifetimes obtained by BC73 and the other experiments is shown in fig. 10. The more recent experiments generally agree on the $D^\pm$ lifetime but BC73 reports a longer D^o lifetime than do the others and hence a lower lifetime ratio.

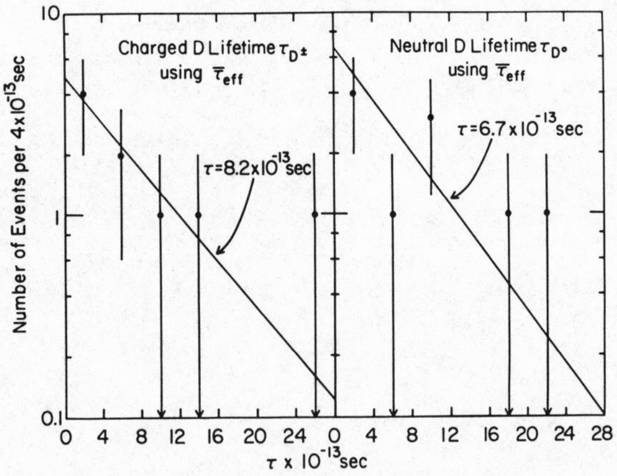

Fig. 9 Lifetime distributions for charged and neutral charmed decays. The curves are derived from Monte Carlo simulations of such variables as $\langle d^{max}\rangle$, $\langle \tau_{eff}\rangle$, $\langle \ell\rangle$, and $\langle \ell_{eff}\rangle$ (see text).

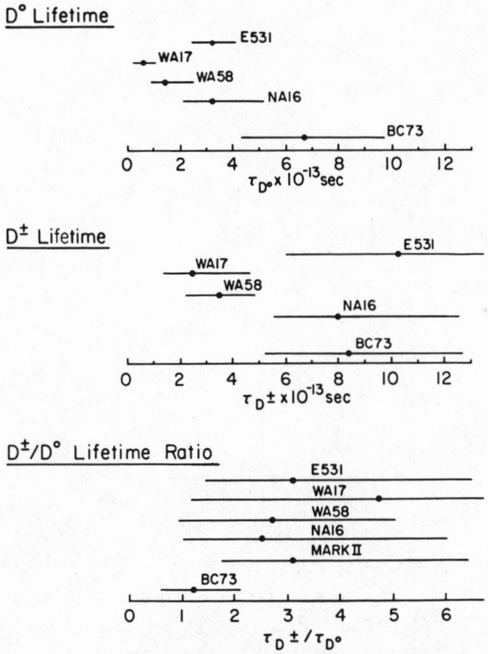

Fig. 10 Comparison of charmed particle lifetime measurements between experiments.

The lifetime ratios from all the experiments, including a result from Mark II, based on the semileptonic D decays, are all consistent with a ratio of 2.5 within 1.5 - 2 standard deviations.

All the charged decays detected by BC73 are consistent with being either $D^{\pm}$ or $F^{\pm}$ and some of the charged decays are also consistent with Λ_c^+. There is, however, no compelling evidence for a large amount of Λ_c^+ production. The momentum spectrum of the D's seem to be more consistent with $D\bar{D}$ production. This is seen in fig. 11, where the visible momentum of the neutral and charged D's is plotted and where two curves representing specific models of $D\bar{D}$ and $D(\Lambda_c^+$ or $\Sigma_c^{++})$ production are also plotted. This is interesting because the CIF experiment[10] using a tagged photon beam with energies ranging from 50 GeV to 200 GeV favored a pair production

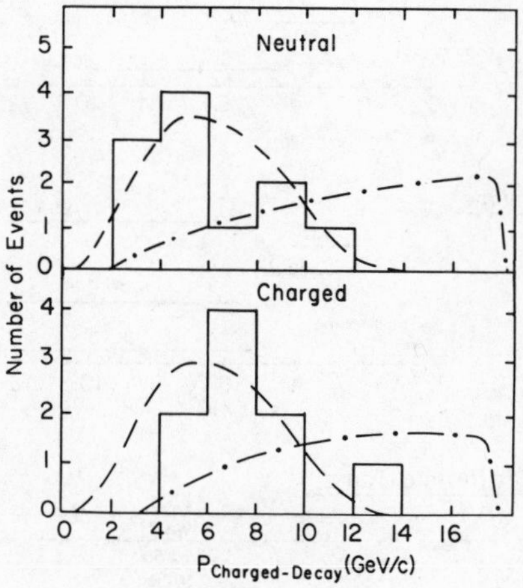

Fig. 11 Visible momentum distribution for charmed particle decays. The dotted curve comes from an associated production model ($\gamma\rho \to \bar{D}^0 \Lambda_c^+$ or $D^- E_c^{++}$) with momentum transfer distribution. The dashed curve results from a pair production model ($\gamma\rho \to D\bar{D}\, N^*$) with e^t momentum transfer distribution.

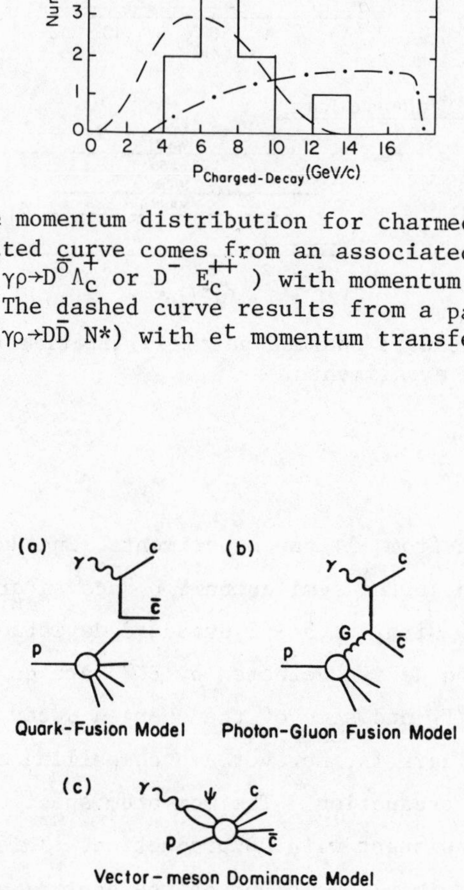

Fig. 12 Three models of charm photoproduction: a. quark-fusion, b. photon-gluon fusion, and c. vector meson dominance.

mechanism, while the WA4 collaboration,[11] with photon energies in the range 40-70 GeV, favored associated production.

Three theoretical models which have been introduced to account for charm photoproduction are shown in fig. 12: the quark fusion model,[12] the photon-gluon fusion model,[13] and the vector-meson dominance model.[14] The model predictions are shown in fig. 13 along with the inclusive charm cross section data from a number of photoproduction experiments ranging in photon energy from ~20 GeV to ~240 GeV. Two experiments, BFP[15] and EMC,[16] are muon beam experiments and the beam consists of virtual photons. The quark fusion model gives too high a cross section at the BC73 energy. The photon-gluon fusion model appears to fit the data reasonably well, but because of the large uncertainties, the data do not yet rule out the predictions of different versions of the model [the naive model has the gluon distribution $xg(x) \sim (1-x)^5$].

In summary, it appears that the expectations of QCD for charmed particles production are satisfied by the data. Charmed particle

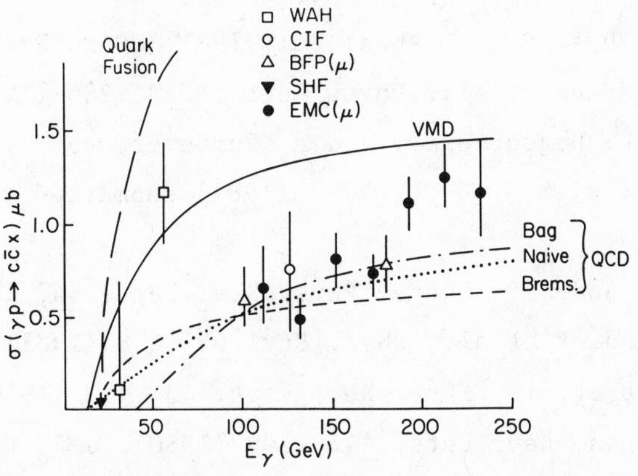

Fig. 13 Charm photoproduction cross section as a function of energy. The curves are predictions of the quark fusion,[12] photon-gluon fusion,[13] and vector meson dominance[14] models.

lifetimes are consistent with naive expectations, and the measured lifetime ratio $\tau_{D^{\pm}}/\tau_{D^{0}}$ is closer now to those expectations than it was two years ago. The photon-gluon fusion model predicts charm photoproduction cross sections. Recent results from BC73 indicate that charmed particle production is predominately pair production rather than associated production. Further advances in charmed particle production undoubtedly await a technological innovation such as silicon microstrip detectors to dramatically increase the number of charmed particle events and better determine their lifetimes and production mechanism.

REFERENCES

1. G. Trilling, Phys. Rept. 75, 57 (1981).
2. D. Allasia et al., Nucl. Phys. B176, 13 (1980).
3. M.I. Adamovich et al., Phys. Lett. 99B, 271 (1981), 89B, 427 (1980), A. Fiorino et al., Lett. Al Nuovo Cim. 30, 166 (1981).
4. N. Ushida et al., Phys. Rev. Lett. 45, 1049, 1053 (1980).
5. M. Aguilar-Benetz et al., "Charm particle production in 360 GeV $\pi^{-}p$ and 360 GeV pp Interactions", paper submitted to European Phys. Soc. Lisbon, July 1981, CERN EP/81, June 15, 1982, B. Adeva et al., Phys. Lett. 102B, 285 (1981).
6. B. Hahn, E. Hugentobler, and E. Ramseyer, Bern preprint.
7. J. Brau et al., SLAC-PUB-2773 (1981) submitted to Nucl. Instr. and Meth.
8. D.W. Duke and J.F. Owens, Phys. Rev. Lett. 44, 1173 (1980).
9. R.H. Schindler et al., Phys. Rev. D24, 78 (1981).
10. M.S. Atiya et al., Phys. Rev. Lett. 43, 414 (1979), P. Avery et al., Phys. Rev. Lett. 44, 1309 (1980), J.J. Russell et al., Phys. Rev. Lett. 46, 799 (1981).

11. F. Richard, talk given at the Int. Symposium on Lepton and Photon Interactions at High Energies, 1979, p. 469, eds. T.B.W. Kirk and H.D.I. Abarbanel; B. d'Almagne, Proc. Int. Conf. on High-Energy Physics, Madison, 1980, p. 221, eds. L. Durand and L.G. Pondrom.
12. F. Halzen and D.M. Scott, Phys. Lett. 72B, 404 (1981).
13. L.M. Jones and H.H. Wyld, Phys. Rev. D17, 759 (1978); J. Babcock, D. Sivers, and S. Wolfram, Phys. Rev. D18, 162 (1978); V.A. Novikov, M.A. Shifman, A.I. Vainshtein, and V.I. Zakharov, Nucl. Phys. B136, 1259 (1978).
14. H. Fritzsch and K.H. Streng, Phys. Lett. 72B, 385 (1978).
15. A.R. Clark et al., Phys. Rev. Lett. 45, 682 (1980).
16. G. Coignet, talk given at European Phys. Soc. Conf., Lisbon, 1981. J.J. Aubert et al., CERN-EP/81 -61,62.

THE WEAK MIXING ANGLE AND ITS RELATION TO THE MASSES M_Z AND M_W*

Emmanuel A. Paschos

Universität Dortmund

Dortmund, West Germany

ABSTRACT

Determinations of the weak mixing angle including one loop radiative corrections are reviewed. The results restrict the masses of the heavy bosons in the ranges $78.9 < M_W < 84.6$ and $90.3 < M_Z < 94.9$ GeV/c^2. Values for $\sin^2\theta_W$ are summarized in figure 7 and compared with ordinary and supersymmetric grand unified theories.

INTRODUCTION

The weak mixing angle is one of the two basic coupling constants of the electroweak theory, whose significance is already appreciated, since

(i) it is a crucial parameter in the prediction of the $W^{\pm}$ and Z^o masses[1,2],

(ii) it is predicted in theories which unify the electroweak and strong interactions[3], and

(iii) it is an important parameter in the prediction[3] of the proton's lifetime.

---

*Supported in part by BMFT

During the past year several groups reported[4-9] new determinations of $\sin^2\theta_W$ which, for the first time, included the photonic corrections and estimates for some of the hadronic model uncertainties. The net effect is a slight modification for the angle and of the masses M_Z and M_W.

Ideally the angle is determined in leptonic reactions which are free of hadronic model ambiguities. Unfortunately, the experimental results for leptonic reactions still carry large errors. From leptonic reactions

$$\sin^2\theta_W = 0.22 \pm 0.08 \quad \text{from } e^+e^- \to \ell^+\ell^- \text{ [10]}$$

and

$$\sin^2\theta_W = 0.25 \begin{matrix} +.0.07 \\ -.0.05 \end{matrix} \quad \text{from } \nu_\mu e^- \to \nu_\mu e^- . \text{ [11]}$$

The experimental situation is better in semileptonic reactions where the errors are between 5 - 10%. The quantities which have been studied[4-6] in greater detail are the ratios

$$D_- = \frac{\sigma(\nu N \to \nu x) - \sigma(\bar{\nu} N \to \bar{\nu} x)}{\sigma(\nu N \to \mu x) - \sigma(\bar{\nu} N \to \bar{\mu} x)} = \frac{1}{2} - \sin^2\theta_W \tag{1}$$

and

$$R_\nu = \frac{\sigma(\nu N \to \nu x)}{\sigma(\nu N \to \mu x)} \tag{2}$$

on isoscalar targets, denoted by N.

For the mixing angle I adopt the definition that it is the coefficient of the component of the neutral current which transforms like the electromagnetic current. This definition does not include any higher order corrections. They will be included explicitly in the calculation of physical observables which are compared with experimental quantities.

In this contribution I review the analysis of several groups and compare their results. Section II describes the corrections in the electroweak theory and discusses the hadronic model ambiguities. Section III includes the analyses of D_- and R_ν. The last section summarizes the conclusions.

ELECTROWEAK CORRECTIONS

The electroweak corrections to neutrino induced reactions have been applied to several processes.[1,2] I will only discuss two topics relevant to the subsequent discussion.* A large one loop correction comes from the self energies of the gauge bosons defined as

$$\pi^i_{\mu\nu}(q^2) = a_i(q^2)\left(q_{\mu\nu} - \frac{q_\mu q_\nu}{q^2}\right) + b_i(q^2)\,\frac{q_\mu q_\nu}{q^2} \quad . \tag{3}$$

One defines renormalized Green's functions as

$$a^R(q^2) \equiv a(q^2) + a_{CT} \quad , \tag{4}$$

with $a(q^2)$ the contribution from Feynman diagrams dimensionally regularized and a_{CT} the counter terms. The counter terms are chosen appropriately in order to cancel the infinities and render $a^R(q^2)$ finite. Renormalized quantities are determined up to arbitrary constants. The constants are fixed by the renormalization conditions

$$a^R_W(M^2_W) = a^R_Z(M^2_Z) = a^R_\gamma(0) = 0 \quad . \tag{5}$$

The renormalized self energies $a^R_W(q^2)/M^2_W$ and $a^R_Z(q^2)/M^2_Z$ are shown in figures 1 and 2. The two curves correspond to Higgs masses of 10 and 100 GeV/c^2. They are very flat for an extended region of q^2 and vanish at the masses M_W and M_Z as indicated by equation (5).

---

*The discussion is based on the on-shell renormalization scheme used extensively by the Dortmund group.[2,4,9,12]

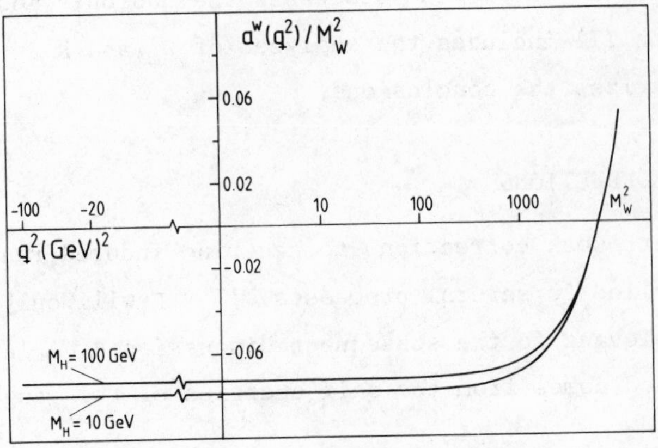

Fig. 1 The renormalized self energy for the $W^{\pm}$ bosons.

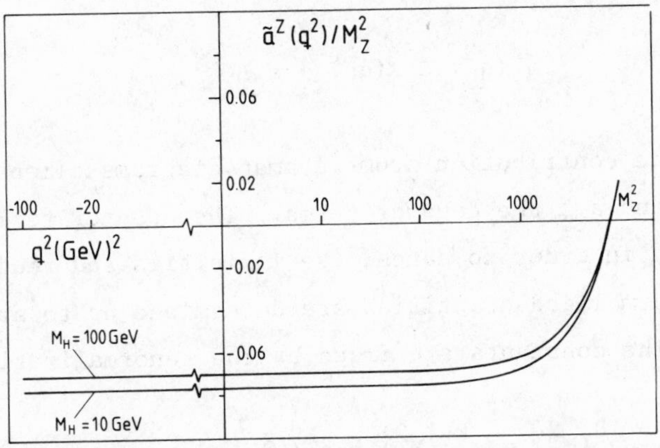

Fig. 2 The renormalized self energy for the Z^o boson.

The curves clearly show that the electroweak corrections are ≈ -7.3% and therefore detectable. This effect was included in the muon lifetime[1,2] and it also accounts for the shifts[1,2] in M_Z and M_W's. In the determination of $\sin^2\theta_W$ from the ratios D_- and R_ν, only the difference of self energies enters and the corrections are indeed much smaller, ≈ 0.3%. Other corrections from fermion self energies, vertices, and box diagrams are also small and are included in the computations.

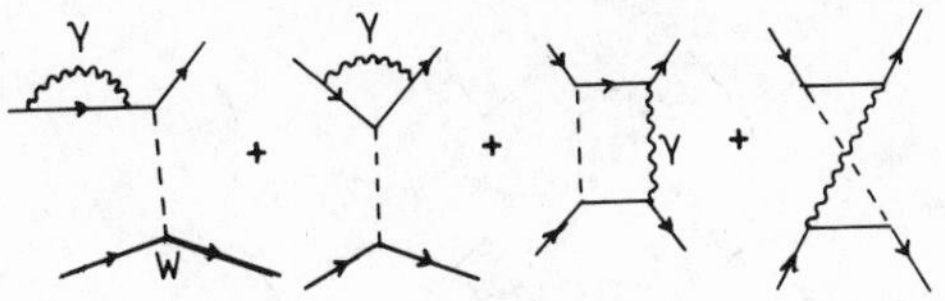

Fig. 3

Several Green's functions receive contributions from diagrams with internal $W^{\pm}$, Z^os, and in addition internal photons. Diagrams with internal photons are handled as follows: (i) the high frequency range of integrations, denoted by $\ell n\Lambda/M_W$, is combined with other diagrams and counter terms to give finite Green's functions, (ii) the low frequency parts, with finite logarithms and all constant terms, are added to the bremsstrahlung terms. I will refer to them together as the photonic correction. A large photonic logarithm arises from the diagrams in figure 3. Their combined effect is

$$\begin{aligned}\Delta_M &= \Delta_\Sigma + \Delta_V + \Delta_B \\ &= \frac{\alpha}{\pi}\{-\frac{1}{4}\sum_i f_i^2 + \frac{1}{2}\sum_{i,j} f_i f_j + \frac{3}{2}\}\,\ell n\,(\frac{M_W^2}{Q^2}) \\ &= \frac{\alpha}{\pi}\,\ell n\,(\frac{M_W^2}{Q^2}) \approx 1.3\%, \text{ for } Q^2 = 20\ (\text{GeV}/c)^2 \quad . \end{aligned} \tag{6}$$

This term occurs only in the charged currents. The summation i runs over charged fermions and ij over the vertices. The logs are again truncated at Q^2, with the rest combined in the bremsstrahlung terms. During the past year several groups[4-7] pointed out that there is a potentially large logarithm shown in equation (6). Its net effect is to lower the mixing angle.

The bremsstrahlung terms are harder to handle since they must be computed explicitly in the kinematic regions of the experiments. In the experiments there is a hadronic energy cut

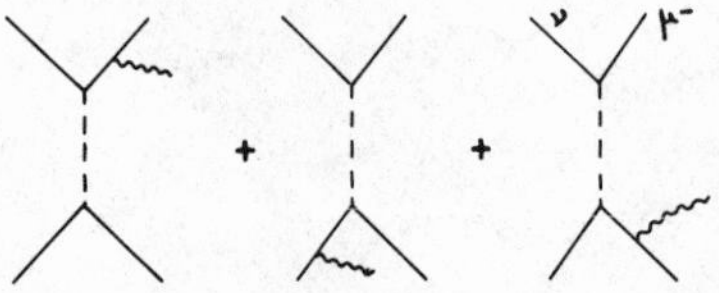

Fig. 4

which excludes the small region of y and a muon energy cut which excludes the high region of y. I describe the bremsstrahlung corrections as developed in references 4,9,13. Bremsstrahlung corrections are larger for the charged currents, where the diagrams in figure 4 contribute together with diagrams containing virtual photons needed in order to cancel the infrared singularities. We describe the effects from bremsstrahlung by $\delta_{cc}(x,y)$ and $\delta_{cc}(y)$ defined as

$$\frac{d\sigma}{dxdy} = \frac{d\sigma_o}{dxdy}\,[1 + \delta_{cc}(x,y)] \tag{7}$$

and

$$\delta_{cc}(y) = \int_o^1 \frac{d\sigma_o}{dxdy}\,\delta_{cc}(x,y)dx \quad , \tag{8}$$

with the subscript zero indicating the Born diagram cross section. Figure 5 shows corrections for two parametrizations of the quark distribution functions. We note that for some values of x and y the corrections are ± 10%. After integration over x the corrections are reduced to those shown in figure 6. The reduction is understood once we realize that the largest correction comes from terms of the form log (Q^2/m_f^2), where m_f is the mass of a fermion. Furthermore, it is evident that by appropriate choices of the region of integration, in figure 6, we can make the bremsstrahlung correction zero. This is consistent with an old theorem,[14] which states that the integrated cross section cannot depend on log $m_{f'}$, where $m_{f'}$ is the mass of a fermion in the final state. Finally, figure 6 shows the dependence of $\delta_{cc}(y)$ on three quark masses:

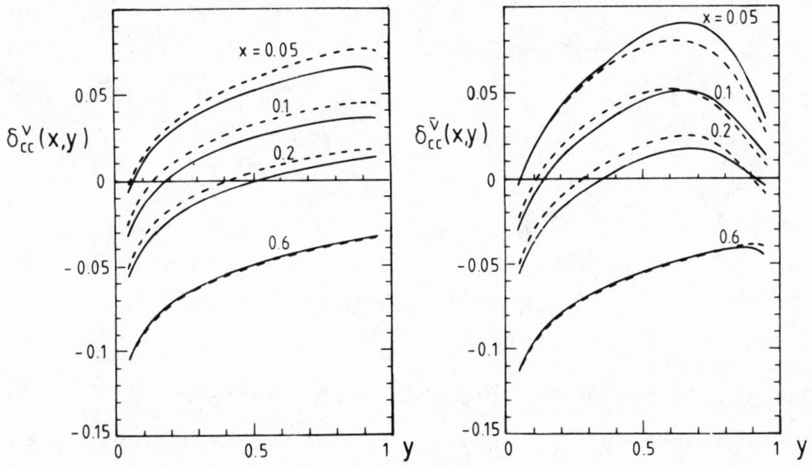

Fig. 5 ——— CDHS
------ Barger/Phillips
Bremsstrahlung correction as function of x and y.

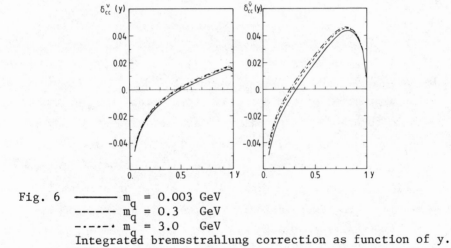

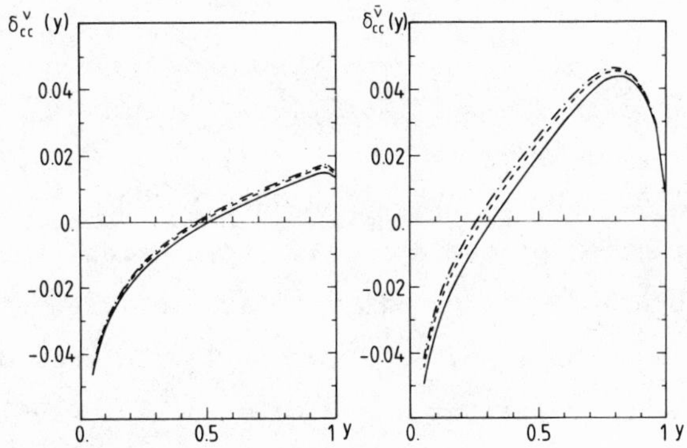

Fig. 6 ——— m_q = 0.003 GeV
------ m_q = 0.3 GeV
-·-·-· m_q = 3.0 GeV
Integrated bremsstrahlung correction as function of y.

m_q = 3, 300, 3000 MeV/c^2. Evidently the dependence on quark masses is very small.

RESULTS

Before discussing specific results it is worthwhile to study the sensitivity of the different ratios. The sum total of the corrections can be combined into the terms δ_- and δ_ν defined by

$$D_- = (1 + \delta_-)\, D_-^o \quad , \tag{9}$$

$$R_\nu = (1 + \delta_\nu)\, R_\nu^o \quad , \tag{10}$$

with the quantities on the left hand sides representing the experimental ratios and D_-^o, R_ν^o denoting the ratios as functions of $\sin^2\theta_W$ computed in the Born approximation. The change of the angle due to the corrections δ_- and δ_ν is

$$\Delta\sin^2\theta_W \approx + \frac{1}{4}\, \delta_- \tag{11}$$

and

$$\Delta\sin^2\theta_W \approx + \frac{27}{17}\, R_\nu\, \delta_\nu \approx + 0.46\, \delta_\nu \quad . \tag{12}$$

Both relations have the remarkable property that the change in $\Delta\sin^2\theta_W$ is a fraction of the δ's. In other words, the determination of the angle through D_- and R_ν is very stable. The stability of D_- is better than that of R_ν by a factor of 2. In the following I discuss each of the relations separately including values for the angle, ambiguities on the hadronic model and the experimental data.

Analysis of the PW relation: Relation (1) was studied extensively. In the limit where the s,c,b,t quarks are neglected and θ_c is set equal to zero, the relation follows from strong isospin invariance as demonstrated in the original paper.[15] In this limit, it is valid independent of any details of the structure functions; it is valid in the presence of arbitrary amounts of scaling violations, and in the presence of arbitrary higher twists, for example. In this limit there are corrections due to a violation of strong isospin invariance[16] as a result of the d-u mass difference. These might be of the order $[(m_d^2 - m_u^2)/q^2]$, which is very small. Thus one is left with corrections from the presence of the heavier quarks which are discussed in detail in a recent article[4] and are found to be small. The main correction arises from the charged currents involving s and c quarks. The total uncertainty from the hadronic model was estimated to bring in a correction for $\sin^2\theta_W$, equal to

$$\Delta\sin^2\theta_W = 0.0035 \text{ to } 0.0045 \ .$$

Note that this correction is positive and increases the angle.

This relation is also attractive from the experimental point of view, because it is independent of the kinematic region so long as the same region is used for both charged and neutral currents. It gives experimentalists the freedom to employ kinematic cuts that reduce the charged current contamination of the neutral current sample. Three experimental groups analysed their data using the PW relation and obtained the values denoted by $\sin^2\theta_W{}^o$ in table 1.

Table 1

Experiment	$\sin^2\theta_W^o$	$\sin^2\theta_W$	Remark
CDHS[17]	0.230 ± 0.016	0.225 ± 0.016	It is necessary to add a correction
CHARM[18]	0.230 ± 0.023	0.225 ± 0.023	
CFRR[19]	0.243 ± 0.016	0.231 ± 0.016	$\Delta\sin^2\theta_W = +\ 0.0035$
Weighted Average		0.227 ± 0.010	from the hadronic model uncertainty

The third column gives values for $\sin^2\theta_W$ including $O(\alpha)$ weak and electromagnetic corrections. The electroweak corrections are from reference 4. However, if I adopt the corrections of Marciano and Sirlin[5], their effect is to decrease $\sin^2\theta_W^o$ by 2% and the above results remain practically the same.* This is a consequence of the remarkable stability of D_-, as indicated by equation (3). Finally the angle as obtained from the D_- ratio carries an experimental error of ± 0.010, which equals the experimental error on the angle as obtained from the R_ν ratio.

Analysis of the R_ν ratio: The data used in the analysis were compiled by Kim et al.[20] in their extensive review on weak neutral currents. Subsequently, several authors adopted their QCD model calculation and deduced values for the angle indicated as $\sin^2\theta_W^o$. In particular the CHARM collaboration repeated the analysis by taking into account the beam spectra and selection criteria of their experiments and obtained the value shown in table 2. The same collaboration also analysed[21] their data using the hadronic energy distributions and I shall return to their results later on. In table 2 I show the values of R_ν reported by the experimental groups groups and values for $\sin^2\theta_W^o$ deduced in ref. 5. The analysis by

Table 2

Experiment	R_ν	$\sin^2\theta_W^o$
CDHS[17]	0.307 ± 0.008	0.230 ± 0.013
CHARM[18]	0.320 ± 0.010	0.220 ± 0.014
BEBC[22]	0.32 ± 0.03	0.217 ± 0.045
CITF[23]	0.28 ± 0.03	0.272 ± 0.055
HPWF[24]	0.30 ± 0.04	0.274 ± 0.075
Kafka et al.[25]	0.30 ± 0.03	

*Private communication W. Marciano.

Table 3

Experiment	$\sin^2\theta_W(-20\text{GeV}^2)$	$\sin^2\theta_W(M_W)$	$\sin^2\theta_W(-20\text{GeV}^2)$
Authors	Marciano-Sirlin[5]	Llewellyn-Smith-Wheater[7]	Wirbel[9]
CDHS	0.217 ± 0.013	0.219	0.222
CHARM	0.211 ± 0.015	0.210	0.215
BEBC	0.203 ± 0.045	0.206	0.208
CITF	0.259 ± 0.055	0.263	0.263
HPWF	0.264 ± 0.075	0.266	0.27
Weighted Average	0.216 ± 0.010	0.217±0.010	0.221 ± 0.010

Kim et al.[20] adopts the quark-parton model and includes QCD corrections. Very recently Glück and Reya[26] argued that higher twist effects might change $\sin^2\theta_W^o$ by as much as ± 0.02, but an explicit computation of higher twist effects is not available. The following discussion of R_ν does not include higher twist effects.

Three groups reported results on the $O(\alpha)$ radiative corrections to R_ν. Their values for the angle are shown in table 3. The errors in the third and fourth columns are the same as in column two. The values in column two and three almost agree with each other.[27] The values by Wirbel are consistently larger by the small amount 0.005. Part of the difference comes from the computation of the bremsstrahlung terms. In their work, Marciano and Sirlin[6] computed the total-neutrino-nucleon cross section, including bremsstrahlung terms, by introducing average values of y. Then they corrected for the $0 \le y \le y_{min}$ region, which is not observed in the experiments. On the other hand, Wirbel[9] integrated the leading-log approximation of the bremsstrahlung terms over the region $y_{min} \le y \le y_{max}$ detected in the experiments. The small difference comes partly from the different methods of including the experimental cuts and partly from the leading-log approximation.

An independent determination of the angle is possible using a more recent analysis[21] of the CHARM collaboration. In this analysis they studied the differential cross sections $d\sigma/dy$ in neutrino and antineutrino reactions. They also included charged current radiative corrections[21] and obtained

$$\sin^2\theta_W^o = 0.222 \pm 0.016 \quad .$$

Since the weak corrections are small and the bremsstrahlung terms were included, we must correct only for the term $\alpha/\pi\ell n(M_W^2/Q^2)$ which reduces the angle by 0.008 to

$$\sin^2\theta_W = 0.214 \pm 0.016 \quad .$$

In summary the radiative corrections to R_ν are reliable and there is good agreement between the groups. Much more uncertain, in my opinion, are the hadronic model ambiguities associated with the quark-parton model and also with higher twists.

The values from other experimental ratios are not as accurate. The electron-deuteron experiment[28] gives

$$\sin^2\theta_W = 0.224 \pm 0.020 \quad .$$

An investigation of the sensitivity of this value to reasonable changes of the quark-parton model parameters varies[30] the central value between 0.207 and 0.227. An analysis of the radiative corrections reports[7] an additional reduction of -0.007.

CONCLUSIONS

1. The values for the mixing angle are summarized in figure 7. The results are consistent with each other, with the neutrino data giving more accurate results.

2. With the angle as an input we can calculate the masses M_W and M_Z

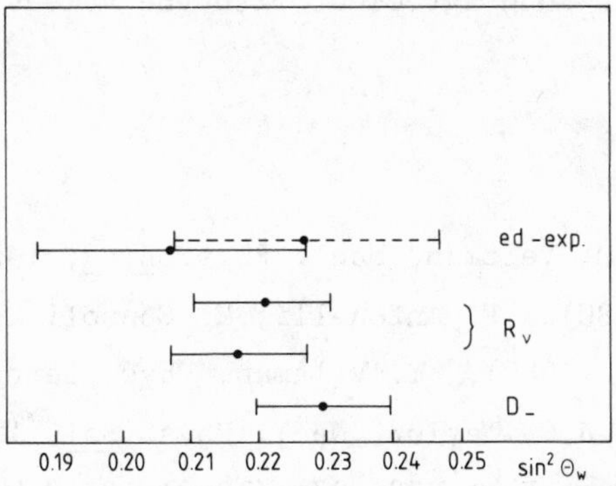

Fig. 7 Summary of values for $\sin^2\theta_W$.

including one loop radiative corrections. From the R_ν ratio

$$M_W = 82.6 \begin{smallmatrix} +\ 2.0 \\ -\ 1.8 \end{smallmatrix}\ GeV/c^2 \text{ and } M_Z = 93.3 \begin{smallmatrix} +\ 1.6 \\ -\ 1.5 \end{smallmatrix}\ GeV/c^2 \ ;$$

from the D_ ratio

$$M_W = 80.6 \begin{smallmatrix} +\ 1.8 \\ -\ 1.7 \end{smallmatrix}\ GeV/c^2 \text{ and } M_Z = 91.7 \pm 1.4\ GeV/c^2 \ .$$

3. The value for the angle predicted in the SU(5) grand unified theory[3]

$$\sin^2\theta_{W|SU(5)} = 0.209 \begin{smallmatrix} +\ 0.003 \\ -\ 0.002 \end{smallmatrix}$$

is in good agreement with the results in figure 7. A more stringent test of SU(5) is provided by the lifetime of the proton which is a sensitive function[29] of the angle. For $\sin^2\theta_W > 0.210$

$$\tau_p \simeq 4 \times 10^{28\pm1}\ \text{yr.}$$

This value is already below the experimental bound.[29] The experiments searching for proton decays have the capability to observe an effect or rule out the SU(5) theory.

4. Supersymmetric grand unified theories predict[29,30] a larger

$\sin^2\theta_W$, which is in good agreement with the values summarized in figure 7.

REFERENCES

1. M. Green and M. Veltman, Nucl. Phys. B169, 137 (1980), ibid. B157, 547 (1980). F. Antonelli, M. Consoli and G. Corbo, Phys. Lett. 91B, 90 (1980). M. Veltman, Phys. Lett. 91B, 95 (1980).
2. D.A. Ross and J.C. Taylor, Nucl. Phys. B51, 125 (1973), R. Sirlin, Phys. Rev. D22, 971 (1980), S. Sakakibara, Phys. Rev. D24, 1149 (1981), W. Marciano and A. Sirlin, Phys. Rev. D22, 2695 (1980).
3. J. Ellis, M.K. Gaillard, D.V. Nanopoulos and S. Rudaz, Nucl. Phys. B176, 61 (1980) and references therein. R. Langacker, Univ. of Penn. preprint, UPR-0170T (1981).
4. E.A. Paschos and M. Wirbel, Nucl. Phys. B197, 189 (1982).
5. R. Sirlin and W. Marciano, Nucl. Phys. B159, 442 (1981).
6. C.H. Llewellyn-Smith and J.F. Wheather, Phys. Lett. 105B, 486 (1981).
7. F. Antonelli and L. Maiani, Nucl. Phys. B186, 269 (1981).
8. S. Dawson, J.S. Hagelin and L. Hall, Harvard preprint HUTP-80/1090.
9. M. Wirbel, Ph.D. Thesis, University of Dortmund (1981).
10. W. Bartel et al. (JADE collaboration) DESY-preprint 81-015 (1981).
11. L. Mo, Proceed. of 15th Rencontre de Moriond, edit. by J. Tran Thanh Van (1981) Vol. II, p. 259.
12. W.R. Murzick, Diplomarbeit, University of Dortmund (December, 1981).
13. A. DeRujula, R. Petronzio and A. Savoy-Navarro, Nucl. Phys. B154, 394 (1979).
14. T. Kinoshita, J. Math. Phys. 3, 650 (1962). T.D. Lee and M. Nauenberg, Phys. Rev. 133, 1549 (1964).
15. E.A. Paschos and L. Wolfenstein, Phys. Rev. D7, 91 (1973).

16. L. Wolfenstein, private communication (1982).
17. C. Geweniger, CDHS collaboration, Proc. Int. Conf. Neutrino 79 (Bergen 1979) p. 392.
18. M. Jonker, CHARM collaboration, Phys. Lett. 99B, 265 (1981).
19. M. Shaevitz (CFRR collaboration), Proceed. of 1981 Int. Conf. in Neutrino Physics and Astrophysics, Vol. I, p. 311.
20. J.E. Kim, P. Langacker, M. Levine and H.H. Williams, Rev. Mod. Phys. 53, 211 (1981).
21. M. Jonker et al., (CHARM collaboration), Phys. Lett. 102B, 67 (1981).
22. BEBC ABCLOS collab., M. Deden et al., Nucl. Phys. B149, 1 (1979).
23. CITF collab., F. Merrit et al., Phys. Rev. D17, 2199 (1979).
24. HPWF collab., P. Wanderer et al., Phys. Rev. D17, 1679 (1978).
25. T. Kafka et al., Stony Brook-Illinois Inst.-Maryland- Tohoku-Tufts collab., preprint (1982).
26. M. Glück and E. Reya, Phys. Rev. Lett. 47, 1104 (1981).
27. The values of Llewellyn-Smith and Wheater are from ref. 6. Shortly after the conference Prof. Dalitz informed me that their values are slightly modified: [CHARM (0.208), CDHS(0.217), BEBC(0.203), CITF(0.257) and HPWF(0.262)].
28. C.Y. Prescott et al., Phys. Lett. 84B, 524 (1979).
29. W. Marciano and G. Senjanovich, BNL-preprint 30398 (1981) and talk by W. Marciano at this conference.
30. See the talk by S. Dimopoulos at this conference.

INTERMEDIATE VECTOR BOSONS AND NEUTRINO COSMOLOGY

Zohreh Parsa

New Jersey Institute of Technology, Newark, NJ 07102
and
Brookhaven National Laboratory, Upton, NY 11973

ABSTRACT

We present decay properties of $W^{\pm}$ and Z^{o}, and give numerical estimates of their decay rates and branching ratios for leptonic, hadronic and higher order rare decays in the standard $SU(3)_c \times SU(2)_L \times U(1)$ model. In addition we discuss bounds on the number of neutrino species and their cosmological implications.

The proposed high luminosity e^+e^-, and pp, accelerators will lead to the production of a large number of $W^{\pm}$ and Z^{o} bosons. Given that fact it will be possible to study the basic properties of these interacting intermediate vector bosons. Hopeful that the masses and decay rates will be precisely measured we will give here a list of predictions for their properties in the standard $SU(3)_c \times SU(2)_L \times U(1)$ model.[1]

At present, the standard model's prediction (including radiative corrections) for the $W^{\pm}$ and Z^{o} masses are[2]

$$m_W = 83.0 \pm 2.4 \text{ GeV}$$
$$m_Z = 93.8 \pm 2.0 \text{ GeV} \tag{1}$$

These estimates include $O(\alpha)$ radiative corrections, which give m_W and m_Z a 5% enhancement over the naive lowest order predictions. Using $m_W = 83$ GeV, $m_Z = 93.8$ GeV, and $s^2 = \sin^2\theta_W = 0.215$, the experimental average, one finds the following predictions for the partial decay rates and branching ratios of the $W^{\pm}$ and Z^0 into leptons and hadrons (including QCD corrections)*, where the muon decay constant $G_\mu = 1.16632 \pm 0.00002 \times 10^{-5}$ GeV^{-2}, and the QCD coupling $\alpha_s(m_W) \simeq \alpha_s(m_Z) \simeq 0.127$ are employed.

$W^{\pm}$ DECAYS

$$\Gamma(W^- \to \ell + \bar{\nu}_\ell) = \frac{G_\mu m_W^3}{6\sqrt{2}\pi} \simeq 0.250 \text{ GeV} \quad \ell = e, \mu, \text{or } \tau \tag{2a}$$

$$\Gamma(W^- \to \text{hadrons}) = \frac{G_\mu m_W^3}{2\sqrt{2}\pi}\left(3 - \frac{3m_t^2}{2m_W^2}\right)\left(1 + \frac{\alpha_s(m_W)}{\pi}\right) \tag{2b}$$

$$\simeq 2.277 \text{ GeV}$$

$$\Gamma(W^- \to \text{all}) \simeq 3.03 \text{ GeV} \tag{2c}$$

and the branching ratio

$$\frac{\Gamma(W^- \to \ell\bar{\nu}_\ell)}{\Gamma(W^- \to \text{all})} \simeq 0.083 \text{ GeV} \quad \ell = e, \mu, \text{or } \tau \tag{2d}$$

Z^0 DECAYS

$$\Gamma(Z^0 \to \ell + \bar{\ell}) = \frac{G_\mu m_Z^3(1-4s^2+8s^4)}{12\sqrt{2}\,\pi} = 0.0920 \text{ GeV} \quad \ell = e, \mu, \text{ or } \tau \tag{3a}$$

---

*We always give rates for W^- decays; the rates for W^+ into charge conjugated final states are the same as these.

$$\Gamma(Z^o \to \nu_\ell+\bar{\nu}_\ell) = \frac{G_\mu m_Z^3}{12\sqrt{2}\pi} \simeq 0.1805 \text{ GeV} \tag{3b}$$

$$\Gamma(Z^o \to \text{hadrons}) = \frac{G\, m_Z^3}{12\sqrt{2}\pi}\left(18-36s^2+40s^4 - 9\frac{m_t^2}{m_Z^2}\right)\left(1+\frac{\alpha_s(m_Z)}{\pi}\right)$$

$$\simeq 2.200 \text{ GeV} \tag{3c}$$

$$\Gamma(Z^o \to \text{all}) \simeq 3.02 \text{ GeV} \tag{3d}$$

and the leptonic branching ratios

$$\frac{(Z^o \to \ell^+\ell^-)}{(Z^o \to \text{all})} = \frac{1-4s^2+8s^4}{24-48s^2+64s^4} \simeq 0.0306 \qquad \ell=e,\mu,\text{or } \tau \tag{3e}$$

$$\frac{(Z^o \to \nu_\ell\bar{\nu}_\ell)}{(Z^o \to \text{all})} = \frac{1}{24-48s^2+64s^4} \simeq 0.0601 \qquad \ell=e,\mu,\text{or } \tau\ . \tag{3f}$$

We obtained the decay widths in (2b) and (3c) by summing over 6 quark flavors (u,c,t,d,s,b). Including leptonic decay modes equations (2c) & (3d) gives the total rates and (2d) & (3e,f) the branching ratios.

It is very important that the branching ratios into electrons and muons be large enough to provide clear experimental signals for $W^\pm$ and Z^o production. One may note that because a value for m_Z that is 5% larger than the naive lowest order estimate was used in the above formulas, our prediction for decay widths, which are proportional to m_Z^3, are generally $\simeq$ 16% larger than the values often quoted in the literature (for a comparable $\sin^2\theta_W$), a sizable difference.

EXOTIC DECAYS

In addition to the leptonic and hadronic rates described above we have studied some of the higher order decays of $W^{\pm}$ and Z^{o}.

We have analyzed the two body radiative decays $W^{\pm} \to P^{\pm} + \gamma$ and $Z^{o} \to P^{o} + \gamma$ where P denotes any pseudoscalar meson. The observation of such exclusive decays would yield useful information regarding strong interaction dynamics, and provide a determination of m_W, m_Z or may lead to the discovery of new pseudoscalar mesons. For the light pseudoscalar mesons, some of our results are:[5,6]

$$\frac{\Gamma(Z^{o} \to \pi^{o} + \gamma)}{\Gamma(Z^{o} \to \text{all})} \simeq 3 \times 10^{-11} \quad , \tag{4a}$$

$$\frac{\Gamma(Z^{o} \to \eta + \gamma)}{\Gamma(Z^{o} \to \text{all})} \simeq 2 \times 10^{-10} \quad , \tag{4b}$$

$$\frac{\Gamma(Z^{o} \to \eta' + \gamma)}{\Gamma(Z^{o} \to \text{all})} \simeq 5 \times 10^{-9} \quad , \tag{4c}$$

$$\frac{\Gamma(W^{\pm} \to \pi^{\pm} + \gamma)}{\Gamma(W^{\pm} \to \text{all})} \simeq 3 \times 10^{-9} \quad , \tag{4d}$$

$$\frac{\Gamma(W^{\pm} \to K^{\pm} + \gamma)}{\Gamma(W^{\pm} \to \text{all})} \simeq 2 \times 10^{-10} \quad . \tag{4e}$$

These branching ratios are probably too small to observe. For very heavy bound state pseudoscalars with mass ≃ 50 ~ 60 GeV we found (roughly)[4]

$$\frac{\Gamma(Z^{o} \to P^{o} + \gamma)}{\Gamma(Z^{o} \to \text{all})} \simeq 10^{-7} \quad , \tag{5a}$$

$$\frac{\Gamma(W^{\pm} \to P^{\pm} + \gamma)}{\Gamma(W^{\pm} \to \text{all})} \simeq 10^{-7} \quad , \tag{5b}$$

somewhat larger but still too small to measure.

In the extended technicolor model[9] (which contains 4 $SU(2)_L$ isodoublets), in the mass range of 10 ~ 40 GeV we have computed the radiative decay rates of the $Z^0 \to P^0 + \gamma$ or $P^3 + \gamma$ and $W^{\pm} \to P^{\pm} + \gamma$ where P^0, P^3, $P^{\pm}$ are very light color singlet pseudo-Goldstone bosons. In this model taking number of technicolors $N_c' = 4$, we find[4,8]

$$\frac{\Gamma(Z^0 \to P^0 + \gamma)}{\Gamma(Z^0 \to \text{all})} \simeq 2 \times 10^{-7} \quad , \tag{6a}$$

$$\frac{\Gamma(Z^0 \to P^3 + \gamma)}{\Gamma(Z^0 \to \text{all})} \simeq 5 \times 10^{-8} \quad , \tag{6b}$$

$$\frac{\Gamma(W^{\pm} \to P^{\pm} + \gamma)}{\Gamma(W^{\pm} \to \text{all})} \simeq 2 \times 10^{-6} \quad . \tag{6c}$$

These branching ratios although still small, are very sensitive to the number of technicolors and isodoublets introduced. Implying that the decay $W^{\pm} \to P^{\pm} + \gamma$ offers the best possibility of being observed at a high luminosity facility such as ISABELLE.

An interesting decay[5] is the process $Z^0 \to \phi^0 + \mu^+ + \mu^-$, where ϕ^0 is the Higgs scalar. The predicted rates are given (for $\varepsilon = m_\phi/m_Z$) by

$$\Gamma(Z^0 \to \phi^0 + \mu + \bar{\mu}) \simeq \frac{\alpha}{\pi \sin^2 2\theta_W} F(\varepsilon) \Gamma(Z^0 \to \mu + \bar{\mu}) \quad , \tag{7}$$

where

$$F(\varepsilon) = \frac{\varepsilon}{\sqrt{4-\varepsilon^2}} (5-2\varepsilon^2+\frac{1}{4}\varepsilon^4)\cos^{-1}\left[\frac{3\varepsilon-\varepsilon^3}{2}\right] - (1-\frac{3}{2}\varepsilon^2+\frac{1}{4}\varepsilon^4)\ell n\varepsilon$$

$$-\frac{1}{24}(1-\varepsilon^2)(47-13\varepsilon^2+2\varepsilon^4) \quad . \tag{8}$$

For $\varepsilon = 0.1$, one finds $F(0.1) \approx 0.69$ which implies a branching ratio

$$\frac{\Gamma(Z^o \to \phi^o+\mu+\bar{\mu})}{\Gamma(Z^o \to \text{all})} = 7 \times 10^{-5} \quad . \tag{9}$$

Unfortunately, the branching ratio decreases very fast with increasing ε. For $\varepsilon = 0.9$, the branching ratio in eq. (9) becomes 7×10^{-10}; much too small for observation.

A somewhat rarer Higgs producing decay is $Z^o \to \phi^o + \gamma$. The rate[6] for this radiative process is given by

$$\frac{\Gamma(Z^o \to \phi^o+\gamma)}{\Gamma(Z^o \to \text{all})} \simeq 2.3 \times 10^{-6}(1-\varepsilon^2)^3(1+0.17\varepsilon^2) \quad . \tag{10}$$

If this mode is observed, it would provide a precise determination of m_ϕ via a measurement of the photon's energy.

NUMBER OF NEUTRINO SPECIES

A much anticipated precise measurement of the Z^o's total decay width, $\Gamma(Z^o \to \text{all})$ will allow one to find out if in addition to the three species ν_e, ν_μ, ν_τ there are other light neutrinos (which may belong to very heavy fourth, fifth, etc., fermions generations). For example if the decays $Z^o \to \nu_i + \bar{\nu}_i$ (where ν_i represents an additional species) occur, by measuring $\Gamma(Z^o \to \text{all})$ and subtracting out the observed partial decay rates, one finds the total number of neutrino species N_ν, from

$$N_\nu \Gamma(Z^o \to \nu+\bar{\nu}) = \Gamma(Z^o \to \text{all}) - \Gamma(Z^o \to \text{hadrons}) - \Gamma(Z^o \to \text{charged leptons}) . \tag{11}$$

Since each additional neutrino species contributes 0.18 GeV to the total width, one finds from eq. (3) that

$$(N_\nu - 3) \times 0.18 \text{ GeV} = \Gamma(Z^o \to \text{all}) - 3.02 \text{ GeV} \quad . \qquad (12)$$

Finding $\Gamma(Z^o \to \text{all}) = 3.02$ GeV will be consistent with $N_\nu = 3$ neutrino species.

NEUTRINO COSMOLOGY

I. Number of Neutrino Species - Big bang Cosmology and the observed abundance of ^4He produced in primordial nucleosynthesis can lead to a significant constraint on the number of types of neutrinos. The argument goes as follows: if all N_ν neutrino species have mass $m_\nu \lesssim 0.3$ MeV and lifetime $\tau > 180$ sec., then each species would be equally abundant in the early evolution of the universe ($t < 180$ sec). Hence, increasing N_ν accelerates the expansion rate for the very early universe which forces the protons and neutrons out of thermal equilibrium at a higher temperature T_c, so the neutron to proton ratio predicted by big bang model

$$\frac{n}{p} = \exp\left[-\frac{(m_n - m_p)c^2}{KT_c}\right] \qquad (13)$$

increases as a function of N_ν. Since almost all the neutrons quickly nucleosynthesize into deuterium, ^3He, ^4He, or ^7Li the primordial abundance of these elements depends sensitively on n/p and therefore on N_ν. Using the presently observed ^4He abundance $\leq 26\%$ of all matter we obtain $N_\nu \leq 3 - 4$. It is interesting to see whether or not a precise measurement of $\Gamma(Z^o \to \text{all})$ supports this bound.

II. Neutrino Mass - According to the big bang theory, neutrinos and photons were in thermal equilibrium in the very early universe. The neutrinos decoupled earlier than the photons and cooled as the universe expanded. The present day temperature, T_ν, of the left over primordial neutrinos is related to the photon blackbody radiation temperature, $T_\gamma \simeq 2.7\ ^oK$, by

$$T_\nu \simeq (4/11)^{1/3}\ T_\gamma \simeq 1.9\ ^oK \quad . \tag{14}$$

This temperature is lower than that of the photons because e^+e^- annihilation heated up the photons after the neutrinos decoupled. Therefore the neutrino number density n_{ν_i} in terms of the photon number density n_γ is expected to be[10]

$$n_{\nu_i} = 3/8\ (T_\nu/T_\gamma)^3\ g_i\ n_\gamma \tag{15}$$

for each type. (The helicity $g_i = 2$ for two component neutrinos and $g_i = 4$ for four component neutrinos). With the present neutrino number density ($n^o_\gamma = 400\ cm^{-3}$)

$$n^o_{\nu_i} \simeq 3/32\ g_i\ n^o_\gamma \quad . \tag{16}$$

The mass density contributed by all distinct neutrino species is

$$\rho_\nu = \Sigma_i\ g_i\ m_{\nu_i}\ (eV) \times 10^{-31}\ g/cm^3 \quad . \tag{17}$$

This formula and the present bound on the total density of the universe (estimated from its age (radioactive dating) and deacceleration)

$$\rho_{total} < 2 \times 10^{-29}\ g/cm^3 \tag{18}$$

leads to the constraint

$$m_{\nu_e} + m_{\nu_\mu} + m_{\nu_\tau} + \ldots = \Sigma\, m_{\nu_i} < 100 \text{ eV}, \quad (g_i=2) \tag{19a}$$

$$< 50 \text{ eV}. \quad (g_i=4) \quad . \tag{20b}$$

It is convenient to express all mass densities in terms of the critical density

$$\rho_c = 5 \times 10^{-30} \text{ g/cm}^3 \quad , \tag{21}$$

which separates those universes which expand forever

$$\rho_{total} \lesssim \rho_c \quad , \tag{22}$$

from those which eventually collapse

$$\rho_{total} > \rho_c \quad . \tag{23}$$

If one believes that the universe is closed (i.e. that it will one day contract to a "big crunch", as shown by eq. (23)) only about 10% of ρ_c can be accounted for by the observed luminous mass in the universe; then where is the rest of the critical mass concentrated? One possible explanation is that relic neutrinos account for the missing mass required to close the universe. Equations (17) and (23) give

$$\Sigma\, m_{\nu_i} > 25 \text{ eV} \quad , \quad (g_i=2) \tag{24a}$$

$$\Sigma\, m_{\nu_i} > 12 \text{ eV} \quad , \quad (g_i=4) \quad . \tag{24b}$$

We conclude from equations (19a) and (24a) that for three species of approximately degenerate neutrinos

$$8 \text{ eV} < m_\nu < 33 \text{ eV} \quad , \tag{25}$$

which is remarkably close to the result of Russian Experiment where they examine the shape of the electron spectrum in tritium β decay $^3H \rightarrow {}^3He + e^- + \bar{\nu}_e$ and find a deviation near the end point which suggests a nonzero mass of $\bar{\nu}_e$ in the range[11]

$$14 \text{ eV} < m_\nu < 46 \text{ eV} \quad .$$

CONCLUSION

In summary we emphasize the need for precise measurements of the fundamental properties of $W^\pm$ and Z^o bosons. Within the Weinberg-Salam framework accurate measurements of the Z^o width would yield a bound for the total number of neutrino species.

Big bang cosmology suggests that there can be three or four distinct neutrino species and that the sum of their masses is less than 100 eV. We conclude that if $\Sigma\, m_{\nu_i} > 25$ eV, then massive neutrinos can provide the "missing mass" necessary to close the universe.

REFERENCES

1. D. Albert, W. Marciano, D. Wyler and Z. Parsa, Nucl. Phys. B166, 460 (1980).
2. W. Marciano, Alberto Sirlin, Proceedings of "ISABELLE Summer Workshop" (1981).
3. W. Marciano, D. Wyler and Z. Parsa, Phys. Rev. Lett. 43, 22 (1979).
4. Z. Parsa, W. Marciano, Proceedings of "Cornell Summer Workshop on Z^o Physics", (1981).
5. J.D. Bjorken, Proceedings of "SLAC Summer Institute", (1976).
6. R. Cahn, M. Chanowitz and N. Fleishon, Phys. Lett. 82B, 113 (1979).
7. L. Arnellos, W. Marciano and Z. Parsa, Nucl. Phys. B196, 365 (1982).

8. Z. Parsa, unpublished.

9. L. Arnellos, W. Marciano, and Z. Parsa, Nucl. Phys. B196, 378 (1982).

10. S. Gershtein and Y. Zeldovich, Sov. Phys.- JETP Lett. 4, 120 (1966); J. McClelland Phys. Rev. Lett. 29, 669 (1972); W. Marciano, Comments Nucl. Part. Phys., Vol. 9, No. 5, (1981).

11. V. Lyubimov et al., ITEP Preprint (Moscow, 1980).

PRECESSIONAL PERIOD AND MODULATION OF RELATIVISTIC RADIAL VELOCITIES IN SS 433

A. Mammano[1,2], F. Ciatti[1], and R. Turolla[3]

Asiago Astrophysical Observatory, University of Padova

Padova, Italy

(presented by F. Ciatti)

SUMMARY

(a) The precessional period of SS 433 is fitted as P = 163.6 ± 0.4 days, with maximum separations computable from the epoch 3669.70 ± 0.06. This period seems stable in 1 part over 10^4, during 1978-81. However improved data may modify this result.

(b) 47 anticorrelated deviations of relativistic radial velocities are recognized. They explain the scatter of data (often double) in a wide strip around the sinusoidal line, as a modulation of the same sinusoidal trend. The period of this modulation should be about 6 days.

1. APPARENT VARIATIONS OF THE PRECESSIONAL PERIOD

Determinations of the precessional period of SS 433 by different authors have been summarized by Newsom and Collins (1981;

[1]Asiago Astrophysical Observatory, University of Padova
[2]University of Messina
[3]International School for Advanced Study, Trieste

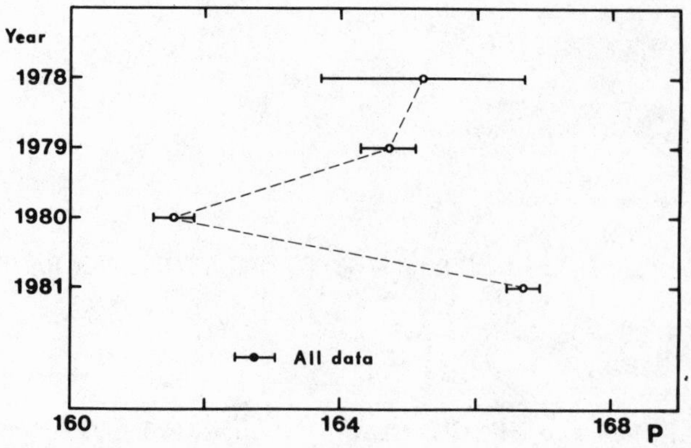

Fig. 1 Fluctuations of the precessional period, as derived with the usual program for data of different years and for the complete sample.

see Table V). Ciatti et al. (1981b) remarked on a numerical decrease of the period, from their previous value $165^{d}.5$ (1978-79) to $164^{d}.7 \pm 0.2$ (1978-80), at the limit allowed by the so called "effective error". This error follows from a perturbation technique which changes the unit weight error σ (say: r.m.s. error) by $\sigma/\sqrt{2n}$, where _n_ is the number of sample data.

Since then a controversy has arisen. A strong decrease in period (-.01 days per day) has been first invoked by Collins and Newsom (1980), and Wagner et al. (1981). Margon (1981) first denied this effect, but later on confirmed the strong decrease (Margon et al., 1981a) over three years, predicting this should be a transient behavior. Four months later, Margon et al. (1981b) reported evidence of an increase of the period during the last 16 months. The last conclusion seems to be confirmed by the analysis performed at Asiago in different time intervals (see Ciatti et al., 1982). Both temporary decreases and increases of the period seem indeed to occur. In this case we should accept strong fluctuations of the precessional period, from the limits of 152^{d} to 172^{d}, as reported in Table 1 and Figure 1. Before assessing the reality of these fluctuations one has to inquire about the epochs of maximum separation. Strong

TABLE 1

Precessional period and epoch in SS 433 in different time intervals, as determined with the usual program JXP.

number of data (blue+red)	year	P	σ_P	E	σ_E	r.m.s.
49 + 48	1978	165.20	1.50	3663.10	1.20	0.0068
		165.25		3663.05		
195 + 178	1979	164.70	0.40	3664.65	1.15	0.0100
93 + 84	1980	161.55	0.30	3674.55	1.45	0.0102
53 + 51	1981	166.70	0.25	3639.60	1.70	0.0097
		166.75		3639.30		
390 + 361	1978/81	162.75	0.30	3668.65	1.05	0.0108
110 + 97	1979 first half	171.20 171.25 171.30	0.60	3651.35 3651.25 3651.15	1.30	0.0083
85 + 81	1979 second half	151.45 151.50 151.55 151.60 151.65 151.70	0.45	3704.00 3703.85 3703.70 3703.55 3703.40 3703.25	1.30	0.0096
46 + 41	1980 first half	164.70 164.75	0.40	3661.70 3661.50	1.55	0.0089
47 + 43	1980 second half	159.75 159.80 159.85	0.35	3683.65 3683.40 3683.15	1.80	0.0099

fluctuations in the epochs were found, anticorrelated with respect to the period variations. We thought it could enventually depend on the inability of our usual computing program to overcome secondary minima in the minimization function.

We employed 390 "blue" and 361 "red" data, from both published Z's and our Asiago observations. Since only one Z value per branch and per spectrum is usually given in the literature, even when more structures are evident in the published spectra (e.g. Margon et al., 1979, 1980; Wagner et al., 1981), we have been obliged to average also our daily Z's for sake of comparison with other authors. We will demonstrate in the following section that this averaging technique implies loss of information, especially with regard to the probable modulation of period about 6 days. Thus we consider our results on the precessional period as preliminary ones, and will defer a final solution at the time when individual Z's per spectrum will be put at our disposal.

2. STABILITY OF THE PRECESSIONAL PERIOD

We have thus processed the same sample data, reported in Table 1, also with a more sophisticated program. The function to be minimized is again the sum of squares of residuals, but new parameters are included. They are: precessional period and epoch, with their corrections, amplitude of the large sinusoidal trend and position of its median line, light time and distance of the emitting regions from the central source, and finally eccentricity. The coefficients of the last three terms were very much smaller than their respective errors and will not be reported in what follows. The following relation represents the shift for the (most accurate) "blue" branch:

$$Z = \underset{(\pm 0.008)}{0.084} - \underset{(\pm 0.005)}{0.003} \sin \phi - \underset{(\pm 0.007)}{0.088} \cos \phi + \underset{(\pm 0.005)}{0.002} \sin 2\phi + \underset{(\pm 0.005)}{0.005} \cos 2\phi \ ,$$

where ϕ is the phase, reckoned from maximum separation. It can be computed with a period of 163.6^d and T_o=3669.7. The covariance matrix is lost; thus the errors are "effective" ones, corresponding to a variation of the unit weight error σ by 0.001, that is of the same order of magnitude as the accuracy of a single observation.

We have then divided the data sample into six independent time subsamples. For each subsample we performed the computation twice, first with and then without smoothing data, that is removing those with (O-C) larger than 0.02 one order of magnitude larger than the possible errors: errors of 0.001-.002 are often quoted in the literature. After smoothing, the unit weight error is reduced from 0.009 to 0.007.

We then fitted the six values with P and $\dot{P}$, and found for the "blue" branch

$$\bar{P} = 163\overset{d}{.}6 - 1.3\ 10^{-4}(d/d) \qquad \pm 0.5 \quad \pm 1\times 10^{-4} \quad .$$

This indicates, at least formally, the stability of the precessional period in one part over 10^4. Due to the presence of possible incorrect data, this result shall be considered as preliminary.

The data of 6 subsamples, treated as a single sample (n=370) give

$$P_{all} = 162\overset{d}{.}6 \pm 0.5 \quad , \quad E_{all} = 3669.70 \pm 0.06 \quad .$$

The original data (n=390) result in slightly larger errors in the subsample periods, but much larger deviations and errors in the epochs. We however obtain with all data together a result very close to the previous one, namely

$$P_{all} = 162\overset{d}{.}8 \pm 0.3 \quad , \quad E_{all} = 3670.05 \pm 0.06 \quad .$$

We also divided the smoothed sample into 29 subsamples, each of them formed with 100 consecutive dates and Z's, and subtracting

Table 2

Results on precessional period from subsamples, with the program of nine parameters.

ORIGINAL INDEPENDENT SAMPLES

Sample	Period	Epoch
A(n=70)	164.2±1.5	3671.4± 3.0
B(70)	162.9 0.3	69.9 2.0
C(70)	163.0 1.0	68.9 2.0
D(70)	162.2 1.6	77.4 11.0
E(70)	162.9 0.3	69.3 0.3
F(40)	162.8 0.4	69.6 3.8
All (n=390)	162.76±0.30	3670.05±0.06
	σ_z(r.m.s.) = 0.009	

SMOOTHED INDEPENDENT SAMPLES

Sample	Period	Epoch
A'(n=70)	163.2±0.5	3669.1±1.2
B'(70)	162.9 0.3	69.8 2.1
C'(70)	163.1 1.2	69.1 3.2
D'(70)	163.4 0.6	69.1 0.9
E'(70)	162.9 0.4	69.4 9.5
F'(20)	162.9 0.6	69.6 6.0
All (n=370)	162.60±0.50	3669.70±0.06
	σ_z(r.m.s.) = 0.007	

SMOOTHED DATA: 100 PER SUBSAMPLE WITH OVERLAPPING OF ± 10

Sample	Period	Epoch
1	162.3±1.0	3667.3±5.9
2	163.3 0.7	68.4 2.8
3	163.0 0.7	69.1 1.7
4	163.2 0.6	69.5 1.5
5	163.8 0.8	70.6 1.6
6	164.7 1.6	69.6 0.2
7	164.0 0.1	69.5 0.5
8	162.8 0.1	68.8 1.6
9	162.8 0.1	69.6 0.7
10	163.0 0.8	70.3 2.7
11	163.2 1.9	74.2 5.2
12	164.6 1.4	69.7 1.8
13	163.5 0.9	69.4 0.2
14	163.4 0.8	69.3 0.4
15	163.8 0.2	69.4 0.4
16	163.4 1.0	70.1 4.6
17	163.2 1.3	70.0 0.8
18	163.2 0.1	70.0 1.0
19	163.2 0.8	69.5 0.3
20	164.2 1.1	69.5 0.7
21	163.5 0.2	70.3 0.7
22	163.7 0.2	70.4 0.1
23	163.2 1.8	69.4 1.6
24	163.5 0.1	69.4 0.8
25	162.9 0.6	69.4 0.1
26	163.4 0.2	69.5 0.2
27	163.0 0.1	69.7 0.6
28	162.8 0.2	70.5 0.3
29	162.9 0.1	69.4 0.5
All (n=370)	162.6±0.5	3669.70±0.06

for each successive subsample the 10 first data while adding the 10 successive ones. These partially overlapped subsamples gave the results reported in Table 2, right side. The overall (n=370) data supply the already reported result:

$$\bar{P}_{all} = 162^{d}.6 \pm 0.5 \ , \quad \bar{E}_{all} = 3669.7 \pm 0.06 \quad (\text{"effective" errors}) \ .$$

We see from Table 2 that each of the 29 subsamples does not allow differences from the average P or E larger than 1 σ. Moreover no systematic trend is inferable from the results in this Table.

We conclude that the period of published averaged Z's seems stable within the obtained effective errors.

The results for the "red" branch are similar, although less reliable because of blending with stationary H_α and A and B telluric bands, as indicated by the larger unit weight error.

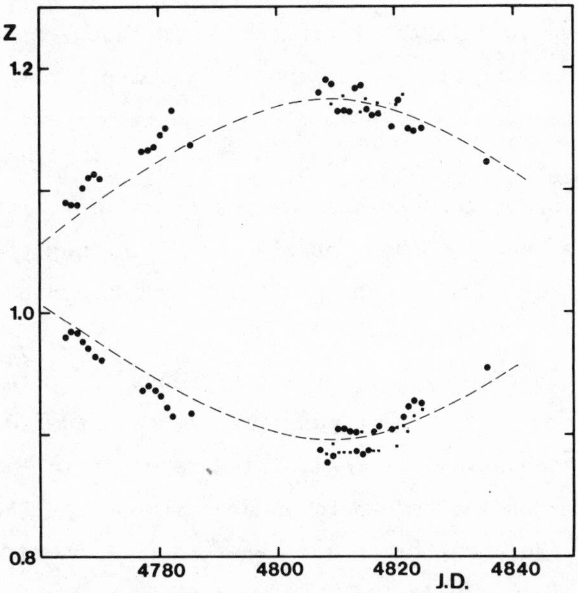

Fig. 2 Observations of moving lines in SS 433 during summer 1981 (sources: Special Astrophysical Observatory USSR and Asiago Observatory).

3. ANTICORRELATED DEVIATIONS FROM SINUSOIDAL TREND

During a coordinated campaign at Asiago Observatory, Special Astrophysical Observatory USSR (6 m telescope), in summer 1981, we have obtained important results. By courtesy of Prof. Kopylov we can report also some of his previous results in advance of publication. The most important are the anticorrelated radial velocity deviations plotted in Figure 2, left part, where no interference of double lines occurs in the two branches. They first demonstrate that the anticorrelation persists twice for about 6 days each, whereas Ciatti et al. (1981a) had shown only one case four days long. Moreover, since spectra are taken one day apart, the light time delay in the system should be less than about this interval, accounting for the inclination of the beams with respect to the line of sight.

Anticorrelation appears also in the right side of the plot, although sometimes disturbed by the presence of double lines.

Stimulated by these findings, we have recognized other slightly less evident anticorrelated deviations, a few of them already independently reported in the literature. In a few cases they have been noticed only on one branch ("red" or "blue"). The epochs of these secondary maxima and/or minima, superimposed on the sinusoidal behavior, are reported in Table 3. It is to be mentioned that in a few cases, in order to recognize minima or maxima of this shorter time trend, we need to assume that strong lines may persist at the same Z up to four days. Such an effect was noticed by Ciatti et al. (1981a).

We rely first upon anticorrelated events free from double lines, in order to attempt to find a possible periodicity from the larger part of the secondary maxima and minima reported in Table 3. Preliminary estimates indicate periods around 6.3 days (6.06, 6.30 and 6.55 days are among best candidates in the literature, see Newsom and Collins, 1980, 1981). The work is in progress.

Our approach is entirely different from that of Newsom and Collins (1981), who found a probable period of $6^{d}.06$ from an

Table 3

SECONDARY MAXIMA AND MINIMA SUPERPOSED TO THE 163^d TREND

Approx. epoch JD 2440000+	Recorded on branches	Sources of data	Approx. epoch JD 2440000+	Recorded on branches	Sources of data
3783	B , R	M	4402	B , R	M
3809.3	B , R	M(note 2)	4405.5	B , R	M, A
3820.0	B	A	4409.5	B , R	M, A
3826	R	A	4412	B , R	M
3833	R	A	4458	(B), R	M
3846	B , R	A, L(note 1)	4465.5	B , R	M, A
3967	B , (R)	A, M	(4468.5)	B , R	M, A
3979.6	B , R	M, S	4487.5	B , R	A, M
3985	B , R	M, A	4490.5	B , R	M, A
(3997)	B , R	A, M	4493.5	B , R	M, A
(4008.3)	B , R	A, M, L	4497	B , R	M
4015.5	B , (R)	A,M,L,S(note 1)	4500	B , R	M
4046	R	M, N	4551	R	M
4069.6	B , R	M, B	4731.9	B , (R)	W(note 4)
(4080)	(B),(R)	M, B	4766	B , R	K
4087	B , R	M, A, Ch	4769.5	B , R	K
(4109.6)	(B),(R)	MU,B,M,A (note 2)	4778	B , R	K
4128	(B), R	B,MU,Ch	4808	B , R	K
4142.3	B , R	M,Ak,Ch	4811.5	B , R	K, A
4169.6	B , R	A, M	4814	B , R	K, A
4185	B , R	A,M,Ch(note 3)	4821	B , R	K, A
(4209)	(B), R	A, M	4893.5	B , R	A
4280.5	B , R	M	(4914.5)	B , R	A
(4342)	B , R	M, A			

Notes: 1 - Noticed by Margon et al. (1979); 2 - Noticed by Margon et al. (1980); 3 - Noticed by Ciatti et al. (1981a); 4 - Noticed by Wagner et al. (1981).

Captions: A - Asiago; B - Bologna; Ch - Chuvaev; Ak - Arhipova et al.; K - Kopylov; L - Liebert et al.; M - Margon et al.; N - Newsom et al.; Mu - Murdin et al.; S - Spinrad; W - Wagner et al.

improvement of the r.m.s. from the fit, when including a short period modulation function. We will defer this job until after we will eventually find a convincing periodicity among the epochs of Table 3.

Both minima and maxima obtained by Wagner et al. (1981) were obtained by averaging double lines according to their respective intensities. As a matter of fact, their first four lower radial velocities appear constant rather than following a sinusoidal trend. Fortunately, their minima and maxima are not much affected if the phenomenon of persistence of lines is actually working.

Thus we think our approach may eventually confirm and improve their result. If a modulation of period around 6 days will be defined, it could explain the observed dispersion of observed data around the sinusoidal line of period 163^d in a wide strip of 100 Å, a situation often entangled by the presence of double lines and their persistence (Mammano et al., 1980, Ciatti et al., 1981a).

We finally stress again the importance of publishing correct JD's and individual Z's per spectrum, together with equivalent widths or, at least, a subjective estimate of intensity.

ACKNOWLEDGEMENTS

A. Mammano is thankful to Prof. L. Rosino for hospitality at Asiago Observatory. We are indebted to Drs. M. Calvani and L. Nobili for discussions and for putting at our disposal computing facilities of the Institute of Physics, University of Padova, and to Mr. M. Franceschi and Alberico Rigoni for valuable use of the Asiago computer and scientific assistance.

REFERENCES

1. F. Ciatti, A. Mammano, A. Vittone: 1981a, Astron. Astrophys. 94, 251.
2. F. Ciatti, A. Mammano, A. Vittone: 1981b, Vistas in Astron. 25, 27.

3. F. Ciatti, A. Mammano, R. Turolla: 1982, in preparation.
4. G.W. Collins, II, G.H. Newsom: 1980, I.A.U. Circular 3547.
5. A. Mammano, F. Ciatti, A. Vittone: 1980, Astron. Astrophys. 85, 14.
6. A. Mammano, R. Margoni, F. Ciatti: 1982, in preparation.
7. B. Margon: 1981, Tenth Texas Symposium on Relativ. Astrophys. (Baltimore 1980).
8. B. Margon, H.C. Ford, S.A. Grandi, R.P.S. Stone: 1979, Astrophys. J. 233, L 63.
9. B. Margon, S.A. Grandi, R.A. Downes: 1980, Astrophys. J. 241, 306.
10. B. Margon, S. Anderson, S. Grandi, R. Downes: 1981a, I.A.U. Circular 3626.
11. B. Margon, S. Anderson, S. Grandi: 1981b, I.A.U. Circular 3649.
12. G.H. Newsom, G.W. Collins, II: 1980, I.A.U. Circular 3459.
13. G.H. Newsom, G.W. Collins, II: 1981, Astron. J. 86, 1250.
14. R.M. Wagner, G.H. Newsom, C.B. Foltz, P.L. Byard: 1981, Astron. J. 86, 1671.

NOTE ADDED IN PROOF:

A period of 6.2851 days has been found by Mammano et al. (1982) for most of the anticorrelated deviations reported in Table 3.

PERIODIC AND SECULAR CHANGES IN SS 433

George W. Collins, II and Gerald H. Newsom

Perkins Observatory, The Ohio State and

Ohio Wesleyan Universities, Columbus and Delaware, Ohio

ABSTRACT

The recent history of SS 433 is reviewed with particular attention being given to the discovery of the periodic phenomena displayed by this object. Several periods ranging from days to months are established as being present in the spectrum of the "moving" lines as well as in other aspects of the emission from the object. In addition evidence for secular change in some of the defining parameters of the system is presented. Although these secular changes may eventually prove to be periodic on a rather long time scale, some interpretation of both the periodic and secular phenomena is possible. It is shown that it is possible to interpret all the known periodic phenomena in terms of a precessing object responding to the time-varying torques that one would expect in a binary system.

I. HISTORY AND OBSERVATIONS

At intervals of about a decade, the recent history of astronomy has been considerably enlivened by the discovery of a new and extra-ordinary class of objects. Two decades ago it was quasars, in the late 1960's and early 1970's it was pulsars and X-ray binary stars. The recent object of interest is SS 433, but in contrast to these other types of highly-publicized objects, SS 433 is the only known member of its class. Attempts to find others have all been fruitless so far, in spite of some rather ingenious detection schemes. If and when a second example is discovered, its similarities and its differences, compared to SS 433 itself, would do much to help understand its strange behavior.

SS 433 received its name by appearing as the 433rd object in the Stephenson-Sanduleak catalogue[1] of stars in the Milky Way with strong Hα emission lines. A year before this publication, however, a weak variable X-ray source had been noted[2] at the same location in the sky with the Ariel-5 satellite. Although the positional coincidence was not immediately noticed, two groups attempted to identify each object in a different region of the electromagnetic spectrum. Clark and Murdin, searching for an optical counterpart of the X-ray source, rediscovered the emission line object SS 433, and noted that it is "richer in emission lines than that for any other compact X-ray object."[3] Seaquist, Gregory and Crane,[4] searching for radio emission from about 20 objects listed as having strong Hα emission in the SS catalogue, found that SS 433 was a highly variable nonthermal radio source in the centimeter wavelength range. (The radio source had previously been catalogued by others,[5,6] but its unusual variability had not been noted.) The attention of Ryle et al.[7] was drawn to the radio source because it is located in the center of a large rather circular region of radio emission[8] which had been identified as a possible supernova remnant.[9] Having drawn attention to itself in three spectral regions, SS 433 soon became the object of detailed spectroscopic study.

The first lines to be identified were prominent lines of H I, He I and He II. However, broad weaker emission lines are also seen at unfamiliar wavelengths. Often these lines consisted of discrete components.[10] It was soon noticed by Mammano, Ciatti and Vittone[10] and Margon[11] that these peculiar features moved across the spectrum! Fabian and Rees[12] proposed that the moving lines originate from Doppler shifts of matter in cool blobs accelerated in two oppositely directed jets. Milgrom[13] interpreted the shifts as periodic variations in Doppler shifts, with a period about 4-6 months; the speed remained constant but the variation in angle between the velocity vectors and the line of sight produced the wavelength changes.

The observational confirmation of the Doppler origin of the shifts, as opposed to such possibilities as large Zeeman splitting,[11] came when several hydrogen and helium lines were found

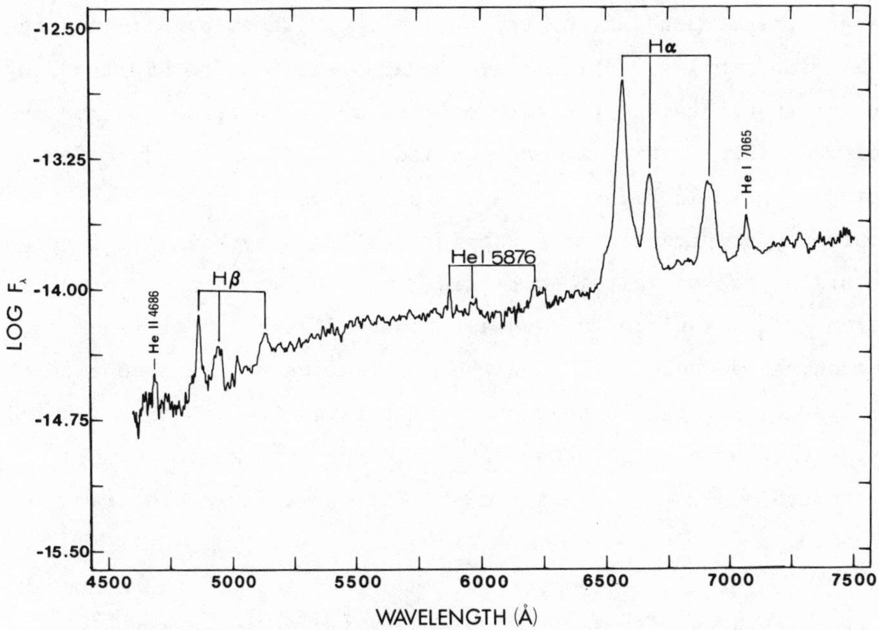

Fig. 1 A Spectrum of SS 433 obtained by Wagner et al.[23] showing the three sets of emission lines of hydrogen and helium: Hα, Hβ and He I λ5876. The left line for each triplet is the "stationary" line, while the other two are the "moving" lines.

by Margon et al.[14] and Liebert et al.[15] to have the same fractional shifts (see Fig. 1). The derived high speeds cause the special relativistic transverse Doppler effect to be significant, and the net red shift in the average of the two jets of Fabian and Rees[12] is now generally used to find the speed, about 0.258 c. Since a transverse Doppler shift is indistinguishable from a gravitational red shift, however, the presence of any gravitational red shift would cause this speed to be overestimated. The periodic behavior of the motions of the lines was also confirmed, although early measurements of this period by various groups often differed by considerably more than their quoted errors, e.g., Ciatti et al.[16] found 165.5±0.16 days and Margon et al.[17] derived 164.0 ± 0.1 days.

The two sets of moving lines, labelled as coming from a "blue beam" and a "red beam," each moved in a fairly pure sine wave with this period, although with considerable night-to-night jitter. The average wavelength for the red beam was displaced 8.3% to a longer wavelength, while the blue beam averaged -1.4%. The amplitude of wavelength excursions for each beam is about 8.5%, so the red and blue beams cross twice in each period. A plot of the data for one cycle is shown in Figure 2. Abell and Margon[18] showed that a simple way to produce these Doppler shifts is by having the velocity vectors of two oppositely-directed jets each move in a conical motion. The displacement and amplitude of each set of lines then depends on the cone angle and the inclination of the cone axis to the line of sight. These two angles turn out to be about 19° and 79°, although which of these angles is the cone angle is not determined a priori. A diagram showing these angles is drawn in Figure 3.

Margon et al.[19] reported evidence that the 19° angle itself varied with an amplitude of 5° and a period of 164 days, but, in common with other observers,[20,21] we find no evidence of any such variation at this period.

While many different ideas have been suggested to yield the

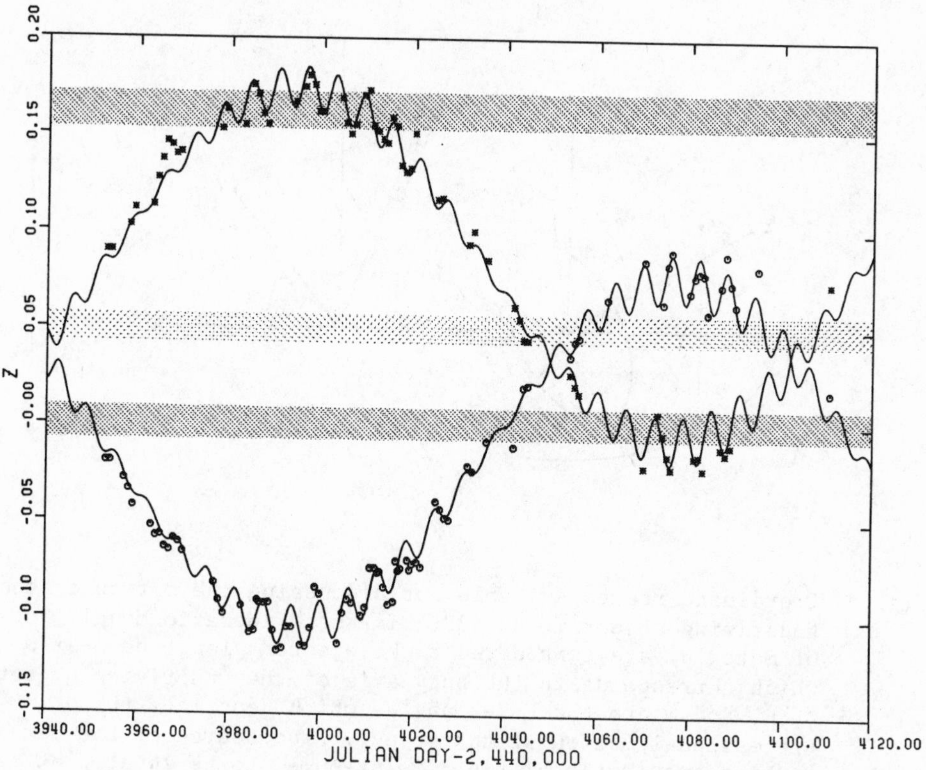

Fig. 2 Wavelength shifts compiled by Collins and Newsom[54] for one full cycle of the "moving" spectral lines. The solid line represents the best least-squares fit to the data and includes the 6-day oscillation as well as the secular changes to the defining parameters. Shaded regions show wavelengths where a moving line is blended with its stationary line, and where moving Hα lines are blended with terrestrial atmospheric absorption bands.

oppositely moving gas streams moving at 1/4 c, some additional observational constraints must be listed.[22]

a) The moving lines, although broad (FWHM of order 50 Å[15,23]), are still much narrower than the amplitude of their Doppler shifts, so the matter is remarkably well collimated to a beam width of order 3°.

b) SS 433 is very energetic. Its absolute magnitude in the V band is about -7,[24,25] and the shape of the continuum agrees with

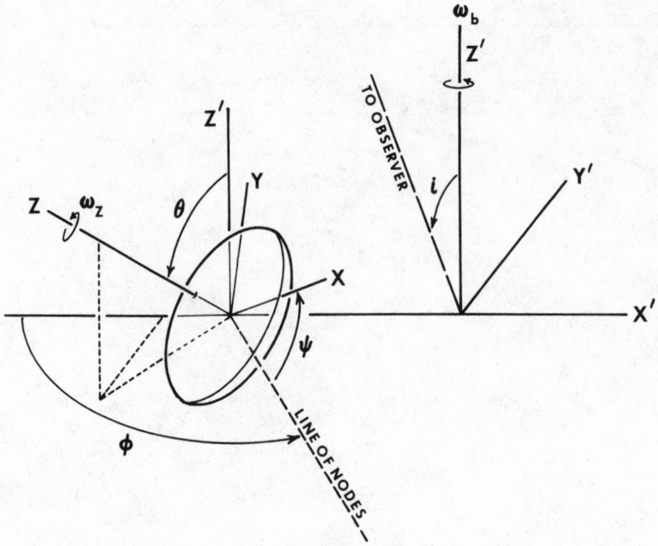

Fig. 3 Coordinate frames suitable for discussing the motion of the underlying object in SS 433. In the "Kinematic Model" jets of material are considered to be ejected along the Z-axis which corresponds to the body axis of the underlying object. ϕ, θ, and ψ are the Euler angles which describes the precessional motion of the object. The source of the perturbing gravitational field is presumed to be located at the origin of X'Y'Z' coordinate frame. We view the object at an angle i to the precessional axis, Z'.

that expected of a highly reddened intrinsically blue star ($T \sim 10^4$-10^5K).[24]

c) The luminosity in each moving Hα line is $\sim 3\times10^{34}$ ergs/sec, or ~ 10 times the solar luminosity.[22] If each Hα photon results simply from recombination of hydrogen in a cooling gas with no repititive ionizations, then a 30 MeV proton is required to obtain each 2 eV photon, and additionally most recombinations will not yield an Hα photon. The 10 $L_\odot$ of photon energy* would then require an input of 10^9 $L_\odot$ as kinetic energy in each beam, an implausibly large value. Hence it seems necessary to arrange for an energy input

*L_0 is the solar luminosity, 4×10^{33} ergs/sec.

into the gas streams to provide multiple excitations or ionizations.

d) Finally, the moving lines come from a relatively cool gas (10^4K).[22] The input of energy must not be too great or else the hydrogen would become all ionized, causing a rapid rise in temperature as the relatively efficient cooling by line emission is lost.

By analogy with other galactic X-ray sources displaying periodic behavior, Collins and Newsom[26] proposed that the object is a binary system. When observational evidence of the binary nature was found,[27] however, the period of orbital motion was much shorter, about 13.1 days. This period showed up as a small variation in wavelength of the peak emission of the "unshifted" lines. While the period is well defined, it is difficult to identify these radial velocities as actual motion of an orbiting object, since the amplitude, phase, and mean velocity are not consistent for all lines. For instance, the mean velocity of the Hβ peak is 223 ± 6 km/s, while that for absorption lines (when they show variations) is -158±9; the variation in the He II λ4686 line has about 2.5 times larger amplitude and occurs early by a quarter phase when compared to Hβ.[28] Hence the motion observed more likely is a result of gas streaming in the binary system, with excitation differing in different parts of the stream.

Once the 13.1-day period was announced, most workers adopted the suggestion first made by Katz[29] and Martin and Rees[30] that the 164-day period results from precession. These two periods, both discovered by wavelength variations in spectral lines, were soon detected by other techniques. Baliunas et al.[31] found indications that the central intentsity of the unshifted Hα line varied in a 13.1-day period, and this variation in the unshifted H and He lines was soon confirmed by several groups.[17,32,33] Broad band photometry showed the integrated brightness of SS 433 also varied in a 13.1-day period,[34,35] but with an amplitude and shape of the light curve which depended on phase of precessional period.[36] Gladyshev et al.[37] used these brightness variations to derive an orbital period of 13.0848 ± 0.0004 days. In addition Kemp[34] found a broad band photometric variation with a 164-day period. This period is also reflected in the broad-band linear polarimetric data.[38]

The unshifted lines also showed effects of the 164-day period; Crampton and Hutchings[32,39] found that the strength of the few absorption lines seen in the spectrum, as well as the profiles of the emission lines, varied with precessional phase. Also the visual and infrared brightness of SS 433 was found to change with the same period.[34,40,41,42]

A completely different manifestation of the 164-day period was discovered by high resolution radio maps of SS 433, both with the Very Large Array[43] and with VLBI techniques.[44] The maps show two jets on opposite sides of the central point source; the jets, with 8-20% linear polarization, oscillate with a 164-day period and 20° amplitude, the same as one possible cone angle of the "kinematic model", about position angles of 100° and 280°. Indications of outward moving condensations are seen on the maps. Although it cannot be proved that this motion represents an extension of the same motion that produces the large Doppler shifts, the connection is highly plausible. If the distance to SS 433 is 5.5 kpc (in reasonable agreement with distances measured by other techniques),[45,46] the outward flow would have the same speed as that required to produce the transverse Doppler shift in the moving lines.

The alignment of the radio jets is significant; X-ray jets, faint filaments of gas seen in red light, and bulges in the apparent supernova remnant W50 all fall along this same axis, as will be discussed later in this session.[47]

Harmonics of the 13.1- and 164-day periods are also seen in the photometric data. Kemp et al.[34] found a brightness variation at 6.54 days, half the orbital period, and Mazeh et al.[42] used a similar technique to find a period of either 6.55 or 6.43 days, although the former would appear to be the more likely. In addition, Margon et al.[19] found indications that this first harmonic period occurs in the peak to continuum ratio of the "stationary" Hα line.

Variations at the first harmonic of the orbital period can be explained as an aspect effect, caused by eclipses or by changes in the projected area of, for example, a tidally distorted member of a close binary system. A second harmonic of any period is harder to explain; hence the origin of a 55-day period (1/3 of 164 days) reported in the visual brightness of SS 433[42] is mysterious.

It must be emphasized in a discussion of periodicities that, whether measuring wavelengths of moving lines or brightness, these periods appear in the midst of considerable scatter or noise in the data. Large optical outbursts have been noted by several groups. Mazeh et al.[42], for example, noted a brightening in B magnitude of 0.76 mag in 3.5 hours, and Wagner et al.[23] found decreases of up to a factor of 25 in the equivalent widths of some of the stationary lines in a few days. The moving lines also vary in brightness,[16,46] but a systematic study of these variations has not been published. The lines appear to move not by a simple motion on a spectrum, but rather by having one component of the line fade away as another component brightens, with each component keeping a fixed wavelength.[33] At S- and X-band radio wavelengths, SS 433 sometimes undergoes rather dramatic flaring episodes for several months, but it can also remain relatively quiet for extended periods.[48] A long-term monitoring program to correlate changes in X-ray, radio and optical brightness and moving and stationary line intensity has not been undertaken, but would be highly desirable.

In view of the often asymmetrical nature of the moving lines, sometimes clearly separated into components, it is not surprising that different observers watching simultaneously report different values for the shift of a line; the definition of effective line center has often been subjective. The residuals from the 5 parameter kinematic model of Abell and Margon[18] were still considerably larger than could be explained by this difficulty, however. To see if some of the scatter could result from a short-period variation in the wavelengths of the moving lines, we searched via a least-squares technique for a significant period. To allow maximum generality, we used a numerical model which allowed the red and blue beams to vary independently with periodic variations in cone angle, azimuth, and speed, with the only restriction being that all these variables must very with the same period. This requires a total of 13 parameters to specify the variation. The significance of the resulting fit to the data was found from the F-statistic, which was computed for an array of test periods.[49] The most significant period was 6.06 days, a number which equals half the period found by adding the angular velocities of the 13.1- and 164-day periods.[49] If we adopted the 19° angle of the kinematic model as the inclination angle and 79° as the cone angle, the 6.06-day variation was dominated by changes in the speed of the emitting regions. Interchanging the values for these angles gave an insignificant variation in speed, with most of the variation in radial velocity resulting from a 3° oscillation of the cone angle about its mean. The red and blue beams displayed variations with opposite phase, indicating that their common axis remained rigid through the 6-day period. We could thus fit the residuals almost as well with a 5 parameter model (phase and amplitude of both cone and azimuth angles, plus period). A plot of derived F-statistic vs period is shown in the upper curve of Figure 4. Many periods on both sides of the 6.06-day peak appear much more significant than the P=.001 level of chance occurrence shown. That these are aliases of the 6.06-day period is demonstrated by subtracting this period from the observed data, as shown in the lower part of Figure 4. A fit to the data for one precessional cycle is shown in Figure 2. The fit to the 6-day period is dominated by a sine curve with 6.06-day period multiplied by the cosine of the phase of the 164-day period. This behavior results from the fact that,

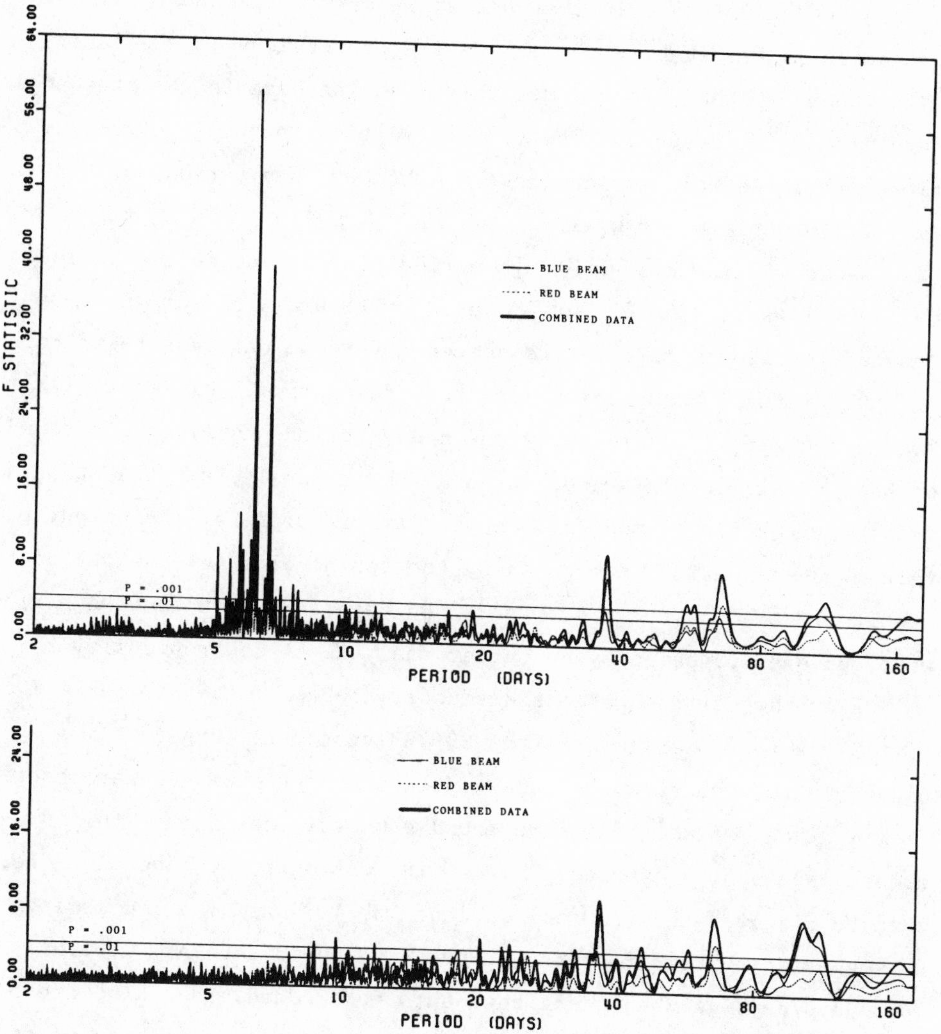

Fig. 4 The F-statistic which measures parameter significance vs period for the residuals of the Kinematic Model fit to the data. The upper graph clearly indicates the presence of the 6.06 day period as well as some alias periods near by. That these periods are indeed aliases is demonstrated by the lower graph where the 6.06-periodicity has been removed from the data and the analysis repeated. There is some suggestion that there are long term variations with periods greater than 35-days. The P = .01 and P =.001 lines denote the levels of 1% and 0.1% chance occurrence.

at the velocity maximum, when the blue beam for example is directed toward us, an increase in the cone angle due to the 6-day period causes an increased blue shift. Half a precessional cycle later, with the same beam giving a red shift, an increase in the cone angle leads to an increased red shift. In addition to the spectral variations, there is some indication that the 6.06-day period may appear in unfiltered light as well.[34]

As if we didn't have enough periods in this object, a period of several weeks (36, 63, or 98 days) in Figure 4 appears to be a real effect, but more data are needed before we can be sure of this.

Our initial announcement of the 6-day period[50] did meet with some disbelief,[51] but recently the 6-day period has also been found by Katz et al.[52]. This group used a Fourier analysis rather than a least-squares fit. However, the Fourier transform of a sine curve with 6.06-day period multiplied by the cosine of a 164-day period results in two peaks, at 6.30 and 5.85 days. Hence these two periods were the ones revealed by a Fourier analysis of the residuals.

A summary of the periods discussed above is given in Table I.

Since the power loss in the SS 433 system is large, it is plausible to expect some of these periods might show secular changes. From a 3-year span of data, we found some evidence for a slow decrease in the 6-day period.[49] We were most surprised shortly afterward to discover that the precessional period was also decreasing, as suggested by Ciatti et al.[21], but at a rapid rate of -0.011 ± .002 days/day.[53] When additional data were added, the value dropped slightly to -0.010 ± .002.[54] If this rapid decrease were to continue, the period would reach zero in 40 years, so obviously it was a trend that required monitoring. As with the 6-day period, our announcement of this decrease also met with some disbelief,[55] but it was subsequently confirmed by Blair[56] and by Margon et al.[57]. With such a large value of $\dot{P}$, it was plausible to expect to find significant higher order derivatives. With data from 1978 through June 1981, we found a marginally significant positive value of $\ddot{P}$,

Table I. Periodic Phenomena Reported for SS 433

Period (days)	Label	Mode(s) of Detection
164 (variable)	Precessional	Wavelengths of "moving lines" Strength of "stationary" absorption lines Profiles of "stationary" emission lines Broad band photometry Position angle of radio-emitting "jets" Amplitude and shape of 13.1-day variation in visual brightness Modulation of 6.06-day period in wavelengths of moving lines Variation in linear polarization
55	1/3 Precessional	Visual brightness
13.1	Orbital	Wavelength of peak of "stationary" lines Central intensity and equivalent widths of "stationary" lines Visual photometry
6.54	1/2 orbital	Broad band visual photometry Peak to continuum ratio for "stationary" Hα
6.06 (variable)	1/2 synodic	Wavelengths of "moving" lines Visual photometry in unfiltered light (?)
6.3,5.8 (variable)	Equivalent to 1/2 synodic	Fourier transform of 6.06-day period

indicating that the rate of decrease in the period was slowing down.[58] More recently, Anderson et al.[59] concluded that the period has begun increasing. Curiously enough, this change in period does not appear to be the main cause of the discrepancies in measured period mentioned earlier. The 165.5 ± 0.16 day period of Ciatti et al.[16] and the 164.0 ± 0.3 day period of Margon et al.[19] both refer to a mean epoch of about J.D. 2443950; yet Anderson et al.[59], using presumably the same data base as that of Margon et al.[19], derived a mean period of about 166.0 ± 0.4 for about this same epoch, in agreement with the value of Ciatti et al.[16] but not of Margon et al.[19]

Since the 6.06-day period is derived from the angular velocities of both the orbital and precessional periods, we expect the decrease in precessional period to lead to a decrease in the 6.06-day period. This effect, suspected by Newsom and Collins,[49] was well established by Wagner et al.[23] Since changes in both precessional and 6-day periods have been independently measured, the rate of change of the orbital period may be determined. Within the errors it appears that the orbital period of the binary has remained constant.[23] Since several of the defining parameters of the kinematic model are clearly not constant, any value assigned to them must include an epoch at which the values are presumed to be instantaneously correct. In Table II we give a summary of our analysis, based on data through June 1981, with the "best" values for the model parameters and their secular changes appropriate for the mean epoch of the data. In the next section we shall provide a basis for understanding these variations - periodic and secular.

II. A DYNAMICAL DESCRIPTION OF SS 433

A. Periodic Phenomena

Having established that the spectra of the moving lines exhibit several periodic and secular characteristics, it is logical to ask if there is a single conceptual framework on which to base an understanding of these phenomena. The kinematical description basically involves two antiparallel velocity vectors which, as they move uniformly, generate a surface of a cone. The angle between the vector and the axis is called the cone angle θ while the axis makes an angle i with the observer's line-of-sight. This simple picture serves well to describe the gross characteristics of the moving line spectrum although it is by no means unique. However, it may

Table II. Values of the Periodic and Secular Parameters Describing SS 433

Parameter	Secular Change
$\gamma = 1.0351 \pm 0.0004$	$\dot{\gamma} = (6 \pm 12)\times 10^{-7}\ \mathrm{Day}^{-1}$
$i = (78.87 \pm 0.11)$ Deg.	$(\dot{i}) = (2 \pm 3)\times 10^{-4}$ Deg/Day
$\theta = (20.05 \pm 0.16)$ Deg.	$\dot{\theta} = (-1.5 \pm 0.5)\times 10^{-3}$ Deg/Day
$P_o = 163.12 \pm 0.01$ Day	$\dot{P}_o = (-0.0087 \pm 0.0006)$ Days/Day
	$\ddot{P}_o = (9.0 \pm 5.6)\times 10^{-6}$ Days/Day
$\phi(3650) = (66.21 \pm 0.64)$ Deg.	
$P_6 = (6.055 \pm .015)$ Days	$\dot{P}_6 = (-3.0 \pm 1.1)\times 10^{-5}$ Days/Day
$P_B = (13.08 \pm 0.07)$ Days[27]	$\dot{P}_B = (-1.4 \pm 3)\times 10^{-5}$ Days/Day

All values refer to an Epoch of J.D.2444167.5

serve as a framework within which to investigate the more subtle behavior of the moving line spectra.

It is generally assumed that the spectra of the "moving" lines reflect the motion of some underlying object. It should be noted that the "moving" lines appear to consist of several components of time-varying strength which seem, once formed, to hold their wavelength constant.[33] However, the mean wavelength of the component ensemble which makes up a moving line does exhibit the smooth single-value motion that one is inclined to associate with the motion of a central source.

As we have seen, SS 433 contains a low-power X-ray binary system with a 13.1 day period. It seems plausible to assume that the source of the moving line emission is a member or associated with one of the members of this binary system. With these two assumptions in mind it is reasonable to ask if the dynamical effects that one expects to be operative within a binary system are sufficient to describe the behavior of the moving lines. There is a general predilection among investigators to associate the large amplitude 164-day variation with the precession of some object.

However, with the exception of precession arising from the General Theory of Relativity, little quantification of the type of precessional effects that may be present is to be found in the literature. In the balance of this paper we shall investigate the extent to which the variation of the moving lines can be understood within the framework of classical mechanics.

If one assumes that the object is kinematically linked to the regions emitting the moving lines and, while subject to time-varying torques from the second member of the binary system, maintains its form in spite of these torques, then one can quantify the subsequent response to those torques. It is worth noting that very little need be assumed concerning the nature of the object. It could be a neutron star, thin or thick disk, or perhaps the envelope of a stellar component of the system. Even the association of the object with either the primary or secondary is arbitrary. With this in mind we offer Figure 3 as a plausible reference frame for the quantification of the object. The angles ψ, ϕ, and θ are the Euler angles as defined by Goldstein.[60]

Since the precessing object is subject to orbital motion by virtue of its being in a binary system, some care must be exercised in developing the Lagrangian for the motion of the object. Utilizing the development of Landau and Lifshitz[61] for the Lagrangian in a noninertial coordinate frame, the Lagrangian and Hamiltonian for the object in terms of the Euler angles become

$$L = \frac{1}{2}I_x(\dot{\theta}^2+\dot{\phi}^2 \sin^2\theta)+\frac{1}{2}I_z(\dot{\psi}+\dot{\phi}\cos\theta)^2+\omega_b(I_x \sin^2\theta+I_z \cos^2\theta)(\dot{\phi}+\frac{1}{2}\omega_b)$$
$$+I_z\omega_b\dot{\psi}\cos\theta - V_o(1-3\sin^2\theta \sin^2\phi) \quad , \tag{1}$$
$$H = \frac{1}{2}I_x[\dot{\theta}^2+(\dot{\phi}+\omega_b)^2 \sin^2\theta] - [I_z\tilde{\omega}\cos\theta+I_x(\dot{\phi}+\omega_b)\sin^2\theta]\omega_b$$
$$+\frac{1}{2}I_z\tilde{\omega}^2 + V_o(1-3\sin^2\theta \sin^2\phi) \ ,$$

where ω_b is the orbital frequency of the binary system. We have made use of the development of Brouwer and Clemence[62] to obtain the potential amplitude

$$V_o = \frac{1}{2}\,\omega_b^2\,\mu\,(I_x - I_z) \quad , \tag{2}$$

where μ is the reduced mass of the system while I_x and I_z are just the moments of inertia about the principle axes of the object. An additional assumption is made here in that the object is assumed to be axisymmetric about z and that the potential is sufficiently approximated by including the J_2 term. The parameter $\tilde{\omega}$ is a constant of the motion resulting from the cyclic nature of ψ.

We note that the body axis of the object will appear to an inertial observer to move about the normal to the binary orbit with an angular frequency $(\dot{\phi}+\omega_b)$ which we shall henceforth denote by ω_o and identify with the precessional frequency.

$$\left.\begin{aligned} \text{Since } \langle\ddot{\theta}_o\rangle &= 0 \\ \langle\omega_o\rangle &= \text{Const.} \\ \langle\psi_o\rangle &= \text{Const.} \end{aligned}\right\} \tag{3}$$

represent a solution to the equation of motion derived from the Lagrangian and averaged over ϕ, it is useful to develop the equation of motion in terms of the variables

$$\left.\begin{aligned} \delta\psi &= \langle\psi_o\rangle - \psi \\ \delta\phi &= \langle\phi_o\rangle - \phi \\ \delta\theta &= \langle\theta\ \rangle - \theta \end{aligned}\right\} \quad . \tag{4}$$

Should the solution to the equation of motion expressed in terms of the variables given by (4) be simply periodic in ϕ then we may expect the solution given by (3) to represent the average behavior of the object. This solution has been called pseudoregular precession by Klein and Sommerfeld.[63] By utilizing the Hamiltonian and the other constant of the motion, after considerable algebra the equations of motion resulting from the Lagrangian reduce to

$$\delta''\theta \sin^2\langle\phi_o\rangle - \delta'\theta\, A \cos 2\langle\phi_o\rangle + \delta\theta[B \sin 2\langle\phi_o\rangle + C \sin 4\langle\phi_o\rangle]$$

$$= Q \sin 2\langle\phi_o\rangle \quad ,$$

where

$$A = [2 - (\langle\omega_o\rangle\omega/_s)(1-q_o)] \cos^2\langle\theta_o\rangle \quad ,$$

$$B = 4 \cos\langle\theta_o\rangle + (\langle\omega_o\rangle/\omega_s)^2(1+q_o) \sin^2\langle\theta_o\rangle \quad ,$$

$$C = (\langle\omega_o\rangle/\omega_s)^2 q_o \quad ,$$

$$Q = (\langle\omega_o\rangle/\omega_s)^2[(\delta\tilde{\omega}/\tilde{\omega})\ (1-q_o) \sin^2\theta_o - 2\delta H/I_x\langle\omega_o\rangle^2] \cot\langle\theta_o\rangle \ , \quad (5)$$

and $\omega_s \equiv \langle\dot{\phi}_o\rangle \quad ,$

$$q_o \equiv 3V_o/I_x\langle\omega_o\rangle^2 \quad .$$

Here we have replaced the independent variable time by $\langle\theta_o\rangle$ and $'$ denotes differentiation with respect to $\langle\phi_o\rangle$. Floquet Theory[64] suggests that the general solution to equation (5) exhibits simply periodic solutions of period π. In the case of equation (5), one can further show that these solutions are stable. Hamilton and Sarazin[65] reach a similar conclusion from a rather general treatment of gravitational spin-orbit and spin-spin coupling in binary systems. Thus we may expect the object to exhibit steady precession characteristic of the pseudoregular solution given by equation (3) modulated by variations in the Euler angles with a frequency $2\omega_s$. This is not a surprising result as $2\omega_s$ is the frequency of the driving torque (whose average value is nonzero) which the object experiences in the binary system. Thus the average value of the torque results in the object being driven to the pseudoregular precession solution while the time variation produces a modulation at frequency $2\omega_s$.

One characteristic of the pseudoregular solution pointed out by Collins, Newsom and Boyd[22] is that the driven precession is retrograde for oblate objects. Thus if one identifies the 164-day periodic variation with the steady retrograde driven precession of frequency ω_o, then the high frequency synodic oscillation $2\omega_s$ would correspond to variations with a period of 6.06 days. Collins, Newsom and Boyd[22] predicted the existence of this period on a somewhat phenomenological basis prior to development of the proof outlined here. With the announcement of the detection of a 6 day periodicity by Mammano et al.[66] and the Fourier equivalent periodicities by Katz et al.[52] we feel that the existence of this period is firmly established. Since the value of the period requires that precession be retrograde, we may take the observational values as confirmatory evidence that the precession is consistent with the forced precession of a slowly spinning oblate object.[22] It should be emphasized that this interpretation is not unique but certainly is strongly suggestive.

B. Secular Phenomena

We have already indicated that some of the canonical parameters of the kinematic description are subject to changes on a time scale greater than the precessional period. It is reasonable to ask if the variations which are observed can be described in terms of the dynamical structure we have suggested. Anyone who has observed a spinning gyroscope slow down as a result of the friction loss of energy and angular momentum knows that as the spin rate decreases the precession rate increases and the gyroscope falls over (cone angle increases). Thus one should not be surprised if an orbiting gyroscope should straighten up (cone angle decreases) and precess more rapidly as spin angular momentum and energy are lost from the system. However, the relation between cone angle and precessional rate changes is not arbitrary but depends on the nature of the gyroscope and perturbing torques. This relationship may be quantified in terms of the 1st order variational equations of motion averaged over phase ϕ. These equations are

$$
\begin{aligned}
&\delta''\langle\psi_o\rangle + \delta''\langle\phi_o\rangle + \delta'\langle\theta_o\rangle\,(\langle\omega_o\rangle/\omega_s)\,\sin\langle\theta_o\rangle = 0\ ,\\
&\delta''\langle\phi_o\rangle + \delta'\langle\theta_o\rangle\,(\langle\omega_o\rangle/\omega_s)(1-q_o)\,\cot\langle\theta_o\rangle = 0\ ,\\
&\delta''\langle\theta_o\rangle - \delta'\langle\phi_o\rangle\,(\langle\omega_o\rangle/\omega_s)(1-q_o)\,\sin\langle\theta_o\rangle\cos\langle\theta_\omega\rangle\ ,\\
&\quad + \delta\langle\theta_o\rangle(\langle\omega_o/\omega_s)^2(1+q_o)\,\sin^2\langle\theta_o\rangle = (\delta\tilde{\omega}/\tilde{\omega})(\langle\omega_o\rangle/\omega_s)^2(1-q_o)\\
&\quad \times \sin\langle\theta_o\rangle\cos\langle\theta_o\rangle\ .
\end{aligned}
\tag{6}
$$

Mindful of the fact that period changes basically represent accelerations, we see that the second of equations (6) bears directly on the relationship between variations in precessional period and cone angle. Indeed, if one regards the binary period as constant we obtain

$$
\dot{P}_o = \langle\dot{\theta}_o\rangle\, P_o(1-q_o)\,\cot\langle\theta_o\rangle\ . \tag{7}
$$

As we have already seen, the value of $\dot{P}_B$ is satisfyingly small and not statistically different from zero, assuring the validity of equation (7). Unfortunately the observed values of $\langle\theta_o\rangle$ and $\dot{P}_o$ require $q_o>0$. For oblate objects $q_o<0$. Therefore it is clear that we cannot accept the intuitively appealing picture of the gyroscope losing spin angular momentum to describe the secular changes which are observed.

If this dynamical description is to remain viable for

understanding the behavior of the moving line spectra of SS 433, only a limited number of possibilities exist for the explanation of the secular variations. Certainly mass exchange in the binary system can result in changes in the dynamical aspects of the system which have not been incorporated into the arguments presented here.

However, within the classical dynamical framework it seems unlikely that plausible mass transfer could produce the very large rate of change (~1%) observed in the period. Thus if one is to avoid very large mass transfer rates which would surely affect the binary period at a detectable level, it would appear that one must look elsewhere for an explanation of the secular effects. While it is certainly possible that a different dynamical picture utilizing General Relativistic effects such as that suggested by Fang LiZhi et al.[67] may explain the secular effects, there is one last classical dynamical effect which is likely to occur in spinning objects which we have not included.

Should the spin axis and body axis be slightly misaligned, then the spin axis will execute a slow rotation about the body axis. This phenomenon, which is often called free precession, would accompany the previously described forced precessional motion of the body axis. The tip of the spin vector would then be expected to trace out a cycloid-like motion on a cone as seen by an inertial observer (see Figure 5).

This would cause the precession rate to alternately run faster and then slower than the pseudoregular precession rate $\langle\omega_o\rangle$, oscillating with a characteristic frequency ω_f. This would be coupled with a similarly periodic variation in the cone angle θ. The observed values of these rates would be approximately

$$\begin{aligned} \omega_{obs} &\cong \langle\omega_o\rangle + \omega_z\, \alpha \cos \omega_f t \quad , \\ \theta_{obs} &\cong \langle\theta_o\rangle + \alpha \sin \omega_f t \quad , \end{aligned} \tag{8}$$

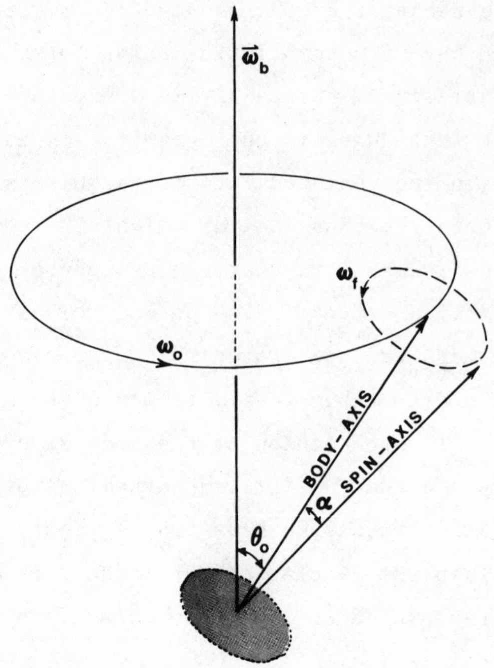

Fig. 5 A schematic representation of the effect of free-body precession ω_f superimposed on the forced precession ω_o. The free-body precession results from a possible misalignment by an angle α between the spin axis and body axis of the object.

where ω_z is the angular velocity of the object about its spin axis, and α is the angle between the spin and body axes. Equation (8) implies that the free precession period is given by

$$P_f = 2\pi\alpha[(P_z/P_o)^2(\dot{P}_o)^2+\langle\dot{\theta}\rangle^2]^{-\frac{1}{2}} \quad , \tag{9}$$

where P_z is the axial spin period.

One normally thinks of free precession occurring on a much shorter time scale than driven precession. Such is certainly the case for the earth. However, if the spin rate is quite low, just the reverse is the case and the free precession period can greatly exceed the driven precession.

We have seen in the previous section that the periodic aspects of the line motion was compatible with driven precession of a very slowly spinning object. If one evaluates equation (9) for small misalignments (i.e. α of the order of a few degrees), and a spin period P_z of the order of the forced precession period P_o we see that one would expect free precession periods of the order of decades. Thus the observed values of $\dot{P}_o$ and $<\dot{\theta}_o>$ are not in conflict with the interpretation of the object experiencing slow free precession in addition to the motion previously discussed. It must be recognized that the total span of modern observations does not even cover one estimated period of the free precession. However, the nature of the variation derived from free precession in the precessional period and cone angle is so specific that one would expect the hypothesis to be confirmed or rejected in the near future.

IV. SUMMARY AND CONCLUSIONS

In this paper we have endeavored to analyze the motion of the 'moving' lines found in the spectra of SS 433. Periodic variations seem to be one of the hallmarks of this object and their nature constitutes one of the primary diagnostic tools for the analysis of the system.

The existence of the periodic phenomena now seems well established. However the nature of the secular variation is still a matter of some controversy. In this work we have tried to provide a basis for describing all kinematic aspects of the moving line spectra of SS 433. We have seen that all presently known properties are consistent with classical dynamical processes which should be present in the system. However, we emphasize that there are General Relativistic effects which would provide a similar explanation for most of the phenomena. It remains to be seen what the relative importance of these effects is when compared to the classical effects.

In our view the dynamical description outlined here has been rather successful. The notion of forced retrograde precession required the existence of a short term periodicity of a precise value. The establishment and subsequent confirmation of this period must be considered a major success for the theory. The secular changes in both the precessional period and the six day period are linked with the dynamical picture in that their independent determination implies an orbital period change which is consistent with estimates of the life-time of the system.[68.69] We have shown that it is possible to understand even these secular changes in terms of dynamical free precession. Although the uncertainty in the secular variations make the verification of this hypothesis tentative at best, the hypothesis is highly testable.

Although we have limited our discussion to a specific aspect of SS 433, namely the moving lines, none of the conclusions are known by us to be in conflict with any other observational aspect of the phenomena. This is not to say that within the context presented here the object stands fully described. There are many other aspects of the phenomena which at present are only dimly understood. However, if one considers the expansion in the state of knowledge of SS 433 during the past two years one is entitled to optimistically believe that this enigma will eventually be unraveled.

We would like to take this opportunity to thank those observers who generously shared their data with us. Particular thanks are due Professor Mammano et al. who first showed that the lines moved and drew our attention to the component structure of the moving lines, and W.P. Blair whose 1981 data established the existence of the secular changes. In addition we are indebted to Professor Remo Ruffini, for without his aid and support this paper would not have been possible.

It is a pleasure to acknowledge Mark Wagner and Ken Rumstay for aid in preparing figures for this paper.

REFERENCES

1. Stephenson, C.B. and Sanduleak, N. Astrophys. J. Suppl., 33, 459-469, 1977.
2. Seward, F.D., Page, C.G., Turner, M.J.L. and Pounds, K.A. Mon. Not. R. Astron. Soc., 175, 39p-46p, 1976.
3. Clark, D.H. and Murdin, P. Nature, 276, 44-45, 1978.
4. Seaquist, E.R., Gregory, P.C. and Crane, P.C. Int. Astron. Union Circ. 3256, 1978.
5. Velusamy, T. and Kundu, M.R. Astron. Astrophys. 32, 375-90, 1974.
6. Clark, D.H., Green, A.J. and Caswell, J.L. Aust. J. Phys. Astrophys. Suppl., No. 37, 75-86, 1975.
7. Ryle, M., Caswell, J.L., Hine, G. and Shakeshaft, J. Nature, 276, 571-573, 1978.
8. Westerhout, G. Bull. Astron. Inst. Neth., 14, No. 488, 215-260, 1958.
9. Holden, D.J. and Caswell, J.L. Mon. Not. R. Astron. Soc., 143, 407-435, 1969.
10. Mammano, A., Ciatti, F. and Vittone, A. Paper submitted to Nature, subsequently published in Astron. Astrophys., 85, 14-19, 1980.
11. Margon, B. in "Ninth Texas Symposium on Relativistic Astrophysics," ed. Ehlers, J., Perry, J. and Walker, M. Ann. N.Y. Acad. Sci., 336, 550-565, 1978.
12. Fabian, A.C. and Rees, M.J. Mon. Not. R. Astron. Soc., 187, 13p-16p, 1979.
13. Milgrom, M. Astron. Astrophys. 76, L3-L6, 1979.
14. Margon, B. et al., Int. Astron. Union Circ. 3345, 1979.
15. Liebert, J., Angel, J.R.P., Hege, E.K., Martin, P.G. and Blair, W.P. Int. Astron. Union Circ. 3345, 1979; Nature, 279, 384-87, 1979.
16. Ciatti, F., Mammano, A. and Vittone, A. Astron. Astrophys. 94, 251-58, 1981.

17. Margon, B., Grandi, S.A., Downes, R., Ford, H.C., Aller, L.H., Plavec, M., Stone, R.P.S., Spinrad, H., Stauffer, J., Burbidge, E.M., Junkkarinen, V.T., Koski, A.T. and Smith, H.E. Bull. American Astron. Soc., 11, 671, 1979.

18. Abell, G.O. and Margon, B. Nature, 279, 701-703, 1979.

19. Margon, B., Grandi, S.A. and Downes, R.A. Astrophys. J., 241, 306-315, 1980.

20. Firmani, C. Paper at SS 433 session, IAU General Assembly, Montreal, August 1979.

21. Ciatti, F., Mammano, A. and Vittone, A. Vistas Astron., 25, 27-30, 1981.

22. Collins, G.W., II, Newsom, G.H. and Boyd, R.N. Astrophys. Space Sci., 76, 417-440, 1981.

23. Wagner, R.M., Newsom, G.H., Foltz, C.B. and Byard, P.L. Astron. J. 86, 1671-78, 1981.

24. Murdin, P., Clark, D.H. and Martin, P.G. Mon. Not. R. Astron. Soc., 193, 135-151, 1980.

25. Dopita, M.A. and Cherepashchuk, A.M. Vistas Astron., 25, 51-59, 1981.

26. Collins, G.W., II, and Newsom, G.H. Nature, 280, 474-475, 1979.

27. Crampton, D., Cowley, A.P. and Hutchings, J.B. Int. Astron. Union Circ. 3388, 1979; Astrophys. J., 235, L131-L135, 1980.

28. Crampton, D. and Hutchings, J.B. Astrophys. J., 251, 604-610, 1981.

29. Katz, J.I., Astrophys. J., 236, L127-L130, 1980.

30. Martin, P.G. and Rees, M.J. Mon. Not. R. Astron. Soc., 189, 19p-22p, 1979.

31. Baliunas, S., Noyes, R., Liller, W. and Tokarz, S. Int. Astron. Union Circ. 3410, 1979.

32. Crampton, D. Bull. American Astron. Soc., 11, 671-2, 1979.

33. Liller, W., Noyes, R., Davis, M., Baliunas, S., Sternberg, A. and Tokarz, S., Bull. American Astron. Soc., 11, 732, 1979.

34. Kemp, J.C., Barbour, M.S., Kemp, G.N. and Hagood, D.M. Vistas Astron., 25, 31-43, 1981.
35. Cherepashchuk, A.M. Mon. Not. R. Astron. Soc., 194, 761-769, 1981.
36. Gladyshev, S.A., Goranskij, V.P., Kurochkin, N.E. and Cherepashchuk, A.M. Astron. Circ. (USSR), No. 1146, 1980.
37. Gladyshev, S.A., Goranskij, V.P., Kurochkin, N.E. and Cherepashchuk, A.M. Astron. Circ. (USSR), No. 1145, 1980.
38. McLean, I.S. and Tapia, S. Vistas Astron., 25, 45-50, 1981.
39. Crampton, D. and Hutchings, J.B. Vistas Astron., 25, 13-21, 1981.
40. Gladyshev, S.A. Astron. Circ. (USSR), No. 1138, 1980.
41. Catchpole, R.M., Glass, I.S., Carter, B.S. and Roberts, G. Nature, 291, 392-394, 1981.
42. Mazeh, T., Leibowitz, E.M. and Lahav, O. Astrophys. Lett., 22, 185-191, 1981.
43. Hjellming, R.M. and Johnston, K.J. Nature, 290, 100-107, 1981; Astrophys. J., 246, L141-L145, 1981.
44. Niell, A.E., Lockhart, T.G. and Preston, R.A. Astrophys. J., 250, 248-253, 1981.
45. van Gorkom, J.H., Goss, W.M. and Shaver, P.A. Astron. Astrophys., 82, L1-L2, 1980.
46. Margon, B., Ford, H.C., Katz, J.L., Kwitter, K.B., Ulrich, R.K., Stone, R.P.S. and Klemola, A. Astrophys. J., 230, L41-L45, 1979.
47. Grindlay, J. Proc. Orbis Scientiae (this volume), 1982.
48. Johnston, K.J., Santini, N.J., Spencer, J.H., Klepczynski, W.J., Kaplan, G.H., Josties, F.J., Angerhofer, P.E., Florkowski, D.R. and Matsakis, D.N. Astron. J., 86, 1377-83, 1981.
49. Newsom, G.H. and Collins, G.W., II. Astron. J., 86, 1250-58, 1981.

50. Newsom, G.H. and Collins, G.W., II. Int. Astron. Union Circ. 3459, 1980.

51. Margon, B., Grandi, S.A. and Downes, R.A. Int. Astron. Union Circ. 3469, 1980.

52. Katz, J.I., Anderson, S.F., Margon, B. and Grandi, S.A. Bull. American Astron. Soc., 13, 801, 1981.

53. Collins, G.W., II, and Newsom, G.H. Int. Astron. Union Circ. 3547, 1980.

54. Collins, G.W., II, and Newsom, G.H. Paper submitted to Nature, subsequently published in Astrophys. Space Sci., 81, 199-208 (1982).

55. Margon, B. Paper given at SS 433 workshop, Tenth Texas Symposium on Relativistic Astrophysics, Baltimore, Dec. 1980.

56. Blair, W.P. private communication, 1981.

57. Margon, B., Anderson, S., Grandi, S. and Downes, R. Int. Astron. Union Circ. 3626, 1981.

58. Collins, G.W., II, and Newsom, G.H. Proceedings of IAU Colloquium 69, Astrophys. Space Sci. Libr., Reidel, 1982 (in press).

59. Anderson, S.F., Margon, B., Grandi, S.A. and Downes, R.A. Bull. American Astron. Soc., 13, 801, 1981.

60. Goldstein, H., 1959 Classical Mechanics, Addison-Wesley Pub. Co. Inc. Reading, London p107, 108.

61. Landau, L.D. and Lifshitz, E.M., 1960, Mechanics, Addison-Wesley Pub. Co. Inc. Reading pp126-129.

62. Brouwer, D. and Clemence, G.M., 1961, Methods of Celestical Mechanics Academic Press, New York, London pp115-124.

63. Klein, F. and Sommerfeld, A. 1965 Über die Theorie des Kreisels Johnson Reprint Corp., New York, p291, 307.

64. Ince, E.L., 1927 Ordinary Differential Equations, Longmans, Green and Co. Ltd. London, New York, Toronto, Bombay, Calcutta and Madrid, pp381-385.

65. Hamilton, A.J.S., and Sarazin, C.L. Mon. Not. R. Astron. Soc. 198, 59-70, 1982.

66. Mammano, A., Ciatti, F., and Turvolla, R., Proc. Orbis Scientiae (this volume), 1982.

67. Fang LiZhi, Ruffini, R., and Stella, L., Vistas Astron., 25, 185-211, 1981.

68. Geldzahler, B.J., Pauls, T. and Salter, C.J. Astron. Astrophys., 84, 237-244, 1980.

69. Seward, F., Grindlay, J., Seaquist, E. and Gilmore, W. Nature, 287, 806-808, 1980.

SUPERSTRINGS*

John H. Schwarz

California Institute of Technology

Pasadena, California 91125

ABSTRACT

Type I superstrings are supersymmetrical open strings whose massless modes belong to the adjoint representation of a Yang-Mills gauge group and singlet closed strings. The interacting theory in ten dimensions is free from ghosts and tachyons and is (at least) one-loop renormalizable. Type II superstrings are supersymmetrical closed strings only. The interacting theory in ten dimensions is also free from ghosts and tachyons and is (at least) one-loop finite. By compactifying six dimensions and letting the radii and Regge slope approach zero, one obtains N = 4 Yang-Mills theory as a limit of theory I and N = 8 supergravity as a limit of theory II. A power-counting argument suggests that N = 4 Yang-Mills theory should be ultraviolet finite to all orders, whereas N = 8 supergravity should have ultraviolet divergences starting at three loops.

*Work supported by the U.S. Department of Energy under Contract No. DE-AC-03-81-ER40050.

This talk is a sequel to one I gave here four years ago.[1] The philosophy, presented at that time and developed originally in collaboration with J. Scherk,[2] is to regard supersymmetrical dual string theory as a candidate theory of fundamental particle interactions. The study of dual strings began in 1968 as an attempt to develop a phenomenological model of hadrons, but over the years it became clear that all the tractable ghost-free theories we could invent contain massless vector and tensor particles. Seeing that these states would not go away, we gradually recognized that these models could not accurately describe hadrons, but that by identifying the massless states as gluons, photon, graviton, and so forth a fundamental unification of interactions including gravity would be a possibility.* This meant that the preferred physical size of the strings was no longer the size of a hadron, but rather a much smaller scale, perhaps the Planck length. Correspondingly, the excited string models describe very high mass states, so that at energies low on this scale one is left with an effective field theory that only involves the massless (before symmetry breaking) string modes. The importance of the full string theory is that it could give a finite or renormalizable theory containing gravity even though the limiting field theory, or any other conventional field theory, may be unable to do so.

During the last two years Michael Green and I have developed a new formalism for describing supersymmetrical strings.[3-6] This formalism, which makes previously-obscure supersymmetry properties manifest, has enabled us to carry out explicit calculations of tree and loop graphs much more easily than was previously possible and to investigate divergences of loop graphs[5] as well as the limit that gives a low-energy effective field theory.[7] Moreover, it

*There probably is a string theory corresponding to the infinite color limit of QCD that is appropriate for hadrons. Such a theory is likely to be quite different from the theories considered here.

has led to the discovery of a second supersymmetrical string theory[6] that appears to be at least as strong a candidate for a unified theory as the original one. Both theories require ten spacetime dimensions (nine space and one time) for their consistency, but 10-D dimensions can form a compact space that would be unobservable at low energies. We define an analytic continuation in D, which is especially useful for regulating infrared divergences in a supersymmetrical fashion and describing the physical case D = 4 as a limit approached from above.

The original ("Veneziano") dual string theory can be described by starting with an action principle for the string coordinates $X^\mu(\sigma,\tau)$.[8] As usual, σ is a spacelike parameter that runs from 0 to π along the string, while τ is a timelike evolution parameter. The Lorentz index μ refers to the physical spacetime, 26-valued for the Veneziano model and 10-valued for the supersymmetrical models. A suitable action is

$$I_V = -\frac{1}{4\pi\alpha'}\int d\sigma d\tau\,\sqrt{-g}\,g^{\alpha\beta}\,\eta_{\mu\nu}\,\partial_\alpha X^\mu\,\partial_\beta X^\nu \quad , \tag{1}$$

where α' is the Regge slope parameter (inverse string tension) and $\eta_{\mu\nu}$ is the Minkowski metric. $g^{\alpha\beta}(\sigma,\tau)$ is an auxiliary two-dimensional metric tensor that implements σ-τ reparametrization invariance as a consequence of the usual mathematics of general relativity. Eliminating $g^{\alpha\beta}$ by means of its equations of motion gives the Nambu-Goto action. A more convenient way of proceeding is to choose a gauge in which $g^{\alpha\beta} \propto \eta^{\alpha\beta}$ and to use the remaining local invariance, as well as the equation of motion for the light-cone component $X^+(\sigma,\tau)$, to set

$$X^+(\sigma,\tau) = x^+ + 2\alpha'\,p^+\tau \quad . \tag{2}$$

The light-cone component X^- is then determined in terms of the other ones, so that the string dynamics resides in the transverse coordinates $X^i(\sigma,\tau)$. This dynamics is described by the "light-cone-gauge" string action

$$I_V^{\ell.c.} = -\frac{1}{4\pi\alpha'}\int d\sigma d\tau\ \eta^{\alpha\beta}\ \partial_\alpha X^i\ \partial_\beta X^i \quad . \tag{3}$$

The supersymmetrical string theories can be described starting from a light-cone-gauge string action that generalizes eq. (3). In retrospect, the reason this result took so long to discover is that it is very difficult (maybe even impossible) to find an appropriate covariant and reparametrization-invariant string action with ten-dimensional supersymmetry from which it may be deduced. (There is a covariant and reparametrization-invariant string action with two-dimensional supersymmetry that gives rise to the old formalism). Instead of looking for one, we jump directly to the analogs of eqs. (2) and (3), which turn out to be all one really needs. The supersymmetrical action involves fermionic degress of freedom $S^{Aa}(\sigma,\tau)$ in addition to $X^i(\sigma,\tau)$. The index A = 1,2 describes a spinor in σ-τ space, whereas a is a 32-valued spinor index. S^{1a} and S^{2a} satisfy simultaneous Majorana and Weyl conditions, as is possible in ten dimensions[9]. In addition to imposing eq. (2), we require that

$$(\gamma^+)^{ab}\ S^{Ab} = 0 \quad . \tag{4}$$

The supersymmetrical light-cone-gauge action is then[6*]

$$I_S^{\ell.c.} = \int d\sigma d\tau\ [-\frac{1}{4\pi\alpha'}\ \partial_\alpha X^i\ \partial^\alpha X^i + \frac{i}{4\pi}\bar{S}\ \gamma^-\rho^\alpha\ \partial_\alpha S] \quad . \tag{5}$$

Setting $\alpha' = \frac{1}{2}$, this action is invariant under the supersymmetry

---

* $\rho^0 = \sigma_2$ and $\rho^1 = i\sigma_1$ are two-dimensional Dirac matrices in a Majorana representation and $\bar{S}^{Aa} = S^{\dagger Bb}(\gamma^0)^{ba}(\rho^0)^{BA}$.

transformations

$$\delta X^i = \frac{1}{\sqrt{p^+}} \bar{\varepsilon}\gamma^i S \quad , \tag{6a}$$

$$\delta S = \frac{1}{\sqrt{p^+}} \gamma_- \gamma_\mu \, \rho \cdot \partial X^\mu \, \varepsilon \quad , \tag{6b}$$

if one assumes periodic boundary conditions, as appropriate for closed strings, and takes ε^{Aa} to be an arbitrary constant Majorana-Weyl Grassmann number. The corresponding generators, obtained by the Noether method, are

$$Q^{Aa} = \frac{i}{\pi\sqrt{p^+}} \int_0^\pi [\gamma_\mu (\rho \cdot \partial X^\mu) \rho^0 S]^{Aa} d\sigma \quad . \tag{7}$$

We can now deduce two distinct supersymmetrical string theories both of which (in contrast to the Veneziano model) are free from tachyons. The first theory consists of open strings and certain closed strings - referred to collectively as <u>type I superstrings</u>. Open-string boundary conditions, together with the equations of motion $\rho \cdot \partial S = 0$ and $\partial \cdot \partial X^i = 0$, lead to the normal-mode expansions

$$X^i = x^i + p^i \tau + i \sum_{n \neq 0} \frac{1}{n} \alpha_n^i \cos n\sigma \, e^{-in\tau} \quad , \tag{8}$$

$$S^{1a} = \sum_{n=-\infty}^{\infty} S_n^a \, e^{-in(\tau-\sigma)} \quad , \tag{9a}$$

$$S^{2a} = \sum_{n=-\infty}^{\infty} S_n^a \, e^{-in(\tau+\sigma)} \quad . \tag{9b}$$

Canonical quantization gives

$$[x^i, p^j] = i\delta^{ij} \quad , \tag{10a}$$

$$[\alpha_m^i, \alpha_n^j] = m\delta_{m+n,0} \, \delta^{ij} \quad , \tag{10b}$$

$$\{S_m^a, \bar{S}_n^b\} = (\gamma^+ h)^{ab}\, \delta_{m+n,0} \quad , \tag{11}$$

$$[\alpha_m^i, S_n^a] = 0 \quad , \tag{12}$$

where h represents a Weyl projection operator $\frac{1}{2}(1\pm\gamma_{11})$. In the case of open-string boundary conditions, eqs. (6a) and (6b) do not describe an invariance, and therefore Q^{1a} and Q^{2a} are not separately conserved, although

$$Q^a = \frac{1}{2}(Q^{1a} + Q^{2a}) = i\sqrt{p^+}\,(\gamma_+ S_0)^a + \frac{i}{\sqrt{p^+}} \sum_{-\infty}^{\infty} (\gamma_i S_{-n})^a\, \alpha_n^i \, , \tag{13}$$

is conserved. Thus type I superstrings have just one supersymmetry in the ten-dimensional sense (corresponding to N = 4 in four dimensions).

The second theory is based on closed strings only - called type II <u>superstrings</u> - and has two supersymmetries in ten dimensions (corresponding to N = 8 in four dimensions). The normal-mode expansions in this case are

$$X^i = x^i + p^i\tau + \frac{i}{2} \sum_{n\neq 0} \frac{1}{n} \left(\alpha_n^i\, e^{-2in(\tau-\sigma)} + \tilde{\alpha}_n^i\, e^{-2in(\tau+\sigma)}\right) \, , \tag{14}$$

$$S^{1a} = \sum_{n=-\infty}^{\infty} S_n^a\, e^{-2in(\tau-\sigma)} \quad , \tag{15a}$$

$$S^{2a} = \sum_{n=-\infty}^{\infty} \tilde{S}_n^a\, e^{-2in(\tau+\sigma)} \quad . \tag{15b}$$

The canonical commutation relations give eqs. (10-12) for the operators with tildes as well as the ones without them. Also

$$[\alpha_m^i, \tilde{\alpha}_n^j] = [\alpha_m^i, \tilde{S}_n^a] = [\tilde{\alpha}_m^i, S_n^a] = \{S_m^a, \tilde{S}_n^b\} = 0 \quad . \tag{16}$$

Substituting eqs. (14) and (15) into eq. (7) gives the conserved supersymmetry charges

$$Q^{1a} = i\sqrt{p^+}\,(\gamma_+ S_0)^a + \frac{2i}{\sqrt{p^+}} \sum_{-\infty}^{\infty} (\gamma_i S_{-n})^a \alpha_n^i \;, \tag{17a}$$

$$Q^{2a} = i\sqrt{p^+}\,(\gamma_+ \tilde{S}_0)^a + \frac{2i}{\sqrt{p^+}} \sum_{-\infty} (\gamma_i \tilde{S}_{-n})^a \tilde{\alpha}_n^i \;. \tag{17b}$$

There are actually two distinct type II theories according to whether S_1 and S_2 (and hence Q_1 and Q_2) have the same or opposite handedness.

An important feature of both type I and type II superstring theories is the existence of a ten-dimensional super-Poincaré algebra. In the case of type I superstrings the Lorentz generators are

$$J^{+-} = \ell^{+-} \;, \tag{18a}$$

$$J^{i+} = \ell^{i+} \;, \tag{18b}$$

$$J^{ij} = \ell^{ij} - i \sum_{1}^{\infty} \frac{1}{n} (\alpha_{-n}^i \alpha_n^j - \alpha_{-n}^j \alpha_n^i) + \frac{i}{8} \sum_{-\infty}^{\infty} \bar{S}_{-n} \gamma^{ij-} S_n \;, \tag{18c}$$

$$J^{i-} = \ell^{i-} - \frac{i}{p^+} \sum_{1}^{\infty} \frac{1}{n} (\alpha_{-n}^i \alpha_n^- - \alpha_{-n}^- \alpha_n^i)$$

$$+ \frac{i}{8p^+} \sum_{-\infty}^{\infty} \bar{S}_{-m} \gamma^{ij-} S_{-n} \alpha_{m+n}^j \;, \tag{18d}$$

where

$$\ell^{\mu\nu} = x^\mu p^\nu - x^\nu p^\mu \tag{19}$$

and

$$\alpha_m^- = \sum_{-\infty}^{\infty} [\tfrac{1}{2} : \alpha_{m-n}^i \alpha_n^i : + \tfrac{1}{4} (n - \tfrac{m}{2}) : \bar{S}_{m-n} \gamma^- S_n :] \;. \tag{20}$$

When evaluated between on-mass-shell states ($\alpha_0^- = p^+ p^-$) $J^{\mu\nu}$, Q^a, and p^μ satisfy a super-Poincaré algebra.[3] The verification that $[J^{i-}, J^{j-}] = 0$, in particular, is especially nontrivial, requiring great care in manipulating the infinite sums as well as the use of Fierz transformation formulas based on eq. (4) as well as the Majorana and Weyl properties of the S's. Similar calculations show that Q^a transforms as a spinor and that the supersymmetry algebra

$$\{Q^a, \bar{Q}^b\} = -2(h\gamma \cdot p)^{ab} \tag{21}$$

is satisfied. The verification of the extended super-Poincaré algebra for type II superstrings is very similar. Eqs. (18c) and (18d) acquire additional contributions from oscillators with tildes, and the requisite physical-state conditions are $\alpha_0^- = \tilde{\alpha}_0^- = \frac{1}{2} p^+ p^-$. The theory I closed-string sector consists of those type II superstrings (in the case that S_1 and S_2 have the same handedness) that are symmetrical under interchange of modes with and without tildes, consistent with the rule $Q^a = \frac{1}{2}(Q^{1a} + Q^{2a})$ found for the open strings.

The description of interactions among type I superstrings can be completely deduced from the vertex operator for the emission of a massless vector state whose momentum is entirely in the minus direction:

$$V_B(\zeta^i, k^-, \tau) = g\, \zeta^i P^i(\tau) e^{-ik^- X^+(\tau)} \quad . \tag{22}$$

Here ζ^i is a transverse polarization vector, X^+ is given in eq. (2), g is a Yang-Mills coupling constant, and

$$P^i(\tau) = \frac{\partial}{\partial\tau} X^i(\sigma=0,\tau) = \sum_{-\infty}^{\infty} \alpha_n^i\, e^{-in\tau} \quad , \tag{23}$$

where we define $\alpha_0^i = p^i$. The vertex with the momentum in an arbitrary orientation is obtained by a Lorentz transformation:

$$V_B(\zeta,k,\tau) = \exp(i \frac{k^j}{k^-} J^{j-}) V_B(\zeta^i,k^-,\tau) \exp(- i \frac{k^j}{k^-} J^{j-}) \quad . \tag{24}$$

The light-cone-gauge formalism enforces the gauge choice $\zeta^- = 0$. All formulas involving vertices are only required to hold up to total τ derivatives, since such derivatives never contribute in amplitude calculations. The only way in which eqs. (22) - (24) differ from the massless vector particle emission formulas of the Veneziano model is in the formula for J^{i-} given in eq. (18d). The emission operator $V_F(u,k,\tau)$ for a massless spinor satisfying the Dirac equation $\gamma \cdot ku = 0$ is given by the supersymmetry relations

$$[\bar{\varepsilon}Q, V_B(\zeta,k,\tau)] = V_F(\tfrac{1}{2}\gamma_{\mu\nu}\zeta^\mu k^\nu \varepsilon,k,\tau) \quad , \qquad (25)$$

$$[\bar{\varepsilon}Q, V_F(u,k,\tau)] = V_B(2\bar{\varepsilon}(\gamma^\mu - \frac{k^\mu}{k^-}\gamma^-)u,k,\tau) \quad . \qquad (26)$$

In particular when the momentum is in the minus direction

$$V_F(u,k^-,\tau) = ig\sqrt{p^+}\,\bar{S}(\tau)\gamma_+ u\, e^{-ik^- X^+(\tau)} \quad , \qquad (27)$$

where

$$\bar{S}(\tau) = \sum_{-\infty}^{\infty} \bar{S}_n e^{-in\tau} \quad . \qquad (28)$$

The general case is given by the analog of eq. (24).

To calculate trees with external massless states we also introduce a propagator $\Delta = \alpha'/L_0$, where

$$L_0 = \alpha' p^2 + N \quad , \qquad (29)$$

$$N = \sum_1^\infty (\alpha_{-n}^i \alpha_n^i + \tfrac{1}{2} n \bar{S}_{-n}\gamma^- S_n) \quad . \qquad (30)$$

Somewhat schematically, the M-particle tree amplitude with arbitrary external massless bosons and fermions is

$$T_M = \Sigma\, tr(\lambda_1\lambda_2\ldots\lambda_M)\langle 1|V(2)\Delta V(3)\Delta\ldots V(M-1)|M\rangle \quad , \qquad (31)$$

where λ_i is an $n \times n$ hermitian matrix giving the U(n) description of the ith particle. Each term in the sum has cyclic symmetry in the M emitted states (this is the "duality" property - a complete proof remains to be constructed), and the sum is over inequivalent permutations. The calculations have been done explicitly for all the four-particle tree amplitudes.[4] In this case eq. (31) is the sum of three terms, $F_{st} + F_{su} + F_{tu}$, where s, t, and u refer to the usual invariant-energy variables whose sum is zero. The result is summarized by the formula

$$F_{st}^{(\text{string theory})} = \frac{\Gamma(1-\alpha' s)\Gamma(1-\alpha' t)}{(1-\alpha' s - \alpha' t)} F_{st}^{(\text{field theory})} . \tag{32}$$

"Field theory" refers to the ten-dimensional supersymmetrical Yang-Mills theory[9,10] given by

$$L = -\frac{1}{4} F^2 + \frac{i}{2} \bar{\psi} \gamma \cdot D\psi \quad , \tag{33}$$

where ψ is an adjoint representation 32-component Majorana-Weyl spinor. Eq. (32) shows that in tree approximation the string-theory amplitude reduce to field-theory ones as $\alpha' \to 0$.

The tree amplitudes for type II superstrings are obtained by a similar analysis. One important difference is that there is no Yang-Mills group, and as a result the analog of eq. (31) involves only a single term that has total symmetry under interchange of the external lines. (This can be understood in terms of the topology of closed-string world sheets). The analog of eq. (32) for four-particle amplitudes with massless external states is

$$T_4^{(\text{string theory})} = \frac{\Gamma(1-\frac{\alpha'}{2} s)\Gamma(1-\frac{\alpha'}{2} t)\Gamma(1-\frac{\alpha'}{2} u)}{\Gamma(1+\frac{\alpha'}{2} s)\Gamma(1+\frac{\alpha'}{2} t)\Gamma(1+\frac{\alpha'}{2} u)} T_4^{(\text{field theory})} . \tag{34}$$

In this case "field theory" refers to extended supergravity in ten dimensions (corresponding to N = 8 supergravity in four dimensions).

The explicit Lagrangian is easily constructed for the case that S_1 and S_2 have opposite handedness by dimensionally reducing 11-dimensional supergravity.[11] Once again we see that the string theory result reduces to the field theory one as $\alpha' \to 0$.

A number of interesting new issues arises in the evaluation of the loop graphs required to implement unitarity perturbatively. There are three topologically distinct types of four-particle one-loop graphs for type I superstrings. They are given, somewhat schematically, by

$$T(1,2,3,4) = \mathrm{tr}(\lambda_1\lambda_2\lambda_3\lambda_4) \int d^{10}p \ \mathrm{tr}\ [V(1)\Delta V(2)\Delta V(3)\Delta V(4)\Delta] \ , \quad (35a)$$

$$T(1,2,3;4) = \mathrm{tr}(\lambda_1\lambda_2\lambda_3)\mathrm{tr}(\lambda_4) \int d^{10}p \ \mathrm{tr}\ [V(1)\Delta V(2)\Delta V(3)\Delta\Omega V(4)\Delta\Omega], \quad (35b)$$

$$T(1,2;3,4) = \mathrm{tr}(\lambda_1\lambda_2)\mathrm{tr}(\lambda_3\lambda_4) \int d^{10}p \ \mathrm{tr}\ [V(1)\Delta V(2)\Delta\Omega V(3)\Delta V(4)\Delta\Omega], \quad (35c)$$

where Ω is the "twist" operator[4] (see eqs. (20) and (30))

$$\Omega = -\ (-1)^N \ e^{-\bar{\alpha}_{-1}} \quad . \quad (36)$$

The method of evaluation consists of first substituting integral representations for the propagators and then performing the momentum integrals, which at this point become gaussians. The evaluation of the traces over nonzero oscillator modes is done by completely standard methods. One utilizes, in particular, the amusing result

$$\mathrm{tr}\ (w^N) = 1 \quad . \quad (37)$$

A calculus for S_0 traces is also required. Introducing transverse vector states $|i\rangle$ satisfying $\langle i|j\rangle = \delta_{ij}$, and spinor states

$$|a\rangle = \frac{i}{8}\ (\gamma_i S_0)^a |i\rangle \quad , \quad (38)$$

it is easy to show using eq. (11) and

$$\frac{1}{8}\,\bar{S}_0\,\gamma^{ij-}S_0|k\rangle = \delta_{ik}|j\rangle - \delta_{jk}|i\rangle \tag{39}$$

that

$$I = |i\rangle\langle i| + |a\rangle(\gamma^-)^{ba}\langle b| \tag{40}$$

acts as an identity operator. Therefore the S_0-space trace of an operator A is

$$\mathrm{tr}(A) = \langle i|A|i\rangle - (\gamma^-)^{ba}\langle b|A|a\rangle \quad . \tag{41}$$

A remarkable consequence of these rules is that the trace of any monomial in S_0 vanishes if it is less than 8th order. As a result one-loop graphs with fewer than four massless external lines vanish identically, while four-particle one-loop graphs are surprisingly simple to evaluate (easier than the trees, in fact).

Evaluating eq. (35a) gives

$$T(1,2,3,4) = K_1\ \mathrm{tr}\ (\lambda_1\lambda_2\lambda_3\lambda_4)\int_0^1 \frac{dq}{q}\int_0^1 \prod_{i=1}^{3}[d\nu_i\,\theta(\nu_{i+1}-\nu_i)]$$
$$\times \prod_{1\le i<j\le 4}[\sin\pi(\nu_j-\nu_i)A_{ji}]^{2\alpha' k_i\cdot k_j} \tag{42}$$

where $\nu_4 = 1$, the kinematical factor K_1 is proportional to $st\,F_{st}^{(\text{field theory})}$ and

$$A_{ji} = \prod_{n=1}^{\infty}(1 - 2q^{2n}\cos 2\pi\,(\nu_j - \nu_i) + q^{4n}) \quad . \tag{43}$$

T(1,2,3,4) has a divergence associated with $q \to 0$ that is proportional to

$$I_1(s,t) = \int_0^1 \prod_{i=1}^{3} [d\nu_i \theta(\nu_{i+1}-\nu_i)] \prod_{1\le i<j\le 4} [\sin\pi(\nu_j-\nu_i)]^{2\alpha' k_i\cdot k_j}$$

$$= -\frac{1}{2\pi\alpha' st} \frac{\partial}{\partial\alpha'} [\frac{\Gamma(1-\alpha' s)\ (1-\alpha' t)}{\Gamma(1-\alpha' s - \alpha' t)}] \quad . \tag{44}$$

In view of eq. (32) and the formula for K_1 this is precisely the relation required to show that to order g^2 the divergence in eq. (42) can be cancelled by a renormalization of α'. Evaluating eq. (35b) gives

$$T(1,2,3;4) = K_1\ tr(\lambda_1\lambda_2\lambda_3)tr(\lambda_4) \int_0^1 \frac{dq}{q} \int_0^1 (\prod_{i=1}^{3} d\nu_i)$$

$$\times [\sin\pi(\nu_2-\nu_1)A_{21}B_{34}]^{-\alpha' s}[\sin\pi(\nu_3-\nu_2)A_{32}B_{14}]^{-\alpha' t}$$

$$\times [\sin\pi(\nu_3-\nu_1)A_{31}B_{24}]^{-\alpha' u} \quad , \tag{45}$$

where

$$B_{ji} = \prod_{n=1}^{\infty} (1-2q^{2n-1} \cos 2\pi(\nu_j-\nu_i) + q^{4n-2}) \quad , \tag{46}$$

and $\nu_4 = 1$. Once again there is a divergence associated with $q \to 0$, this time proportional to

$$J(s,t) = \int_0^1 d\nu_1 d\nu_2 d\nu_3 \left[\frac{\sin\pi(\nu_2-\nu_1)}{\sin\pi(\nu_3-\nu_1)}\right]^{-\alpha' s} \left[\frac{\sin\pi(\nu_3-\nu_2)}{\sin\pi(\nu_3-\nu_1)}\right]^{-\alpha' t}$$

$$= -\frac{2}{\pi^2} \{\frac{\Gamma(-\alpha' s)\Gamma(-\alpha' t)}{\Gamma(1-\alpha' s-\alpha' t)} + \frac{\Gamma(-\alpha' s)\Gamma(-\alpha' u)}{\Gamma(1-\alpha' s-\alpha' u)} + \frac{\Gamma(-\alpha' t)\Gamma(-\alpha' u)}{\Gamma(1-\alpha' t-\alpha' u)}\}. \tag{47}$$

This divergence has exactly the form that can be cancelled to order g^2 by a renormalization of the wave function of particle #4 when it is a singlet of the SU(n) algebra. For eq. (35c) one obtains

$$T(1,2;3,4) = K_1 \, tr(\lambda_1\lambda_2)tr(\lambda_3\lambda_4) \int_0^1 \frac{dq}{q} q^{-\alpha' s/2}$$

$$\times \int_0^1 \left(\prod_{i=1}^{3} d\nu_i\right)[4\sin\pi(\nu_2-\nu_1)\sin\pi(\nu_4-\nu_3)A_{21}A_{43}]^{-\alpha' s}$$

$$(B_{32}B_{41})^{-\alpha' t} (B_{31}B_{42})^{-\alpha' u} \quad . \tag{48}$$

An expansion about $q = 0$ reveals s-channel poles with masses given by $\alpha' s = 0, 4, 8, \ldots$. They precisely correspond to SU(n) singlet closed-string "bound states". Thus the four-particle graphs demonstrate that all divergences are satisfactorily accounted for at the one-loop level, and type I superstrings are well-behaved to this order.

The analysis of loops for type II superstrings is quite similar. Just as for type I superstrings, the one-loop amplitude vanishes when there are fewer than four massless external lines. The four-particle one-loop amplitude is given by the single term

$$T_4^{(loop)} = K_2 \int d^2\tau (\mathrm{Im}\,\tau)^{-2} F(\tau) \quad , \tag{49}$$

where the kinematical factor K_2 is proportional to s t u times the tree approximation amplitude of the extended supergravity field theory. The function $F(\tau)$ is given by

$$F(\tau) = (\mathrm{Im}\,\tau)^{-3} \int \left(\prod_{i=1}^{3} d^3\nu_i\right)$$

$$\times \prod_{1\leq i<j\leq 4} \left(\exp\left[\frac{-\pi(\mathrm{Im}\,\nu_{ji})^2}{\mathrm{Im}\,\tau}\right]|\theta_1(\nu_{ji}|\tau)|\right)^{2\alpha' k_i\cdot k_j} \quad , \tag{50}$$

where $\nu_{ji} = \nu_j - \nu_i$, $\nu_4 = \tau$, and the integration limits are

$$0 \leq \mathrm{Im}\,\nu_i \leq \mathrm{Im}\,\tau \; ; \quad -\frac{1}{2} \leq \mathrm{Re}\,\nu_i \leq \frac{1}{2} \qquad i = 1,2,3 \quad . \tag{51}$$

$F(\tau)$ is automorphic, invariant under the modular group

$$\tau \rightarrow \frac{a\tau + b}{c\tau + d} \quad , \tag{52}$$

where a,b,c,d are integers satisfying ad - bc = 1. Accordingly the τ integration in eq. (49) is restricted to the fundamental region

$$-\frac{1}{2} \leq \text{Re}\ \tau \leq \frac{1}{2}\ , \quad \text{Im}\ \tau \geq 0,\ |\tau| \geq 1 \quad . \tag{53}$$

Since the loop amplitude in eq. (49) is completely free from divergences (interpreted by analytic continuation where necessary), no renormalizations are required for type II superstrings at the one-loop level!

In order to make contact with four-dimensional physics it is necessary to suppress six of the spatial dimensions. A consistent possibility is to take 10-D dimensions to be circles of radius R (the radii may be different, but that is a trivial detail). Whether or not this is the only topological possibility consistent with the existence of spinors and the absence of a cosmological constant is not yet clear. It is certainly the easiest one to analyze. The compactification is expected to make possible the introduction of consistent supersymmetry breaking along the lines discussed in ref.[12], although this has not yet been worked out. In the simpler case without symmetry breaking the consequences of the compactification have been deduced.[7] The massless external particles have vanishing momentum components in the compactified dimensions, and their wave functions can be decomposed into spin multiplets appropriate to D dimensions if one so desires. That is the whole story for the tree amplitudes. In the case of loops one must also take account of the fact that for internal lines the momentum components corresponding to compactified dimensions are quantized. For open-string loop amplitudes such as in eqs. (42), (45), and (48), this modification is achieved by inserting the additional factor $[F_1(a,q)]^{10-D}$, where

$$F_1(a,q) = \sum_{N=-\infty}^{\infty} q^{N^2/2a^2} , \tag{54}$$

and a is the dimensionless ratio $\sqrt{\alpha'}/R$. The inclusion of this factor does not affect the slope-parameter and wave-function renormalizations required to cancel the divergences in eqs. (42) and (45), but it does affect the spectrum of closed-string poles in eq. (48). The new poles may be interpreted as states for which the string is wound around the circular compactified dimensions.[13] If it is wrapped N_i times around the ith circular dimension, the mass-squared receives a contribution $\frac{1}{\alpha'}(N_i/a)^2$.

The modification to closed-string loop amplitudes due to compactification must allow both for the discretization of internal momentum components and for possible windings around circular dimensions. These effects are accounted for by inserting into eq. (49) the finite-radius factor $[F_2(a,\tau)]^{10-D}$, where

$$F_2(a,\tau) = a\sqrt{\text{Im}\,\tau} \sum_{MN} \exp[-2\pi i MN\,\text{Re}\tau - \pi(a^2M^2+N^2/a^2)\,\text{Im}\,\tau] . \tag{55}$$

It is both important and true that $F_2(a,\tau)$ is invariant under the modular group described in eq. (52).

We are now in a position to study the field-theory limits of the loop amplitudes. By allowing $R \to 0$ at the same time that $\alpha' \to 0$, we can obtain N = 4 Yang-Mills theory as a limit of theory I and N = 8 supergravity as a limit of theory II. In defining the limit one should absorb powers of R in the definitions of the coupling constants and wave functions so as to give them the dimensionality appropriate to D-dimensional spacetime. Doing this, the limit of the tree amplitudes is completely straightforward, not sensitive to the order in which the limits are performed. The analysis of the loop amplitudes is more sensitive since either $R \to 0$ or $\alpha' \to 0$ separately is a singular limit. However, one can obtain the desired limiting theories by taking the limits at the same time. In the case of type I superstrings, the massless closed-string sector (supergravity) decouples if the limit is taken so that

$$\frac{1}{\alpha'} R^{\frac{10-D}{2}} \to 0 \quad . \tag{56}$$

The analysis is technically most straightforward if one holds $a = \sqrt{\alpha'}/R$ fixed in the limit.

Taking the limit $\alpha' \to 0$, at fixed a, eq. (42) with the finite-radius factor $[F_1(a,q)]^{10-D}$ inserted, becomes[7]

$$T(1,2,3,4) \sim K_1\, g_D^2\, \pi^{\gamma-1}\Gamma(-\gamma) \int_0^1 \Big(\prod_{i=1}^{4} d\eta_i\Big)\delta(1-\Sigma\eta_i)(\eta_1\eta_3 s+\eta_2\eta_4 t)^{\gamma}$$

$$= K_1\, g_D^2\, c(\gamma)[I_\gamma(s,t) + I_\gamma(t,s)] \quad , \tag{57}$$

where $\gamma = \frac{1}{2}$ D-4 and

$$c(\gamma) = \frac{1}{4}\left(\frac{\pi}{4}\right)^{\gamma+1/2} [\sin\pi\gamma\ \Gamma(\gamma+5/2)]^{-1} \quad , \tag{58}$$

$$I_\gamma(s,t) = t^{\gamma+1} \int_0^1 \frac{(1-x)^{\gamma+1}dx}{sx-t(1-x)} \quad . \tag{59}$$

Dependence on the parameter a has dropped out in the limit. This result is <u>finite</u> for 4 < D < 8, the upper limit representing the onset of ultraviolet divergences (at one loop) and the lower limit representing the onset of infrared divergences. The analysis assumes an analytic continuation in D, or equivalently the number of compactified dimensions. It maintains the full supersymmetry of the theory, which is mostly contained in the structure of the factor K_1, for all D. The infrared singularity is proportional to $(D-4)^{-2}$, which is characteristic of a Yang-Mills theory at one loop. The asymptotic analysis of eqs. (45) and (48) gives essentially identical results. If the kinematical factor K_1 should be common to the multiloop graphs as well, as we suspect is required by supersymmetry, then we can use power-counting considerations to anticipate the dimension at which UV divergences should begin. The result is that we expect N = 4 Yang-Mills theory to be UV finite at ℓ loops for $D < 4+4/\ell$. Thus it should be UV finite to all orders in four dimensions, corresponding to the vanishing of the β function.

The limit of theory II that gives N = 8 supergravity may be analyzed in a similar fashion. Letting $\alpha' \to 0$ at fixed a in eq. (49), with the finite-radius factor $[F_2(a,q)]^{10-D}$ inserted, we obtain[7]

$$T_4^{(\text{loop})} \sim \kappa_D^2 K_2\, c(\gamma)[I_\gamma(s,t)+I_\gamma(t,s) + I_\gamma(s,u)+I_\gamma(u,s)+I_\gamma(t,u)+I_\gamma(u,t)] \quad , \tag{60}$$

with c and I_γ as given in eqs. (58) and (59). Once again the result is finite for $4 < D < 8$. The only significant difference between the Yang-Mills and supergravity one-loop results is that the symmetrization of terms in eq. (60) softens the infrared divergence in the supergravity case to one proportional to $(D-4)^{-1}$. If, as before, we assume the universality of the factor K_2 and predict the onset of UV divergences in multiloop graphs by a power-counting argument, we are led to conclude that N = 8 supergravity should be UV finite at ℓ loops for $D < 2 + 6/\ell$. If this is correct, then UV divergences should begin to appear in four dimensions at the three-loop level.

In conclusion, type I superstrings appear to give a renormalizable theory that reduces to a finite N = 4 Yang-Mills theory at low energy, whereas type II superstrings are likely to give a finite theory that reduces to a singular N = 8 supergravity theory at low energy. Explicit calculations of amplitudes are not as difficult as might have been anticipated, which provides some optimism about the prospects for verifying these speculations at the multiloop level. Calculating the string-theory amplitudes and extracting their α', $R \to 0$ limits may even be the easiest methods of obtaining explicit results for the limiting field theories.

Note added: After presenting this work I realized that U(n) gauge groups lead to inconsistencies and must be rejected. It turns out, however, that SO(n) or Sp(2n) groups can be used without inconsistency. Their use requires that nonorientable diagrams be included.

REFERENCES

1. J.H. Schwarz in "New Frontiers in High-Energy Physics," Proc. Orbis Scientiae, 1978, ed. A. Perlmutter and L.F. Scott (Plenum, New York, 1978) p. 431.
2. J. Scherk and J.H. Schwarz, Nucl. Phys. B81 (1974) 118; Phys. Letters 57B (1975) 463.
3. M.B. Green and J.H. Schwarz, Nucl. Phys. B181 (1981) 502.
4. M.B. Green and J.H. Schwarz, Caltech preprint, CALT-68-872.
5. M.B. Green and J.H. Schwarz, Caltech preprint, CALT-68-873.
6. M.B. Green and J.H. Schwarz, Caltech preprint, CALT-68-874, to be published in Phys. Letters B.
7. M.B. Green, J.H. Schwarz, and L. Brink, Caltech preprint, CALT-68-880.
8. Dual Models, ed. M. Jacob (North-Holland, Amsterdam 1974); J. Scherk, Rev. Mod. Phys. 47 (1975) 123.
9. F. Gliozzi, J. Scherk, and D. Olive, Nucl. Phys. B122 (1977) 253.
10. L. Brink, J.H. Schwarz, and J. Scherk, Nucl. Phys. B121 (1977) 77.
11. E. Cremmer, B. Julia, and J. Scherk, Phys. Lett. 76B (1978) 409.

12. J. Scherk and J.H. Schwarz, Phys. Lett. 82B (1979) 60; Nucl. Phys. B153 (1979) 61;
E. Cremmer, J. Scherk, and J.H. Schwarz, Phys. Lett. 84B (1979) 83.

13. E. Cremmer and J. Scherk, Nucl. Phys. B103 (1976) 399.

A GEOMETRICAL APPROACH TO QUANTUM FIELD THEORY

F. R. Ore, Jr. and P. van Nieuwenhuizen

State University of New York at Stony Brook

Stony Brook, Long Island, New York 11794

(presented by P. van Nieuwenhuizen)

1. INTRODUCTION

It is well-known that unitarity and renormalizability of gauge theories are most conveniently proved by using BRST (Becchi-Rouet-Stora, and independently, Tyutin) invariance.[1] From the invariance of the effective quantum action

$$L(\text{qu}) = L(\text{classical}) + L(\text{gauge fixing}) + L(\text{ghost}) \tag{1}$$

under the global BRST transformations one can derive Ward identities for the connected Green's functions which allow a simple proof of unitarity,[2] and Ward identities for the one-particle irreducible Green's functions which allow a simple proof of renormalizability.[3] Thus BRST transformations are of great practical use.

Most physicists are of the opinion that classical gauge field theory has a beautiful geometrical basis, but that the introduction of L(gauge fixing), needed to define propagators for perturbation theory†

†Working with lattices avoids the need to add an L(gauge-fixing) but leads to another loss: that of manifest invariance under continuous global Lorentz transformations.

is a necessary evil because it destroys the classical gauge invariance of the action and thus the geometry associated with it, while also the choice of L (gauge-fixing) is arbitrary. Below we want to show that there exists an elegant geometry for L(qu), based on the same BRST transformations which are so useful for practical purposes. Moreover, in this geometrical approach the quantum artifacts are no longer arbitrary.

There exist, in fact, in addition to BRST transformations, also anti-BRST transformations,[4] and by requiring L(qu) to be invariant under both, we will find that L(gauge-fixing) is unique. This will be our first result:[5] the standard gauge-fixing terms as well as the ensuing L(ghost) of Yang-Mills theory, gravity and supergravity[6] all follow from the same geometrical principle. One might call this the unification of all gauge-fixing terms.

The constant parameters of both BRST transformations are anticommuting. Now it is well known that for real commuting spinor fields or real anticommuting boson fields the action is a total derivative.[7] However, for complex fields this is not so (which explains, by the way, why Faddeev-Popov ghosts are complex) and hence one can invent a new gauge theory: local BRST invariance. This will be our second result.[5] The gauge field of local BRST symmetry is a complex anticommuting vector field. It violates the spin-statistics theorem, and might be just what is needed[8] to cancel a similar anticommuting vector field in the electroweak theories based on the <u>simple</u> supergroup SU(2|1).[9]

Our third result is a new class of gauge algebras. An example is provided by supergravity, where the structure constants of the gauge algebra depend on the fields which represent the algebra ("structure functions"). In more generality, we will ask whether one can define such systems abstractly, and we will present generalized Jacobi identities and make contact with certain motions of group-theory and differential geometry, due to Cartan. This is only the beginning of a potentially interesting new field in group theory; for example, we do not know anything about a classification of such algebras.

We should point out that our approach is totally different from that by Ne'eman and Thierry-Mieg.[9] These authors consider a supergroup (namely SU(2|1)) and a group manifold with coordinates x^μ (ordinary space) and y^α (the (super) group coordinates). The left-vierbeins of Cartan are $A = A_\mu dx^\mu + C_\alpha dy^\alpha$ and the right-vierbeins are $A^*_\mu dx^\mu + C^*_\alpha dy^\alpha$. All fields are (super)algebra valued, $A = T_a A^a$. They identify fields as follows. The ghosts and antighosts are one-forms (which explains their opposite statistics) corresponding to dy^α and even generators T_e. The odd generators T_o yield one-forms $h^o_\alpha dy^\alpha$ and $h^{o*}_\alpha dy^\alpha$ which are the Higgs fields, but also spin statistics violating vector fields ξ^o_μ and ξ^{o*}_μ. Finally, a reality condition is imposed: $A^e_\mu = A^{e*}_\mu$ and h^o and h^{o*}. The gauge-fixing terms are not discussed, and the discussion is based on Yang-Mills theories. We start below with the gauge fixing terms and consider only completely general theories.

Some recent work on the renormalizability of gauge theories based on anti-BRST symmetry can be found in ref. (17). In ref. (18), the BRST techniques are applied to theories in which the gauge fixing terms are quadratic in gauge fields, and violate even the global symmetry of the classical action.

2. GEOMETRY OF GAUGE FIXING

As an example of the systems we will discuss, consider the ordinary Yang-Mills systems with Lorentz gauge fixing term and standard Faddeev-Popov ghosts. In order that the BRST algebra is nilpotent, one needs also real (commuting) Nielsen-Kallosh ghosts $\vec{d}$.[10] The action reads

$$L = -\frac{1}{4}\vec{G}_{\mu\nu}(\vec{A})^2 + \frac{1}{2}\vec{d}^2 - \vec{d}\,\partial.\vec{A} + \vec{C}^*\,\partial.D(\vec{A})\vec{C} \quad . \tag{2}$$

Clearly, in this example $\vec{d}$ is nonpropagating, and eliminating it by its algebraic field equation $\vec{d} = \partial.\vec{A}$, we retrieve the Lorentz gauge fixing term $-\frac{1}{2}(\partial.\vec{A})^2$. However, in supergravity these Nielsen-Kallosh ghosts are propagating and contribute, for example, to the axial anomaly.[11]

The action in (2) is invariant under the following BRST and anti-BRST transformations

$$\begin{aligned}
&\delta\vec{A}_\mu = D_\mu(\vec{C}\Lambda) \quad , \quad && \partial\vec{A}_\mu = D_\mu(\vec{C}^*\zeta) \quad , \\
&\delta\vec{C} = -\frac{1}{2}\vec{C}\wedge\vec{C}\Lambda \quad , \quad && \delta\vec{C}^* = -\frac{1}{2}C^*\wedge C^*\zeta \quad , \\
&\delta\vec{C}^* = -\vec{d}\Lambda \quad , \quad && \delta\vec{C} = \vec{d}\zeta - \vec{C}\wedge\vec{C}^*\zeta \quad , \\
&\delta\vec{d} = 0 \quad , \quad && \delta\vec{d} = -\vec{d}\wedge\vec{C}^*\zeta
\end{aligned} \tag{3}$$

where $D_\mu\vec{C}\Lambda = \partial_\mu\vec{C}\Lambda - \vec{A}_\mu \wedge \vec{C}\Lambda$ and $(\vec{C}\wedge\vec{C})^a = \varepsilon^{abc}\, C^b C^c$. The proof is simple. For the BRST transformations, the classical Yang-Mills action is separately invariant because $\delta\vec{A}_\mu$ are classical gauge transformations (with parameter $\vec{C}\Lambda$). The variations of $\vec{d}$ and $\vec{C}^*$ in the last two terms cancel because

$$-\vec{d}\cdot\partial_\mu D_\mu(\vec{C}\Lambda) - \Lambda\,\vec{d}\,\partial_\mu D_\mu\, C = 0 \quad , \tag{4}$$

since Λ is anticommuting and each term is bosonic (the variation of an action is bosonic since the action is bosonic). Finally, the variation of $D_\mu\vec{C}$ vanishes, too (which at the same time proves the nilpotency of $\delta(\Lambda)$ on $\vec{A}_\mu$). In order to demonstrate certain manipulations which one needs repeatedly we prove the last result explicitly.

$$\begin{aligned}
&\delta\,\partial_\mu\vec{C} = -\frac{1}{2}\,\partial_\mu(\vec{C}\wedge\vec{C}\Lambda) = -\,\partial_\mu\vec{C}\wedge\vec{C}\Lambda \quad , \\
&\delta\,\vec{A}_\mu\wedge C = (D_\mu\vec{C}\Lambda)\wedge\vec{C} - \frac{1}{2}\,\vec{A}_\mu\wedge(\vec{C}\wedge\vec{C}\Lambda) \quad .
\end{aligned} \tag{5}$$

Now $(D_\mu \vec{C}\Lambda)_\wedge \vec{C} = -(D_\mu \vec{C})_\wedge \vec{C}\Lambda$ and from the "Jacobi identity"

$$(\vec{A}_\mu{}_\wedge \vec{C})_\wedge \vec{C} - (\vec{C}_\wedge \vec{A}_\mu)_\wedge \vec{C} + (\vec{C}_\wedge \vec{C})_\wedge \vec{A}_\mu = 0 \quad , \tag{6}$$

we see that $(\vec{C}_\wedge \vec{A}_\mu)_\wedge \vec{C}$ equals $\frac{1}{2}(\vec{C}_\wedge \vec{C})_\wedge \vec{A}_\mu$ and the proof is easily completed.

From the transformations in (3) we can extract the extended BRST algebra

$$[\delta(\Lambda_1), \delta(\Lambda_2)] = [\delta(\zeta_1), \delta(\zeta_2)] = [\delta(\Lambda), \delta(\zeta)] = 0 \ . \tag{7}$$

Because Λ_i and ζ_i are anticommuting constant Lorentz-scalars these commutators are equivalent to nilpotency

$$\delta(\alpha\Lambda_1 + \beta\zeta_1)\ \delta(\alpha\Lambda_2 + \beta\zeta_2) = 0 \tag{8}$$

for arbitrary α and β. In particular, for $\alpha = 0$ or $\beta = 0$ one recovers the well-known nilpotency rules

$$\delta(\Lambda_1)\ \delta(\Lambda_2) = \delta(\zeta_1)\ \delta(\zeta_2) = 0 \quad . \tag{9}$$

The need for $\vec{d}$ is clear: replacing $\vec{d}$ by its field equation $\partial.\vec{A}$, (9) evaluated on $\vec{C}^*$ vanishes only on-shell, being proportional to the antighost field equation $\partial_\mu D_\mu \vec{C}$. Hence $\vec{d}$ plays a role similar to the auxiliary fields in the supersymmetry algebra off-shell, and this is a hint to apply supersymmetry techniques to BRST symmetry. Other hints are the anticommuting nature of the parameters and the fact that all fields in (2,3) are in the same adjoint representation.[12]

Let us now consider the general case. We will use the supercondensed notation of DeWitt,[13] where the classical gauge fields are denoted collectively by ϕ^i and i stands for all labels: internal indices as well as spacetime points. Faddeev-Popov ghosts and antighosts are denoted by C^{α} and $C^{*\alpha}$, respectively, and Nielsen-Kallosh ghosts by d^{α}. The classical gauge transformation read

$$\delta\phi^{i} = R^{i}_{\ \alpha}(\phi)\ \xi^{\alpha} \quad . \tag{10}$$

We will assume that the classical gauge algebra closes. This means that the commutator of two gauge transformations $\delta(\xi^{\alpha})$ and $\delta(\eta^{\alpha})$ is again a sum of gauge transformations, with structure functions $f^{\alpha}_{\ \beta\gamma}$ (which may depend on the classical gauge fields ϕ^{i}). The latter are defined by

$$[\delta(\eta^{\alpha}),\ \delta(\xi^{\alpha})] = \delta(f^{\alpha}_{\ \beta\gamma}\ \eta^{\gamma}\xi^{\beta}) \quad . \tag{11}$$

A lesson learned from supergravity is to consider the algebra first, and independently from any action. It is straight forward to show that the most general transformation rules satisfying (7), are unique, and given by

$$\begin{aligned}
\delta\phi^{i} &= R^{i}_{\ \alpha}\ C^{\alpha}\Lambda \quad , & \delta\phi^{i} &= R^{i}_{\ \alpha}\ C^{*\alpha}\zeta \quad , \\
\delta C^{\alpha} &= -\frac{1}{2}\ f^{\alpha}_{\ \beta\gamma} C^{\gamma}\Lambda C^{\beta} \quad , & \delta C^{*\alpha} &= -\frac{1}{2}\ f^{\alpha}_{\ \beta\gamma} C^{*\gamma}\zeta C^{*\beta} \quad , \\
\delta C^{*\alpha} &= -\ d^{\alpha}\Lambda \quad , & \delta C^{\alpha} &= -\ f^{\alpha}_{\ \beta\gamma} C^{*\gamma}\zeta C^{\beta} + d^{\alpha}\zeta \quad , \\
\delta d^{\alpha} &= 0 \quad , & \delta d^{\alpha} &= -\ f^{\alpha}_{\ \beta\gamma} C^{*\gamma}\zeta d^{\beta} \\
& & &\quad -\frac{1}{2}\ (-)^{\beta+\gamma}\ f^{\alpha}_{\ \beta\gamma,\delta} C^{\delta} C^{*\gamma}\zeta C^{*\beta} \ .
\end{aligned} \tag{12}$$

A few remarks are in order. Indices are contracted in north-easterly[12] fashion. For example, in $f^{\alpha}_{\ \beta\gamma}$ and $C^{\gamma}\Lambda$ the γ contraction acts as a commuting object, so that we can pull C^{β} till it sits between the β and γ in $f^{\alpha}_{\ \beta\gamma}$ and yields another valid contraction. But in the last term in (12) the δ-contraction yields an anticommuting object because we have C^{δ} instead of $C^{\delta}\zeta$, and this causes the extra sign.

Because we only want structure functions and fields in the transformation rules, all indices of C^{α}, $C^{*\alpha}$ and d^{α} must be superscripts. This is important for what follows and differs from the conventions of previous authors (including us).

The proof of (12) is not particularly difficult. One writes the most general expression compatible with statistics, dimensions and spins (the latter meaning that "all indices match"). Then one requires (9), at which point one still has some free parameters, and C and C^{*} appear symmetrically. Finally, the requirement of (7) fixes the freedom and breaks the C, C^{*} symmetry "spontaneously": the result in (12) is only ambiguous in so far that one might have interchanged C and C^{*}.

Having obtained the transformation rules, our next aim is to find an action $L(\mathrm{qu})$, invariant under these transformations. This is very simple, and summarizes all previous particular models and discussions. Usually one considers actions of the form

$$L(\mathrm{qu}) = L(\mathrm{class}) + \frac{1}{2}\,(-)^{\alpha}F^{\alpha}\gamma_{\alpha\beta}F^{\beta} + C^{*\alpha}\gamma_{\alpha\beta}\,\frac{\delta}{\delta\Lambda}\,F^{\beta}\ . \qquad (13)$$

The term $\frac{1}{2}\,(-)^{\alpha}\,F^{\alpha}\gamma_{\alpha\beta}F^{\beta}$ is the gauge fixing term, and by varying F^{β} one finds in the usual way the Faddeev-Popov action in the last term. If $\gamma_{\alpha\beta}$ is field-dependent (as in some gauges of supergravity) there are extra terms which will be reobtained below. Because $C^{*\alpha}$ always commutes with $\gamma_{\alpha\beta}F^{\beta}$ there is no extra sign $(-)^{\alpha}$ needed in the last term in (13).

In the Yang-Mills case we needed the Nielsen-Kallosh ghosts d^{α}, so we expect instead of (13) the following gauge fixing term

$$L\ (\text{gauge fixing}) = -\,\frac{1}{2}\,(-)^{\alpha}d^{\alpha}\gamma_{\alpha\beta}d^{\beta} - (-)^{\alpha}F^{\alpha}\gamma_{\alpha\beta}d^{\beta}\ . \qquad (14)$$

One can now at once write down the most general action which is

BRST invariant and contains (14).

$$L\text{ (qu)} = L\text{ (class)} + \frac{\delta}{\delta\Lambda}\,[C^{*\beta}\gamma_{\beta\alpha}\,(F^{\alpha} + \frac{1}{2}\,d^{\alpha})] \quad , \tag{15}$$

where $^{\delta}/\delta\Lambda$ means: perform a $\delta(\Lambda)$ variation, and remove Λ from the right. Clearly, the $C^{*\beta}$ variation in (15) produces (14), and the F^{α} variation the Faddeev-Popov action, but when $\gamma_{\alpha\beta}$ is field-dependent, the extra terms also at once follow.

The action in (15) is unique if one considers the general case, but in particular cases there may be extra terms invariant under BRST transformations. For example: if $\gamma_{\alpha\beta}$ is field-independent, one may add to (15) a term $d^{\alpha}\gamma_{\alpha\beta}d^{\beta}$. However, our strategy is the opposite: we consider only the most general cases, and allow only the results which are completely general. Thus (15) is unique.

The result in (15) states that the most general BRST invariant action is a sum of a classical-gauge invariant term plus a total BRST derivative. This at once suggests the solution for the most general extended BRST invariant action

$$L\text{ (qu)} = L\text{ (class)} + \frac{\delta}{\delta\Lambda}\,\frac{\delta}{\delta\zeta}\,(\ldots.) \quad . \tag{16}$$

To find the dots in (16) we must write the expression between parentheses in (15) as a $\delta/\delta\zeta$ derivative. Writing

$$L\text{ (qu)} = L\text{ (class)} + \frac{\delta}{\delta\Lambda}\,\frac{\delta}{\delta\zeta}\,(\frac{1}{2}\,C^{*\beta}\gamma_{\beta\alpha}C^{\alpha}+F) \quad , \tag{17}$$

where F is arbitrary at this point, we see that $\delta(\zeta)$ on C^{α} reproduces the term $\frac{1}{2}\,C^{*\beta}\gamma_{\beta\alpha}d^{\alpha}$ in (15). The other term in (15) can only come from the $\delta/\delta\zeta$ variation of F in (17). Hence

$$F_{\beta} = F,_{\beta} \tag{18}$$

where $F_{\beta} \equiv \gamma_{\beta\alpha}F^{\alpha}$ and $\delta(\zeta)\ F \equiv F,_{\beta}C^{*\beta}\ \zeta$. All gauge fixing terms F_{β}

are derived from the same scalar function F!!

The question is now: what is F? For Yang-Mills theory $F = \vec{A}.\vec{A}$ and for gravity and supergravity $F = (e^m{}_\mu \eta^{\mu n} - m \leftrightarrow n)^2$ reproduce all desired gauge fixing terms. (Namely: Lorentz, deDonder, and $\gamma \cdot \psi$). Hence the general rule we propose is

$$F = g_{ij}\, \phi^j \phi^i \quad , \tag{19}$$

where g_{ij} is a "suitable" metric on the infinite dimensional manifold spanned by the ϕ^i. We have not found a good geometrical determination of g_{ij}, but suspect that it has something to do with an invariant metric. We would like to find a g_{ij} which is invariant under global gauge transformations of the representations contragredient to ϕ^i. That would probably solve an outstanding problem: a background field formalism for supergravity in ordinary spacetime.

3. NEW LOCAL ALGEBRA

The attentive reader may have noticed that we did not elaborate on the details of the proof that (12) satisfies (7). We postponed a discussion because this leads into our next subject: a discussion of _general_ algebras with structure functions which depend on the fields which represent the algebra.

To begin with, the nilpotency on ϕ^i is simple: it just defines $\delta(\Lambda)\, C^\alpha$ and $\delta(\zeta)\, C^{*\alpha}$. But nilpotency of $\delta(\Lambda)$ transformations on C^α (and of $\delta(\zeta)$ transformations on $C^{*\alpha}$) is only simple when $f^\alpha{}_{\beta\gamma}$ are constant: then it is just the Jacobi identity. When $f^\alpha{}_{\beta\gamma}$ are field-dependent, one finds that one needs the following _modified_ Jacobi _identity_ involving triple commutators of classical gauge transformations

$$[\delta(\eta), [\delta(\zeta), \delta(\zeta)]] + \text{cyclic terms} \equiv 0 \quad . \tag{20}$$

Evaluation, using (11) leads to

$$f^{\delta}{}_{\gamma\varepsilon} f^{\varepsilon}{}_{\alpha\beta} - f^{\delta}{}_{\gamma\alpha,\beta} + \text{supercyclic terms in } \alpha\beta\gamma \equiv 0 \; . \tag{21}$$

New are the terms with $f^{\delta}{}_{\gamma\alpha,\beta}$. (Supercyclic means: cyclic plus an extra sign whenever two fermionic indices pass each other). The first interesting problem is to find a classification of such algebraic systems. Note that in ordinary Lie algebras the Jacobi identities are equivalent to $f^{\delta}{}_{\gamma\alpha;\beta} = 0$ with $f^{\mu}{}_{\nu\rho}$ as connections in the covariant derivatives, but that for our algebras this is not true.

Another point we would like to make is that it is not difficult to transform ordinary Lie algebras into the new algebras. For example, in general relativity one usually has in terms of holomic parameters

$$[\delta(\xi^{\mu}),\, \delta(\eta^{\mu})] = \delta[\eta^{\nu}\partial_{\nu}\xi^{\mu} - \xi^{\nu}\partial_{\nu}\eta^{\mu}] \; . \tag{23}$$

However, considering the tetrad fields e^{μ}_{m} as a representation, and defining $\xi^{m} \equiv \xi^{\mu} e^{m}_{\mu}$ to be local parameters, one finds

$$[\delta(\xi^{m}),\, \delta(\eta^{m})] = \tag{24}$$

$$\delta[\eta^{n} e^{\mu}_{n} \partial_{\mu} \xi^{m} + e^{m}_{\mu}(\partial_{\nu} e^{\mu}_{k}) e^{\nu}_{n} \eta^{n} \xi^{k} - \xi \leftrightarrow \eta] \quad ,$$

and clearly structure functions are produced. The question is thus: which theories with structure functions cannot be transformed (by such acceptable transformations as in the example) to theories with structure constants? Can one find a new basis for ξ^{α} and ϕ^{i} in supergravity such that the $f^{\alpha}{}_{\beta\gamma}$ become field-independent?

The Jacobian of $\delta(\Lambda)$ transformations is equal to the Jacobian of $\delta(\zeta)$ transformations, and equals unity if

$$(-)^{\alpha j} R^{j}{}_{\alpha,j} + (-)^{\beta} f^{\beta}{}_{\beta\alpha} = 0 \quad , \tag{25}$$

where ,j means right-differentiation w.r.t. ϕ^j. For physics this criterion is important for the derivation of Ward identities. Mathematically it restricts representations. If the ϕ^i form a Lie-algebra valued vector field, and transform as covariant derivatives in the adjoint representation, then this criterion is automatically satisfied. However, in supergravity it is satisfied too, although there the ϕ^i are not of this kind.[6]

We found an action which was invariant under extended BRST transformation, (a sufficient condition) but what would have happened if we started from

$$L(\text{qu}) = L(\text{class}) + (-)^{\alpha} \frac{1}{2} d^{\alpha}\gamma_{\alpha\beta}F^{\beta} + C^{*\alpha}\gamma_{\alpha\beta}F^{\beta}{}_{,\gamma}C^{\gamma} , \quad \gamma_{\alpha\beta} \text{ constant,} \tag{26}$$

and required it to be invariant under $\delta(\zeta)$ transformations (the necessary condition)? One finds two conditions

$$\begin{aligned} &F_{\alpha,\beta} - (-)^{\alpha\beta}F_{\beta,\alpha} = F_{\gamma}f^{\gamma}{}_{\alpha\beta} , \quad F_{\alpha} \equiv F^{\delta}\gamma_{\delta\alpha} , \\ &(-)^{\alpha+\beta} d^{\beta}d^{\alpha}f_{\alpha\beta\gamma} = 0 , \quad f_{\alpha\beta\gamma} \equiv \gamma_{\alpha\delta}f^{\delta}{}_{\beta\gamma} . \end{aligned} \tag{27}$$

These first conditions look like the Cartan-Maurer equations, except that one needs two F's on the right hand side. On the other hand, $F_{\alpha,\beta}$ is a variational derivative, not an ordinary derivative. If one would expand $F_{\alpha} = F_i R^i{}_{\alpha}$, then with (11) one finds

$$F_{i,j} - (-)^{ij} F_{j,i} = 0 \quad \text{(ordinary derivatives)} ,$$

whose solution is $F_i = F_{,i}$ and hence $F_{\alpha} = F_{,\alpha}$ is also a necessary criterion. The integrability conditions are precisely the generalized Jacobi identities in (21).

One can, in fact, define covariant derivatives,

$$F_{\alpha;\beta} \equiv F_{\alpha,\beta} - \frac{1}{2} F_\gamma f^\gamma{}_{\alpha\beta} \tag{29}$$

in terms of which $F_{\alpha;\beta} - (-)^{\alpha+\beta} F_{\beta;\alpha} = 0$. Closure of the gauge algebra becomes $R^i{}_{\alpha;\beta} - (-)^{\alpha+\beta} R^i{}_{\beta;\alpha} = 0$. These derivatives are strongly reminiscent of Cartan's (0) connection, which is also proportional to 1/2 times the structure constants, and which leads to vanishing torsion on the group manifold but to nonvanishing curvature.[14] Assuming right-invariant frames f^μ_m satisfying the Cartan-Maurer equations, we can compute the curvature in the standard way, taking as connection $w^m_{\mu n} = \lambda f^k_\mu f^m{}_{kn}$ and λ constant. One finds, not surprisingly, an extension of Cartan's result. Taking for simplicity only bosonic indices,

$$R_{mn}{}^a{}_b = (\lambda^2 f^a{}_{m\ell} f^\ell{}_{bn} + \lambda f^a{}_{bm,n} - m(\leftrightarrow)n) - \lambda f^a{}_{bc} f^c{}_{mn} \quad . \tag{30}$$

As a check one may verify that for $\lambda = 0$ or 1 (Cartan's (-) and (+) connections) the curvature would vanish if one had structure constants. For $\lambda = \frac{1}{2}$, there is no torsion, hence the cyclic identity for the curvature should hold. In fact, being precisely proportional to the generalized Jacobi identities in (21), the cyclic identity does hold. Note that this does not imply that the Ricci tensor is symmetric because $R_{mn}{}^b{}_b \neq 0$.

One of our central unsolved problems is the choice of $\gamma_{\alpha\beta}$. It should be the extension of the Killing form to the case of structure functions. The second condition (27) tells us that.[15] But how should one define the extension? One possibility is to use commutators. This is standard in ordinary Lie algebras

$$\begin{aligned} &[\delta(a^\alpha)\ [\delta(b^\alpha),\ \delta(C^\alpha)]] + a(\leftrightarrow)\ b = \delta[G^\alpha(a,b,c)] \quad , \\ &\gamma_{\alpha\beta} a^\beta b^\alpha = \delta/_{\delta c^\gamma}\ G^\gamma\ (a,b,c) \quad , \end{aligned} \tag{31}$$

and one finds extra terms due to the field-dependence of $f^\alpha{}_{\beta\gamma}$,

$$\gamma_{\alpha\beta} = (-)^{(\alpha+1)\delta} f^{\delta}{}_{\alpha\varepsilon} f^{\varepsilon}{}_{\delta\beta} - (-)^{\delta} f^{\delta}{}_{\delta\alpha,\beta} + (-)^{\alpha\beta} (\alpha(\leftrightarrow)\beta) . \quad (32)$$

However, one could also drop the terms with $(-)^{\delta} f^{\delta}{}_{\delta\alpha,\beta}$. Another, less convincing choice would be $\gamma_{mn} = R_{m\ell}{}^{\ell}{}_{n}$, or its symmetric part. We leave this problem to future research.

4. SUPERSPACE AND LOCAL BRST INVARIANCE

In this section we will rederive the BRST transformation rules, but now using the formalism of superspace. Although the BRST formalism holds also for nonsupersymmetric theories, the superspace formalism, originally derived for applications to supersymmetric theories, is actually also very convenient for extended BRST symmetry.[19] In fact, the manipulations below are the same as one uses over and over again in superspace supersymmetry, and reading the section below might also be useful for learning these techniques.

The superspace we consider below has four bosonic coordinates (the usual x^{μ}) and two fermionic coordinates (corresponding to the two extended BRST symmetries). It has nothing to do with the chiral superspace of Ogievetski and Sokatchev[6] (which has a left-handed spinor and complex x^{μ}), because the fermionic coordinates are assumed to be inert under Lorentz transformations.

The reason we consider superspace is that it allows one to extend in a simple way the global BRST symmetry into a local invariance. Again this extension is a reflection of the transition from supersymmetry to supergravity. In the local case, our parameters will become functions of x^{μ}, but for the present we take them to be constant.

To begin, we define a vector ghost field $C_a = (C^{*\alpha}, C^{\alpha})$ and a vector parameter $\zeta^a = (\zeta, \Lambda)$, a = 1,2, such that the transformation rule for ϕ^i is

$$\delta(\zeta)\phi^i = R^i{}_{\alpha} C^{\alpha}{}_a \zeta^a \quad . \qquad (4.1)$$

Thus, we unite into a single transformation the two BRST transformations of Eqs. (2.12). We next assume that $C^\alpha{}_a$ transforms as

$$\delta(\zeta)C^\alpha{}_a = C^\alpha{}_{ab}\,\zeta^b \quad , \tag{4.2}$$

with the $C^\alpha{}_{ab}$ to be determined. Nilpotency of $\delta(\zeta)$, which is simply expressed as $[\delta(\zeta), \delta(\zeta')] = 0$, when written on ϕ^i yields the condition

$$C^\alpha{}_{(ab)} = (-1)^\beta \frac{1}{2} f^\alpha{}_{\beta\gamma}\, C^\gamma{}_b\, C^\beta{}_a \quad , \tag{4.3}$$

but fails to constrain $C^\alpha{}_{[ab]}$. Here, $A_{(ab)} \equiv \frac{1}{2}(A_{ab} + A_{ba})$ and $A_{[ab]} \equiv \frac{1}{2}(A_{ab} - A_{ba})$. Since our ζ^a space is two dimensional, we can write any antisymmetric tensor as $A_{[ab]} = \varepsilon_{ab}A$, $A = \frac{1}{2}\varepsilon^{ab}A_{[ab]}$, where ε_{ab} is the Levi-Civita tensor in two-dimensions: $\varepsilon_{12} = \varepsilon^{12} = 1$. We will define raising and lowering of indices by ε^{ab} and ε_{ab} as follows: $A^a \equiv A_b\varepsilon^{ba}$, $A_a \equiv \varepsilon_{ab}A^b$, so that ε_{ab} plays the role of metric tensor (the metric is antisymmetric because the coordinates are fermionic). Thus $A_{[ab]} = -\frac{1}{2}\varepsilon_{ab}A_c{}^c$. A simple mnemonic for these operations is that indices are contracted in a northeasterly direction, i.e. from lower left to upper right.

Returning to our problem, $C^\alpha{}_{[ab]}$ may be set equal to $-b^\alpha\varepsilon_{ab}$, where b^α is a new field (in fact, b^α will be related to the b^α field of section 2.) The most general transformation for $C^\alpha{}_a$ which respects nilpotency on ϕ^i is therefore

$$\delta(\zeta)C^\alpha{}_a = -\frac{1}{2} f^\alpha{}_{\beta\gamma}C^\gamma{}_b\zeta^b C^\beta{}_a - b^\alpha\varepsilon_{ab}\zeta^b \quad . \tag{4.4}$$

The transformation rule for b^α is still undetermined; we shall fix it by requiring nilpotency on $C^\alpha{}_a$. Thus, let

$$\delta(\zeta)b^\alpha = b^\alpha{}_a\zeta^a \quad , \tag{4.5}$$

and demand that $[\delta(\zeta'), \delta(\zeta)]C^{\alpha}{}_{a} = 0$. In this commutator one finds four terms quadratic in the structure constants. Two of them are equal, and eliminating the other two by means of the Jacobi identities, the sum of the original two is multiplied by a factor $\frac{1}{2}$, and one arrives at

$$\tilde{b}^{\alpha}{}_{c}\varepsilon_{ab} + \tilde{b}^{\alpha}{}_{b}\varepsilon_{ac} = \{ -\frac{1}{4}(-1)^{\gamma} f^{\alpha}{}_{\beta\varepsilon} f^{\varepsilon}{}_{\gamma\delta}$$

$$+ \frac{(-1)^{\gamma}}{4}(-1)^{\beta\delta}[f^{\alpha}{}_{\beta\gamma,\delta}(-1)^{\beta\delta} - f^{\alpha}{}_{\gamma\delta,\beta}(-1)^{\beta\gamma}$$

$$+ f^{\alpha}{}_{\delta\beta,\gamma}(-1)^{\gamma\delta}]\, C^{\delta}{}_{c} C^{\gamma}{}_{b} C^{\beta}{}_{a} \quad ,$$

$$\tilde{b}^{\alpha}{}_{a} \equiv b^{\alpha}{}_{a} - \frac{1}{2} f^{\alpha}{}_{\beta\gamma} b^{\gamma} C^{\beta}{}_{\alpha} \quad . \tag{4.6}$$

It is not necessarily true that (4.6) can be satisfied; it is only integrable if the totally symmetric part in abc of the tensor on the right-hand-side vanishes. It does vanish, though, as a consequence of the generalized Jacobi identity, so we can proceed to solve (4.6). Multiplying by ε^{ab} we find

$$\tilde{b}^{\alpha}{}_{a} = -\frac{1}{12}(-1)^{\gamma} f^{\alpha}{}_{\beta\varepsilon} f^{\varepsilon}{}_{\gamma\delta} C^{\delta}{}_{a} C^{\gamma}{}_{b} C^{\beta b}$$
$$+ \frac{1}{b}(-)^{\beta} f^{\alpha}{}_{\delta\beta,\gamma} C^{\gamma}{}_{b} C^{\beta b} C^{\delta}{}_{a} \quad , \tag{4.7}$$

$$\delta(\zeta) b^{\alpha} = -\frac{1}{2} f^{\alpha}{}_{\beta\gamma} C^{\gamma}{}_{a} \zeta^{a} b^{\beta}$$
$$- \frac{1}{12}(-)^{\beta} f^{\alpha}{}_{\beta\varepsilon} f^{\varepsilon}{}_{\gamma\delta} C^{\delta}{}_{a} \zeta^{a} C^{\gamma}{}_{b} C^{\beta b} \tag{4.8}$$
$$+ \frac{1}{6}(-1)^{\beta} f^{\alpha}{}_{\delta\beta,\gamma} C^{\gamma}{}_{b} C^{\beta b} C^{\delta}{}_{a} \zeta^{a} \quad .$$

There is no freedom left in the system; therefore, nilpotency on b^{α}

must occur with just the set of transformations we already possess. To demonstrate that it obtains is a nontrivial exercise which we do not present here. To accomplish it one must make repeated use of

$$f^{\alpha}{}_{\beta\gamma,\delta\varepsilon} - (-)^{\delta\varepsilon} f^{\alpha}{}_{\beta\gamma,\varepsilon\delta} = f^{\alpha}{}_{\beta\gamma,\zeta} f^{\zeta}{}_{\delta\varepsilon} \quad , \tag{4.9a}$$

which is nothing but closure of the gauge algebra applied to $f^{\alpha}{}_{\beta\gamma}(\phi)$, and the identity

$$C^{\alpha\beta\gamma}_{(ab)c} = C^{\alpha\beta\gamma}_{(abc)} + \frac{1}{6}\varepsilon_{bc} C^{\alpha\beta\gamma}_{dae}\varepsilon^{de} + \frac{1}{6} C^{\alpha\beta\gamma}_{dbe}\varepsilon^{de}\varepsilon_{ac} + \frac{1}{6}\varepsilon_{bc} C^{\alpha\beta\gamma}_{ade}\varepsilon^{de} + \frac{1}{6}\varepsilon_{ac} C^{\alpha\beta\gamma}_{bde}\varepsilon^{de} \quad , \tag{4.9b}$$

which is rather like the Fierz identities so essential to calculations in supersymmetry, where $C^{\alpha\beta\gamma}_{abc} \equiv C^{\alpha}{}_{a} C^{\beta}{}_{b} C^{\gamma}{}_{c}$, and $A_{(abc)}$ is the totally symmetric part of A_{abc} (our definition includes the 1/3!). Thus, one finds a consistent, nilpotent algebra

$$\delta(\zeta)\phi^{i} = R^{i}{}_{\alpha} C^{\alpha}{}_{a}\zeta^{a} \quad ,$$

$$\delta(\zeta)C^{\alpha}{}_{a} = -\frac{1}{2} f^{\alpha}{}_{\beta\gamma} C^{\gamma}{}_{b}\zeta^{b} C^{\beta}{}_{a} - b^{\alpha}\varepsilon_{ab}\zeta^{b} \quad ,$$

$$\delta(\zeta)b^{\alpha} = -\frac{1}{2} f^{\alpha}{}_{\beta\gamma} C^{\gamma}{}_{a}\zeta^{a} b^{\beta} - \frac{1}{12}(-1)^{\beta} f^{\alpha}{}_{\beta\varepsilon} f^{\varepsilon}{}_{\gamma\delta} C^{\delta}{}_{a}\zeta^{a} C^{\gamma}{}_{b} C^{\beta b} + \frac{1}{6}(-1)^{\beta} f^{\alpha}{}_{\delta\beta,\gamma} C^{\gamma}{}_{b} C^{\beta b} C^{\delta}{}_{a}\zeta^{a} \quad . \tag{4.10}$$

It will be useful to know also the second variation of these fields. We present without proof the formulas

$$\delta(\zeta')\delta(\zeta)\phi^i = [\tfrac{1}{2}(-1)^{\alpha+1}R^i{}_{\alpha,\beta}C^\beta{}_a C^{\alpha a} - R^i{}_\alpha b^\alpha]\varepsilon_{cd}\zeta'^d\zeta^c \quad ,$$

$$\delta(\zeta')\delta(\zeta)C^\alpha{}_a = [f^\alpha{}_{\beta\gamma}b^\gamma C^\beta{}_a + \tfrac{1}{3}(-1)^\beta f^\alpha{}_{\delta\beta,\gamma}C^\gamma{}_b C^{\beta b}C^\delta{}_a$$

$$- \tfrac{1}{6}(-1)^\gamma f^\alpha{}_{\beta\varepsilon}f^\varepsilon{}_{\gamma\delta}C^\delta{}_a C^\gamma{}_b C^{\beta b}]\varepsilon_{cd}\zeta'^d\zeta^c \quad ,$$

$$\delta(\zeta')\delta(\zeta)b^\alpha = 0 \quad . \tag{4.11}$$

It is interesting to note that the commutator $[\delta(\zeta'),\delta(\zeta)]$ on these fields now trivially and manifestly vanishes, because $\delta(\zeta')\delta(\zeta)$ is proportional to $\varepsilon_{ab}\zeta'^b\zeta^a$, which is symmetric in $\zeta \leftrightarrow \zeta'$. Finally, let us make contact with the results of section 2. The field b^α we have defined here is given in terms of d^α by

$$b^\alpha = d^\alpha + \tfrac{1}{2}(-1)^\beta f^\alpha{}_{\beta\gamma}C^\gamma C^{*\beta} \quad . \tag{4.12}$$

The algebra (4.10) and second variations (4.11) can be realized in a six-dimensional superspace formed by adding to the usual Minkowski four-space two fermionic coordinates θ^a, $a = 1,2$, which correspond to the parameters ζ^a in the sense that the transformation with parameter ζ^a represents in superspace a translation of the θ^a coordinate. That is, if we associate with each field a superfield which depends on x^μ and θ^a then the variation of the original field with respect to ζ^a will be associated with the derivative with respect to θ^a of the superfield.

We define, then, superfields $\Phi^i(x,\theta)$, $\chi^\alpha{}_a(x,\theta)$ and $B^\alpha(x,\theta)$ which are associated with $\phi^i(x)$, $C^\alpha{}_a(x)$, $b^\alpha(x)$ in this way. Since the two θ^a are fermionic, the product of three of them always vanishes, and as a result any superfield may be expanded in a finite power series in θ^a; indeed, the power series has at most three terms. Thus, we write

$$\Phi^i(x,\theta) = \phi^i(x) + \phi^i{}_a(x)\theta^a + \frac{1}{2}\tilde{\phi}^i(x)\theta_a\theta^a ,$$

$$\chi^\alpha{}_a(x,\theta) = C^\alpha{}_a(x) + C^\alpha{}_{ab}(x)\theta^b + \frac{1}{2}\tilde{C}^\alpha{}_a(x)\theta_b\theta^b ,$$

$$B^\alpha(x,\theta) = b^\alpha(x) + b^\alpha{}_a(x)\theta^a + \frac{1}{2}\tilde{b}^\alpha(x)\theta_a\theta^a , \tag{4.13}$$

where the $\theta^a \to 0$ limit must give the ordinary Minkowski space fields. Using the formula for (global) translations

$$\delta(\zeta)\Phi(x,\theta) \equiv \Phi(x,\theta+\zeta) - \Phi(x,\theta)$$
$$= \frac{\partial}{\partial\theta^a}\Phi(x,\theta)\zeta^a , \tag{4.14}$$

(evidently, our $\partial/\partial\theta$ derivatives are right-derivatives) we may solve for the various functions appearing in (4.13), if we identify $\delta(\zeta)$ with the transformation (4.10). Equation (4.14) must be satisfied order by order in θ^a. To zeroth order, one has

$$\phi^i{}_a(x) = R^i{}_\alpha C^\alpha{}_a ,$$

$$C_{ab}(x) = \frac{1}{2} f^\alpha{}_{\beta\gamma} C^\gamma{}_b C^\beta{}_a (-1)^\beta - b^\alpha \varepsilon_{ab} ,$$

$$b^\alpha{}_a(x) = \frac{1}{2} f^\alpha{}_{\beta\gamma} b^\gamma C^\beta{}_a$$
$$- \frac{1}{12}(-1)^\gamma f^\alpha{}_{\beta\varepsilon} f^\varepsilon{}_{\gamma\delta} C^\delta{}_a C^\gamma{}_b C^{\beta b}$$
$$+ \frac{1}{6}(-1)^\beta f^\alpha{}_{\delta\beta,\gamma} C^\gamma{}_b C^{\beta b} C^\delta{}_a , \tag{4.15}$$

for example. Continuing in this way, one derives

$$\Phi^i(x,\theta) = \phi^i(x) + R^i{}_\alpha C^\alpha{}_a\theta^a + \frac{1}{2}\left[\frac{1}{2}(-1)^{\alpha+1} R^i{}_{\alpha,\beta} C^\beta{}_a C^{\alpha a} - R^i{}_\alpha b^\alpha\right]\theta_b\theta^b ,$$

$Ds^3 = M^3$

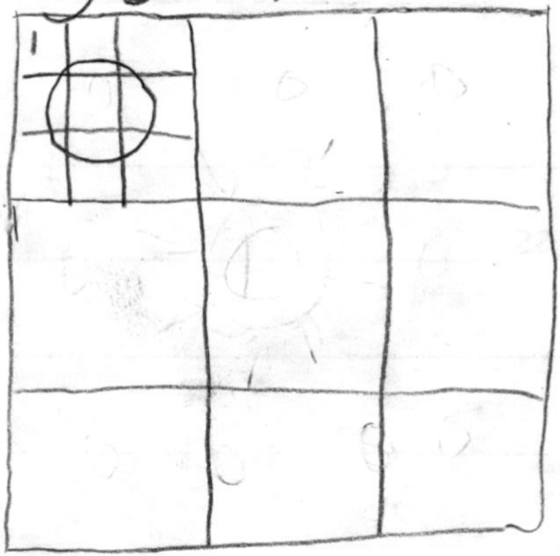

$G = (Fs^3)$ 9 TONS OF E^-

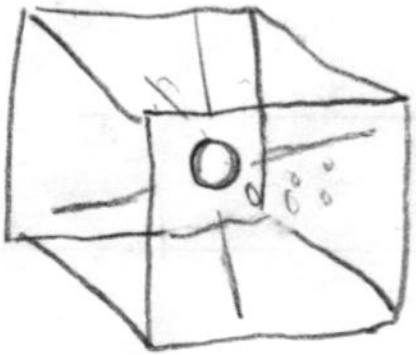

$A^3 \quad B^1 \quad M$

ACCT. NO. ____________________

DATE ____________________

NAME ____

ADDRESS ____

____

____

____

____

$$\chi^{\alpha}{}_{a}(x,\theta) = C^{\alpha}{}_{a}(x) - [\tfrac{1}{2}f^{\alpha}{}_{\beta\gamma}C^{\gamma}{}_{b}C^{\beta}{}_{a}(-1)^{\beta+1} + b^{\alpha}\varepsilon_{ab}]\theta^{b}$$
$$+ \frac{1}{2}[f^{\alpha}{}_{\beta\gamma}b^{\gamma}C^{\beta}{}_{a} + \frac{1}{3}(-1)^{\beta}f^{\alpha}{}_{\delta\beta,\gamma}C^{\gamma}{}_{b}C^{\beta b}C^{\delta}{}_{a}$$
$$- \frac{1}{6}(-1)^{\gamma}f^{\alpha}{}_{\beta\varepsilon}f^{\varepsilon}{}_{\gamma\delta}C^{\delta}{}_{a}C^{\gamma}{}_{b}C^{\beta b}]\theta_{c}\theta^{c} \quad ,$$

$$B^{\alpha}(x,\theta) = b^{\alpha}(x) + \frac{1}{2}[f^{\alpha}{}_{\beta\gamma}b^{\gamma}C^{\beta}{}_{a}$$
$$+ \frac{1}{3}(-1)^{\beta}f^{\alpha}{}_{\delta\beta,\gamma}C^{\gamma}{}_{b}C^{\beta b}C^{\delta}{}_{a}$$
$$- \frac{1}{6}(-1)^{\gamma}f^{\alpha}{}_{\beta\varepsilon}f^{\varepsilon}{}_{\gamma\delta}C^{\delta}{}_{a}C^{\gamma}{}_{b}C^{\beta b}]\theta^{a} \quad . \tag{4.16}$$

We shall make use of Eqs. (4.16) in what follows. The Lagrangian of the quantum theory constructed using these fields is given by Eq. (2.17). We shall rewrite this Lagrangian using some techniques of superspace. Since integration over Grassmann variables obeys the rules [ref. (6), appendix]

$$\int d\theta = 0 \quad ,$$

$$\int \theta\, d\theta = 1 \quad , \tag{4.17}$$

we have

$$L^{(0)} = L_{c\ell}(\phi) + \; d^{2}\theta[F(\Phi^{i}) - \frac{1}{4}\chi^{\alpha}{}_{a}\gamma_{\alpha\beta}\chi^{\beta b}] \quad . \tag{4.18}$$

Because of (4.17), the integration yields a nonzero result only for the term of a given superfield which is quadratic in θ; thus, integration is equivalent to differentiation, and (4.18) is equal to (2.17).

Using (4.18) we can now consider the problem of constructing a locally BRST-invariant system. To do so we shall first specialize to the case of a Yang-Mills theory. Although what we do is applicable in a modified form to more general gauge theories, the details for the latter are somewhat cumbersome and tend to obscure the nature of the effect we wish to demonstrate. Thus from now on we use fields $\phi^i = A_\mu{}^\alpha$ for which the gauge transformation rule is

$$\delta A^\alpha_\mu = D_\mu{}^\alpha{}_\beta(A)\xi^\beta \quad , \tag{4.19}$$

with $D_\mu{}^\alpha{}_\beta(A) = (\partial_\mu\delta^\alpha{}_\beta + gf^\alpha{}_{\beta\gamma})A^\gamma_\mu$, and the structure constants $f^\alpha{}_{\beta\gamma}$ are truly constant, as is the metric $\gamma_{\alpha\beta}$.

To find the correct transformation rule for fields under a local BRST transformation with parameter $\zeta^a(x)$ we note that the only global transformation of Eqs. (4.10) which involves derivatives of a field is that on ϕ^i. Therefore, provided that $\zeta^a(x)$ only appears coupled with $C^\alpha{}_a(x)$ to form a gauge parameter $\xi^\alpha(x) = C^\alpha{}_a(x)\zeta^a(x)$, the classical Lagrangian will still be invariant under the local transformation. Thus, let

$$\begin{aligned}\delta A^\alpha_\mu &= D_\mu{}^\alpha{}_\beta(C^\beta{}_a\zeta^a) \\ &= (D_\mu{}^\alpha{}_\beta C^\beta{}_a)\zeta^a + C^\alpha{}_a\partial_\mu\zeta^a \quad .\end{aligned} \tag{4.20}$$

Enlarging this into a superfield equation and combining it with the corresponding ghost equation we have

$$\begin{aligned}\delta\Phi^\alpha_\mu &= \frac{\partial}{\partial\theta^a}\,\Phi^\alpha_\mu(x,\theta)\zeta^a(x) + \chi^\alpha{}_a(x,\theta)\partial_\mu\zeta^a(x) \quad , \\ \delta\chi^\alpha{}_a &= \frac{\partial}{\partial\theta^b}\,\chi^\alpha{}_a(x,\theta)\zeta^b(x) \quad ,\end{aligned} \tag{4.21}$$

where $\Phi^\alpha_\mu(x,\theta)$ is the superfield corresponding to $A^\alpha_\mu(x)$, and we have used (4.14). Equation (4.21) is in fact a generalization of (4.14) which can be seen as follows. Let us define a six-vector superfield $\Phi^\alpha_A(x,\theta)$ by

$$\Phi^{\alpha}_{A}(x,\theta) = (\Phi^{\alpha}_{\mu}(x,\theta),\ \chi^{\alpha}{}_{a}(x,\theta)) \quad . \tag{4.22}$$

Then a general coordinate transformation in the θ^a variable of the form $\theta^a \to \theta^a + \zeta^a(x)$ changes $\Phi^{\alpha}_{A}(x,\theta)$ by

$$\delta\Phi^{\alpha}_{A}(x,\theta) = \partial_a\Phi^{\alpha}_{A}(x,\theta)\zeta^a(x) + \Phi^{\alpha}_{a}(x,\theta)\partial_A\zeta^a(x) \quad . \tag{4.23}$$

Decomposing (4.23) into its four-and two-space components yields Eqs. (4.21).

Let us now vary (4.18) with respect to (4.21). As we stressed, $L_{c\ell}(A)$ is already invariant, so we have

$$\begin{aligned}\delta L^{(o)} &= \int d^2\theta \ [\frac{\delta F}{\delta\Phi^{\alpha}_{\mu}} \chi^{\alpha}{}_{a}\ \partial_{\mu}\ \zeta^a(x) \\ &\quad + \frac{\partial}{\partial\theta^a}(F - \frac{1}{4}\chi^{\alpha}{}_{a}\gamma_{\alpha\beta}\chi^{\beta a})\zeta^a(x)] \\ &= \int d^2\theta \ \frac{F}{\delta\Phi^{\alpha}_{\mu}}\chi^{\alpha}{}_{a}\partial_{\mu}\zeta^a(x) \quad ,\end{aligned} \tag{4.24}$$

where the total derivative can be dropped, thanks to Eqs. (4.17). To cancel (4.24) we add to $L^{(0)}$ a term involving a new field $a^a_{\mu}(x)$, a Grassmann-valued vector field transforming as

$$\delta a^a_{\mu}(x) = \partial_{\mu}\zeta^a(x) \quad . \tag{4.25}$$

That is, let

$$\begin{aligned}L^{(1)} &= L^{(0)} + \Delta L^{(1)} \quad , \\ \Delta L^{(1)} &= - \int d^2\theta \ \frac{\delta F}{\delta\Phi^{\alpha}_{\mu}}\chi^{\alpha}{}_{a}a^a_{\mu}(x) \quad .\end{aligned} \tag{4.26}$$

Then, the variation of a^a_{μ} in (4.26) precisely cancels that of $L^{(0)}$. However, the process does not end here, because we must also vary the other fields in $\Delta L^{(1)}$

$$\delta L^{(1)} = -\int d^2\theta \frac{\delta^2 F}{\delta\Phi^\beta_\nu \delta\Phi^\alpha_\mu} \chi^\beta{}_b \partial_\nu \zeta^b \chi^\alpha{}_a a^a{}_\mu \quad , \tag{4.27}$$

where we have dropped a total derivative as in (4.24). To cancel (4.27) we add another term to $L^{(0)}$:

$$L^{(2)} = L^{(1)} + \Delta L^{(2)} \quad ,$$

$$\Delta L^{(2)} = \frac{1}{2} \int d^2\theta \frac{\delta^2 F}{\delta\Phi^\beta_\nu \Phi^\alpha_\mu} \chi^\beta{}_b a^b{}_\nu \chi^\alpha{}_a a^a{}_\mu \quad , \tag{4.28}$$

and repeat the process. The pattern is now obvious, and we derive the local BRST-invariant Lagrangian

$$\begin{aligned} L &= \lim_{n\to\infty} L^{(n)} \\ &= \int d^2\theta \, \{-\frac{1}{4} \chi^\alpha{}_a \gamma_{\alpha\beta} \chi^{\beta a} + F(\Phi) \exp[-\frac{\overleftarrow{\delta}}{\delta\Phi^\alpha_\mu} \chi^\alpha{}_a a^a{}_\mu]\} \\ &= \int d^2\theta \, \{F(\Phi^\alpha_\mu - \chi^\alpha{}_a a^a{}_\mu) - \frac{1}{4} \chi^\alpha{}_a \gamma_{\alpha\beta} \chi^{\beta a}\} \; . \end{aligned} \tag{4.29}$$

The effect of requiring local invariance is therefore to shift the gauge superfield by $-\chi^\alpha{}_a a^a{}_\mu$. In terms of four-space fields, this entails the replacements

$$A^\alpha_\mu \to A^\alpha_\mu - C^\alpha{}_a A^a{}_\mu \equiv A^{\alpha\,\mathrm{cov}}_\mu \quad ,$$

$$(D_\mu C_a)^\alpha \to (D_\mu C_a)^\alpha + (b^\alpha \varepsilon_{ab} + \frac{1}{2} f^\alpha{}_{\beta\gamma} C^\gamma{}_a C^\beta{}_b) A^b_\mu \equiv (D^{\mathrm{cov}}_\mu C_a)^\alpha \quad ,$$

$$\begin{aligned} (D_\mu b)^\alpha &\to (D_\mu b)^\alpha + \frac{1}{2} [f^\alpha{}_{\beta\gamma} b^\gamma C^\beta{}_a + \frac{1}{6} f^\alpha{}_{\beta\varepsilon} f^\varepsilon{}_{\gamma\delta} C^\beta{}_b C^{\gamma b} C^\delta{}_a] A^a_\mu \\ &\equiv (D^{\mathrm{cov}}_\mu b)^\alpha \quad , \end{aligned}$$

and so defines the local-BRST-covariant derivatives, that is, the derivatives which transform without derivatives on the parameter.

We can, of course also give a^a_μ a BRST-invariant kinetic Lagrangian of the form

$$L(a) = -\frac{1}{4\kappa^2} f_{\mu\nu a} f^{\mu\nu a} \quad ,$$
$$f^a_{\mu\nu} = \partial_\mu a^a_\nu - \partial_\nu a^a_\mu \quad , \tag{4.31}$$

where κ must be a constant with dimensions of length owing to the fact that ζ^a has dimensions of length. It is possible that κ has some connection with the gravitational constant of the same name, since both arise from local symmetries; however, it should be remembered that in the present case the symmetry is not one of physical fields but rather of fields which appear only in a functional integral.

We do not know if the field a^a_μ, with unphysical statistics and noncanonical dimensions, will be of any use in the quantization of gauge theories, but it may help to illuminate somewhat the connection between the physical gauge field and the unphysical ghost from which it sprang. If one were to apply path-integral methods to this system, one would have to fix the local BRST invariance, introduce new ghosts etc. One might even repeat this process an arbitrary, even infinite number of times. What the limiting theory would be, we don't know. The gauge field of local BRST invariance violates the spin-statistics theorem, as we discussed.

ACKNOWLEDGEMENT

This work is supported in part by the N.S.F. Grant #PHY81-09110.

REFERENCES

1. C. Becchi, A. Rouet and R. Stora, Ann. of Phys. 98, 98 (1976), I.V. Tyutin, inst. report FIAN 39 (1975).
2. G. 't Hooft, Nucl. Phys. B33, 173 (1971), G. Sterman, P.K. Townsend and P. van Nieuwenhuizen, Phys. Rev. D17, 1501 (1978).
3. B. Lee, Lectures les Houches 1975; J. Zinn-Justin, Bonn Lectures, 1975; C. Itzykson and J.B. Zuber, Quantum field theory, 1980, McGraw Hill.
4. G. Curci and R. Ferrari, Phys. Lett. 63B, 51 (1976), idem, Nuov. Cim. 30A, 155 (1975) and Nuov. Cim. 32A, 151 (1976) and Nuov. Cim. 35A, 1, 273 (1976) and erratum Nuov. Cim. 47A, 555 (1978).
5. F.R. Ore Jr. and P. van Nieuwenhuizen, Stony Brook preprint ITP-SB-82-11.
6. For a review of quantum supergravity, see P. van Nieuwenhuizen, Physics Report 68, 189 (1981) section 2.
7. The resulting problems for quantum field theory were discussed in the article by the author in Unification of the Fundamental Particle Interactions, S. Ferrara, J. Ellis and P. van Nieuwenhuizen, editors, Plenum Press, 1980.
8. We thank H. Pagels for this suggestion.
9. Y. Ne'eman, Phys. Lett. 81B, 190 (1979).
10. N.K. Nielsen, Nucl. Phys. B140, 499 (1978), R. Kallosh, Nucl. Phys. B141, 141 (1978).
11. N.K. Nielsen, M.T. Grisaru, H. Römer and P. van Nieuwenhuizen, Nucl. Phys. B140, 477 (1978). In the unweighted gauge $\gamma\cdot\psi = 0$, there are no Nielsen-Kallosh ghosts, and one can use topological methods for pure helicity $\pm$ 3/2, see S.M. Christensen and M.J. Duff, Phys. Lett. 76B, 571 (1978).
12. R. Delbourgo and P.D. Jarvis, preprint April 1981.
13. B.S. DeWitt, Phys. Rev. 162, 1195 (1967).
14. E. Cartan, Qeuvres completes, Gauthier-Villars, 1952, volume 2, pg. 673.

15. The proof goes as follows for ordinary Lie algebras: $f_{\alpha\beta\gamma} - f_{\beta\alpha\gamma} = 0$ means that $\gamma_{\alpha\beta}$ is an invariant tensor of the coadjoint representation. If R^{α} transforms in the adjoint representation, $R^{\alpha}\gamma_{\alpha\beta}R^{\beta}$ is a scalar. But in the direct product $R \otimes R$ one finds only one scalar, as one may verify for SU(n) using Young tableaux.

16. Y. Ne'eman and J. Thierry-Mieg, Texas preprint 1981.

17. L. Baulieu and J. Thierry Mieg, Columbia preprint CU-TP-196, L. Alvarez-Gaume and L. Baulieu, HUTP-81/A058.

18. A. Das and M.A. Namazie, Phys. Lett. 99B, 463 (1981), A. Das, Phys. Rev. D to be published.

19. A superspace formalism for ordinary BRST symmetry was first proposed by S. Ferrara, S. Piguet and M. Schweda, Nucl. Phys. B119, 493 (1977). See also K. Fujikawa, Progr. Theor. Phys. 59, 2045 (1977). For extended BRST symmetry, a superspace approach was first presented by L. Bonora and M. Tonin, Phys. Lett. 98B, 48 (1981). An extensive discussion can be found in a recent preprint by R. Delbourgo and P.D. Jarvis.

GLUEBALLS - A STATUS REPORT*

Daniel L. Scharre

Stanford University

Stanford, California 94305

ABSTRACT

The current experimental status of glueballs is reviewed. The possibility that the $\iota(1440)$ and $\theta(1640)$ are glueballs is examined.

I. INTRODUCTION

It is expected from quantum chromodynamics (QCD) that glueballs,[1-4] bound states which contain gluons but no valence quarks, should exist. To date, no conclusive evidence for glueballs has been presented. After a brief review of the expected properties and experimental signatures of glueballs, I will discuss the status of some glueball candidate states.

Bound states of both two gluons and three gluons are expected to exist. The gluon selfcoupling implies that these states will not be pure two-gluon (gg) or three-gluon (ggg) states. However, the admixture of ggg component in the lowest-lying gg states (which are probably the states of immediate experimental interest) is expected to be small and will be ignored in the subsequent discussion.

*Work supported by the Department of Energy, contract DE-AC03-76SF00515.

Table I

gg Quantum Numbers and Masses

State	J^{PC}	Mass[a]
ground	0^{++}, 2^{++}	960 Mev
first excited	0^{-+}, (1^{-+}), 2^{-+}	1290 Mev

[a] Ref. 2.

gg states are required to have even charge conjugation parity (C-parity) while ggg states are allowed to have either even or odd C-parity depending on whether the SU(3) coupling is antisymmetric or symmetric. Except for the lowest lying states, the spin-parity (J^P) classification of glueballs is a function of the model used to construct the states. Table I gives the spin-parities of the lowest states.[1] (The $J^{PC} = 1^{-+}$ state is not allowed for massless gluons by Yang's theorem.[5]) For the quantum numbers of higher excited states and ggg states, one should refer to the literature.[1]

Most theoretical predictions place the masses of the lowest lying glueballs between approximately 1 and 2 GeV.[1-3] As an example, the masses predicted by the naive bag model without intergluon interactions[2] are given in Table I. In general, it is expected that hyperfine splittings due to intergluon interactions will push the masses of the spin 2 states up relative to the masses of the spin 0 states.[3]

Based on OZI suppression arguments, it is expected that glueball widths are relatively narrow compared to quark state ($q\bar{q}$) widths. This can be seen by comparison of the quark-line diagrams for glueball decay and OZI suppressed quark state decay (e.g., $\phi \to \rho\pi$). Since the glueball decay diagram is half the quark state diagram, the glueball width is expected to be suppressed by the square root of a typical OZI suppression factor. Thus, glueball

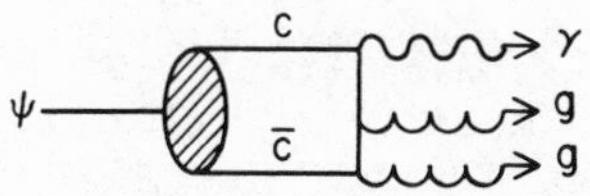

Fig. 1 Leading order diagram for $\psi \to \gamma + X$.

widths are expected to be about an order of magnitude smaller than nonsuppressed quark state widths.

It is clear that unambiguous identification of a glueball will not be easy. Observation of a state in a "glueball-favored" channel, i.e., a channel in which an intermediate state of two or more gluons is produced, is the first indication that a state is a glueball. The classic glueball-favored channel is the process $\psi \to \gamma + X$. The leading-order diagram for this process[6] is shown in Fig. 1. The inclusive branching ratio for this process is predicted to be approximately 8%. Experimental confirmation of this direct-photon component in ψ decays with approximately the correct branching ratio has been made by the Mark II.[7] Thus, if gg states exist in the kinematically allowed mass range for this process, they are expected to be produced here. The particular states which are likely to be produced are those with $J^P = 2^+$, 0^-, and 0^+ as predicted by a spin-parity analysis of the γgg final state.[8]

A second process which is a likely channel for production of glueballs is $\pi^- p \to \phi\phi n$.[9] This OZI suppressed process is shown in

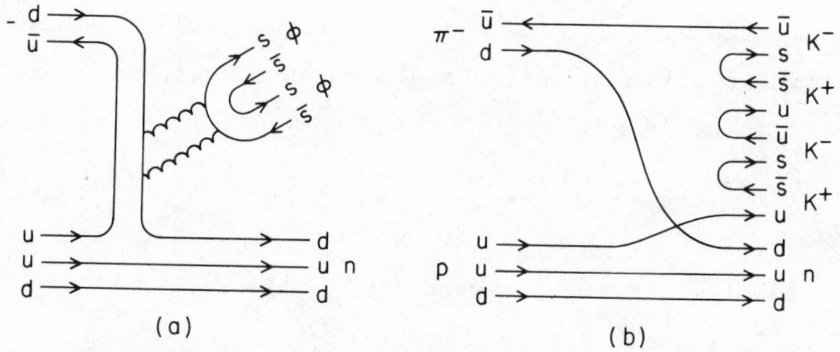

Fig. 2 Diagrams for (a) $\pi^- p \to \phi\phi n$ and (b) $\pi^- p \to K^+K^-K^+K^- n$.

Fig. 2 compared with the OZI allowed process $\pi^- p \to K^+K^-K^+K^- n$. Thus, one does not expect strong $\phi\phi$ production relative to the $K^+K^-K^+K^- n$ background unless there are one or more glueballs which couple to $\phi\phi$ in the intermediate state.

As there are glueball-favored channels, there are also glueball disfavored channels. Since the coupling of gluons to photons is expected to be small, states which are observed in $\gamma\gamma$ interactions are not likely to be glueballs. In addition, most states produced in standard hadronic processes (e.g., πp or Kp) are expected to be ordinary quark states.

After observation of a new state, an important clue to its identification as a glueball is the difficulty in accommodating the state as a member of a standard $q\bar{q}$ SU(3) nonet. For example, one might find that there are three isoscalar states with the same quantum numbers, and hence one of them would be a candidate for a glueball state.

Decay modes of a state also give important clues as to its identification. Glueballs are unitary singlets and hence they couple to all quark flavors equally. On the other hand, quark states couple in a non-OZI violating manner. Thus, glueballs and quark states with the same quantum numbers have different sets of allowed and forbidden decay modes and different relations can be derived among the allowed decay modes.[4] Unfortunately, the situation might be confused by mixing between glueballs and nearby quark states which have the same quantum numbers. However, no matter how strong the mixing between glueballs and quark states, a glueball will manifest itself as an extra state which cannot be accounted for in the quark model.

II. ι(1440)

The ι(1440) was first observed in ψ radiative transitions by the Mark II.[10] It was seen in the decay

$$\psi \rightarrow \gamma K_S K^{\pm} \pi^{\mp} \quad , \tag{1}$$

and was originally identified as the E(1420) as this meson has a mass, width, and decay mode similar to the ι. The ι was subsequently observed by the Crystal Ball[11] in the decay

$$\psi \rightarrow \gamma K^+ K^- \pi^0 \quad . \tag{2}$$

The mass, width, and branching ratio measurements from the two experiments are consistent and are listed in Table II. Figures 3 and 4 show the $K\bar{K}\pi$ invariant mass distributions for events consistent with (1) and (2) from the Mark II and Crystal Ball experiments. The "δ cut" requirement (i.e., the requirement that the $K\bar{K}$ mass be low) enhances the peaks as shown by the shaded regions.

Table II

ι(1440) Parameters

Parameter	Experimental Measurement	
	Mark II	Crystal Ball
M (MeV)	$1440 {}^{+10}_{-15}$	$1440 {}^{+20}_{-15}$
Γ (MeV)	$50 {}^{+30}_{-20}$	$55 {}^{+20}_{-30}$
$B(\psi \rightarrow \gamma\iota) \times B(\iota \rightarrow K\bar{K}\pi)$ [a]	$(4.3 \pm 1.7) \times 10^{-3}$ [b]	$(4.0 \pm 1.2) \times 10^{-3}$
C	+	+
J^P		0^-

[a] I = 0 was assumed in the isospin correction

[b] This product branching ratio has been corrected by me to account for the efficiency correction required under the spin 0 hypothesis.

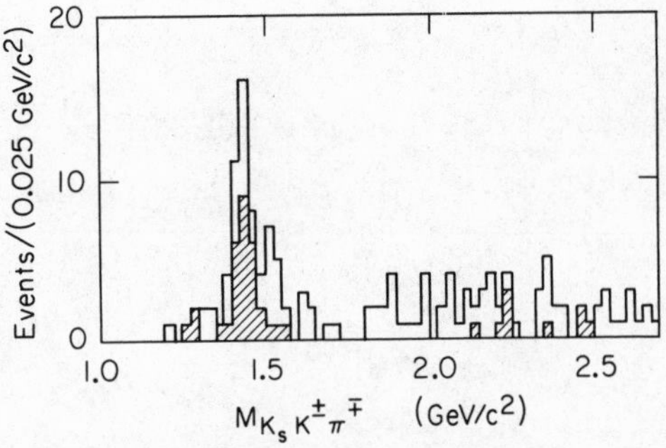

Fig. 3 $K_S K^\pm \pi^\mp$ invariant mass for events consistent with $\psi \to \gamma K_S K^\pm \pi^\mp$ in the Mark II. $M_{K\bar{K}} < 1.050$ GeV for events in the shaded region.

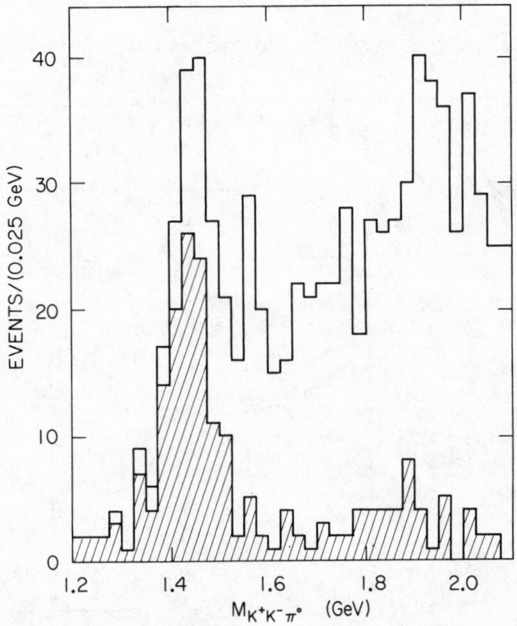

Fig. 4 $K^+K^-\pi^0$ invariant mass for events consistent with $\psi \to \gamma K^+K^-\pi^0$ from the Crystal Ball. $M_{K\bar{K}} < 1.125$ GeV for events in the shaded region.

The spin-parity of the ι was determined to be 0^- from a partial-wave analysis of the $K^+K^-\pi^0$ system in (2) by the Crystal Ball.[11] Five partial waves were included in the analysis:

$$K\bar{K}\pi \text{ flat} \quad ,$$
$$K^*\bar{K} + \text{c.c.} \ (J^P = 0^-) \quad ,$$
$$K^*\bar{K} + \text{c.c.} \ (J^P = 1^+) \quad ,$$
$$\delta\pi \ (J^P = 0^-) \quad ,$$
$$\delta\pi \ (J^P = 1^+) \quad .$$

Only three partial waves contributed significantly to the likelihood in the analysis. Their contributions are shown in Fig. 5 as functions of $K^+K^-\pi^0$ invariant mass. The $K^*\bar{K}$ + c.c. ($J^P = 1^+$)

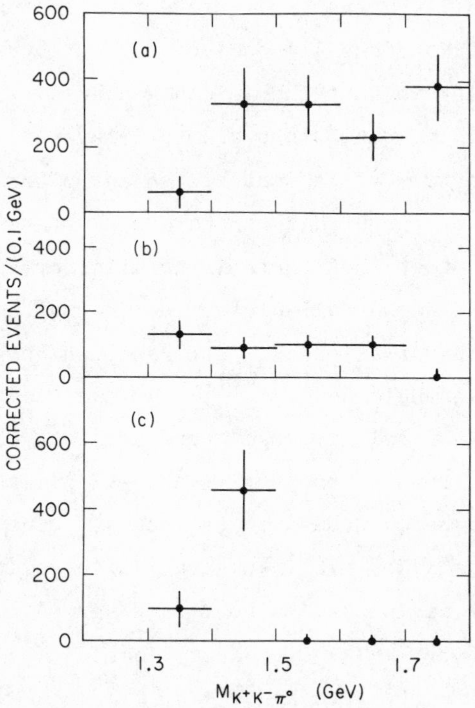

Fig. 5 Partial-wave contributions (corrected for efficiency) as functions of $K^+K^-\pi^0$ mass for (a) $K\bar{K}\pi$ flat, (b) K*K + c.c. ($J^P = 1^+$), and (c) $\delta\pi(J^P = 0^-)$.

partial wave [see Fig. 5(b)] is fairly independent of mass and shows no evidence for resonance structure. [This is the partial wave in which one would expect to see the E(1420).] On the other hand, the $\delta\pi$ ($J^P = 0^-$) partial wave [see Fig. 5(c)] shows a significant peak at the mass of the ι. There is no evidence for $\delta\pi$ production off resonance.

From the time of the original observation of the ι by the Mark II, there has been much controversy over the interpretation of this state.[3,11-15] Much of the confusion resulted from the assumption that the ι and the E(1420) were the same state. Since the spin-parity of the E had been measured[16] to be 1^+, it was difficult to understand why this state would be produced so strongly in ψ radiative decays since the two-gluon system is not allowed to couple to $J^P = 1^+$ states. The measurement of the spin-parity of the ι has established that this state is different from the E, but its interpretation is still subject to controversy.

One possibility is that the ι is a glueball. It is produced with a large branching ratio in a glueball-favored channel. The branching ratio for $\psi \to \gamma\iota$ is larger than any other known radiative branching ratio from the ψ to a noncharmonium state. Equally interesting is the fact that there is no evidence for ι production in πp interactions, a glueball-disfavored channel. (Possible ι production in $\bar{p}p$ annihilations at rest has been observed, but the evidence is not overwhelming.[13,14,17]) An additional factor supporting this hypothesis is that there exist at least three isoscalar, $J^P = 0^-$ mesons, the η, η', and ι. All three of these states cannot be pure $q\bar{q}$ ground state mesons. Finally, the $\delta\pi$ decay mode is allowed for a $J^{PC} = 0^{-+}$ unitary singlet state whereas the $K^*\bar{K}$ + c.c. decay mode is not.[4] The ι is observed to decay into $\delta\pi$ with no evidence for a $K^*\bar{K}$ + c.c. decay mode.[10,11] Note that the E(1420) decays primarily into $K^*\bar{K}$ + c.c.[16]

Although there appears to be no problem with the identification of the ι as a glueball, there is also the possibility that the ι is a radially-excited quark state. However, there are difficulties with this interpretation. Transitions to radially-excited states are expected to have smaller branching ratios than transitions to ground states with the same quantum numbers. Thus, it is expected that $B(\psi \to \gamma\iota)$ would be smaller than $B(\psi \to \gamma\eta')$, which is in disagreement with the experimentally determined branching ratios. However, it might be possible to understand the experimental numbers if the ground and excited states are mixed significantly.[18]

There is the additional problem of understanding the $\zeta(1275)$[13] which has been observed in

$$\pi^- p \to \eta\pi^+\pi^- n \tag{3}$$

at 8.45 GeV/c.[19] The ζ has a mass of 1275 ± 15 MeV, a width of 70 ± 15 MeV, and is observed to decay into $\eta\pi\pi$. The spin-parity has been determined to be 0^- from a partial-wave analysis. Cohen and Lipkin,[20] based on a model in which the η' is a mixture of ground state and radially-excited state wave functions, predict a pseudoscalar meson with mass near 1280 MeV. Thus, it is logical to associate the ζ with this radially-excited state. It can be shown that the ι is not the partner of the ζ.[13,14]

First, the ζ is not observed in $\psi \to \gamma K\bar{K}\pi$ (see Figs. 3 and 4) or in $\psi \to \gamma\eta\pi\pi$.[10,15] ζ production is clearly smaller than ι production in $K\bar{K}\pi$, and smaller than η' production in $\eta\pi\pi$. Thus, $B(\psi \to \gamma\zeta) \ll B(\psi \to \gamma\iota)$. Since the two-gluon system couples only to SU(3) singlet states, the ι must be mostly singlet and the ζ mostly octet if they are members of the same nonet. One then expects[14]

$$\frac{\sigma(\pi^- p \to \iota n)}{\sigma(\pi^- p \to \zeta n)} \approx 2 \quad . \tag{4}$$

In (3), there is possible evidence for an additional $J^P = 0^-$ state near 1400 MeV.[19] However, the enhancement is not really consistent with the ι as the mass appears to be too low and the state appears in the $\varepsilon\eta$ (rather than the $\delta\pi$) partial wave. Even if it were the ι, the cross section is approximately five times smaller than the prediction of Eq. (4). The process

$$\pi^- p \to K_S K^\pm \pi^\mp n \tag{5}$$

at 3.95 GeV/c has been studied also.[16] The spin-parity of the E(1420) was determined from a partial-wave analysis of this data. In the $J^P = 0^-$ partial wave, there is at best a one standard deviation enhancement at the mass of the ι. Even if one were to assume that this was evidence for ι production, the cross section would again be in serious disagreement with Eq. (4). [Since the ζ is not observed in (5), the calibration between (3) and (5) is made by comparing all cross sections with D(1285) production which is observed in both channels.] Despite this arguement, there is still the possibility that the ζ is not a radially-excited state and the ι is, but then one would have the equally difficult problem of interpreting the ζ.

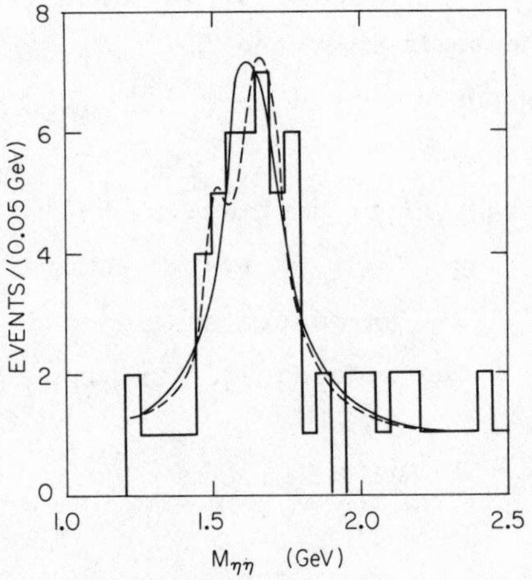

Fig. 6 $\eta\eta$ invariant mass distribution for events consistent with $\psi \to \gamma\eta\eta$ from the Crystal Ball. Solid (dashed) curve represents fit to one (two) Breit-Wigner resonance(s) plus flat background.

Table III

$\theta(1640)$ Parameters

Parameter	One-resonance Fit	Two-resonance Fit
M (MeV)	1640 ± 50	1670 ± 50
Γ (MeV)	$220\ ^{+100}_{-70}$	160 ± 80
$B(\psi \rightarrow \gamma\theta) \times B(\theta \rightarrow \eta\eta)$	$(4.9 \pm 1.7) \times 10^{-4}$	$(3.8 \pm 1.6) \times 10^{-4}$
$B(\psi \rightarrow \gamma f') \times B(f' \rightarrow \eta\eta)$		$(0.9 \pm 0.9) \times 10^{-4}$

III. $\theta(1640)$

The $\theta(1640)$ was recently observed by the Crystal Ball in the decay process[11,21]

$$\psi \rightarrow \gamma\eta\eta \quad . \tag{6}$$

Figure 6 shows the $\eta\eta$ invariant mass distribution for events consistent with (6). The solid curve represents a fit to the mass distribution for a single Breit-Wigner resonance plus a flat background. The parameters of the θ are listed in Table III.

Because of the limited statistics, it is not possible to establish that the $\eta\eta$ peak consists of only one resonance. The $f'(1515)$ has mass $M = 1516 \pm 12$ MeV, width $\Gamma = 67 \pm 10$ MeV, and $J^{PC} = 2^{++}$. No $\eta\eta$ decay mode has been observed, but the upper limit on the branching ratio is 50%.[22] The results of a two-resonance fit which includes an f' contribution are also listed in Table III. The dashed curve in Fig. 6 shows the fit to the mass distribution.

The spin-parity of the θ is favored to be 2^+ as determined from a maximum likelihood fit[11,21] to the angular distribution[23] $W(\theta_\gamma, \theta_\eta, \phi_\eta)$ for the process

$$\psi \rightarrow \gamma\theta, \quad \theta \rightarrow \eta\eta \quad . \tag{7}$$

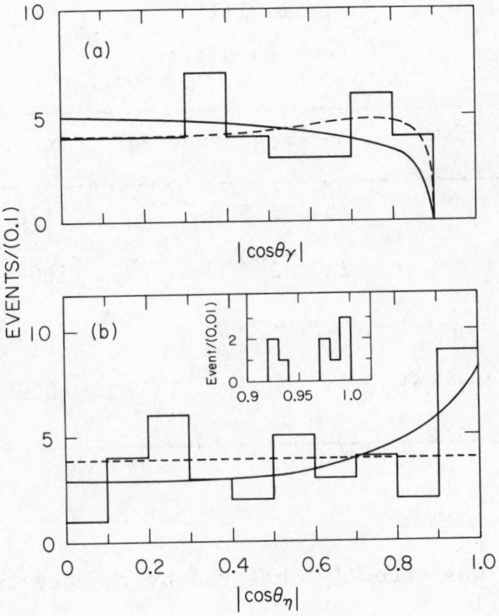

Fig. 7 (a) $|\cos\theta_\gamma|$ and (b) $|\cos\theta_\eta|$ projections for $\psi \to \gamma\theta$, $\theta \to \eta\eta$. Solid (dashed) curves represent best fits for spin 2 (spin 0). Insert shows $|\cos\theta_\eta|$ distribution for $0.9 \leq |\cos\theta_\eta| \leq 1.0$.

(The relative probability of the spin 0 hypothesis is 5%.) θ_γ is the polar angle of the γ with respect to the beam direction; θ_η and ϕ_η are the polar and azimuthal angles of one of the η's with respect to the γ in the rest frame of the θ. Figure 7 shows the $|\cos\theta_\gamma|$ and $|\cos\theta_\eta|$ projections for (7) compared to the best fit curves for spin 0 and spin 2.

As the θ has the quantum numbers expected for the gg ground state, it is a likely candidate for a glueball. Like the ι, the θ is observed only in a glueball-favored channel. Although the measured branching ratio for (7) is somewhat small (nearly a factor of ten smaller than the corresponding ι branching ratio), it is expected that $B(\theta \to \eta\eta)$ is fairly small and hence $B(\psi \to \gamma\theta)$ is relatively large.

There are some possible problems with the identification of

the θ as a glueball. If it is assumed that both the ι and the θ are glueballs, one has the peculiar situation of a gg ground state with a higher mass than the first excited state. However, as mentioned previously, intergluon interactions are expected to push up the spin 2 masses relative to the spin 0 masses. Various calculations have predicted hyperfine splittings which are in agreement with the experimental measurements.[3] Also, the width of the θ is larger than one might naively expect. However, the number of allowed decay modes is large and the width is not so unreasonable when compared with the width of the ι.

The most serious problem is the nonobservation of a $\pi\pi$ decay mode of the θ. Figure 8 shows the $\pi^+\pi^-$ invariant mass distribution for $\psi \to \gamma\pi^+\pi^-$ from the Mark II[11,15] and Fig. 9 shows the $\pi^0\pi^0$ invariant mass distribution for $\psi \to \gamma\pi^0\pi^0$ from the Crystal Ball.[11,24] The 90% confidence level upper limits for $B(\psi \to \gamma\theta) \times B(\theta \to \pi\pi)$ are 6×10^{-4} from the Crystal Ball[11,21] and 2×10^{-4} from the Mark II. (Isospin correction factors have been applied in both calculations.) Based on the postulated unitary singlet nature of the θ, one expects $B(\theta \to \pi\pi) = 3B(\theta \to \eta\eta)$ (without phase space corrections). From the measurement of $B(\psi \to \gamma\theta) \times B(\theta \to \eta\eta)$, one predicts $B(\psi \to \gamma\theta) \times B(\theta \to \pi\pi) = (6.0 \pm 2.1) \times 10^{-3}$ (where p^{2L+1} phase space corrections have been applied.) The Mark II limit is a factor of 30 times smaller than the expected branching ratio. Even if it is assumed that the θ is spin 0, the Mark II limit is still an order of magnitude smaller than expected.

Before ruling out completely the glueball interpretation of the θ, let me comment on the naiveté of the unitary symmetry calculation. The relation between the $\pi\pi$ and $\eta\eta$ branching ratios is based on the assumption that only the diagram shown in Fig. 10 contributes in the decay and that there is an equal probability of pulling any $q\bar{q}$ pair out of the vacuum. The most serious shortcoming of this calculation is that there is no allowance for multibody decays. It is likely (because of phase space) that the all-pion decay modes

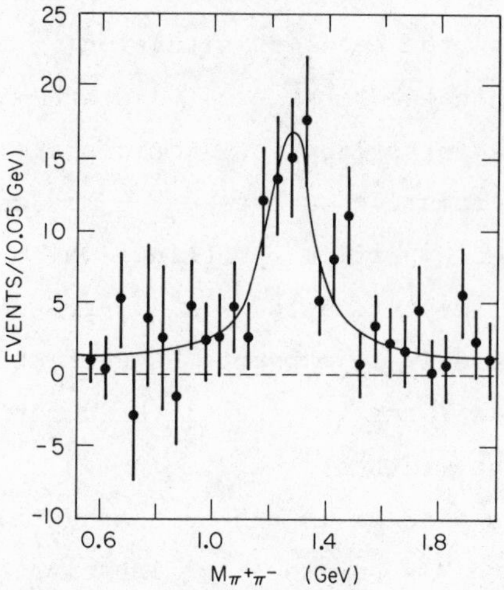

Fig. 8 $\pi^+\pi^-$ invariant mass distribution for events consistent with $\psi \to \gamma\pi^+\pi^-$ after subtraction for feeddown from $\psi\to\rho\pi$ from the Mark II. Solid curve shows fit to f(1270) plus background.

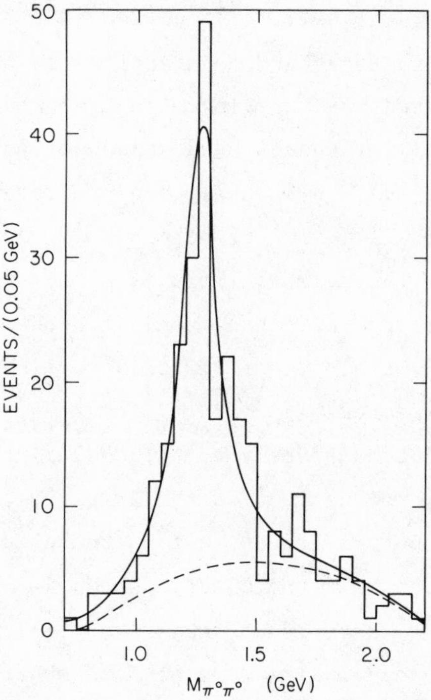

Fig. 9 $\pi^0\pi^0$ invariant mass distribution for events consistent with $\psi\to\gamma\pi^0\pi^0$ from the Crystal Ball. Solid curve shows fit to f(1270 plus background. Dashed curve shows background contribution.

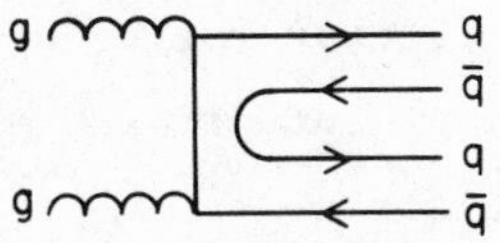

Fig. 10 Diagram for gg decay.

would favor 4π decays over ππ decays. [See, for example, the relative branching ratios for ρ'(1600) and g(1700) decays.[22]] The ηη decays are less likely to have additional pions.

The possibility that the θ is a radially-excited state seems rather unlikely. The mass is too close to the f(1270) and f'(1515) masses, but see Ref. 18 for a possible explanation. In addition, the branching ratio for $\psi \to \gamma\theta$ is probably larger than expected for a radial excitation, assuming that $B(\theta \to \eta\eta)$ is a reasonable (i.e., small) number. If it were to turn out that $B(\theta \to \eta\eta)$ is large ($\gtrsim$30%) so that $B(\psi \to \gamma\theta)$ is reasonable, it would be difficult to understand why a normal quark state would have such a large ηη branching ratio.

An exotic possibility is that the θ is a $q\bar{q}q\bar{q}$ state.[25] In general, it is expected that such states "fall apart" into two $q\bar{q}$ states (as in Fig. 11) and hence have widths of the same order as their masses.[26] Such states would not be observable experimentally. An exception occurs in the case of a $q\bar{q}q\bar{q}$ state that is below threshold for all kinematically allowed fall apart modes, in which case the state will be relatively narrow. A state with which the θ might be identified is $(1/\sqrt{2})[u\bar{u} + d\bar{d}]s\bar{s}$.[14] This state has no kinematically allowed fall apart modes. The prominent two-body decay modes are ηη and $K\bar{K}$ with $B(\theta \to K\bar{K}) \approx 2B(\theta \to \eta\eta)$. $\theta \to \pi\pi$ is

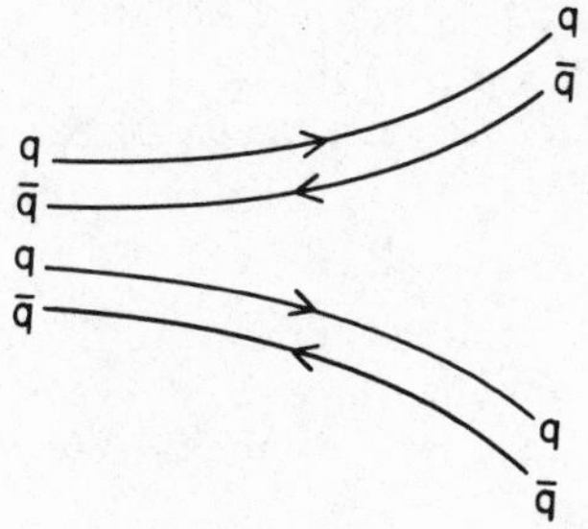

Fig. 11 Diagram for $q\bar{q}q\bar{q}$ decay.

not allowed. Thus, this interpretation of the θ not only explains the lack of a $\pi\pi$ decay mode, but it also places $B(\psi \to \gamma\theta)$ in the expected range for a $q\bar{q}q\bar{q}$ state.

IV. $\psi \to V + X$

It is of interest to compare the process $\psi \to \gamma + X$ with the process $\psi \to V + X$, where V is an isoscalar vector meson (i.e., ω

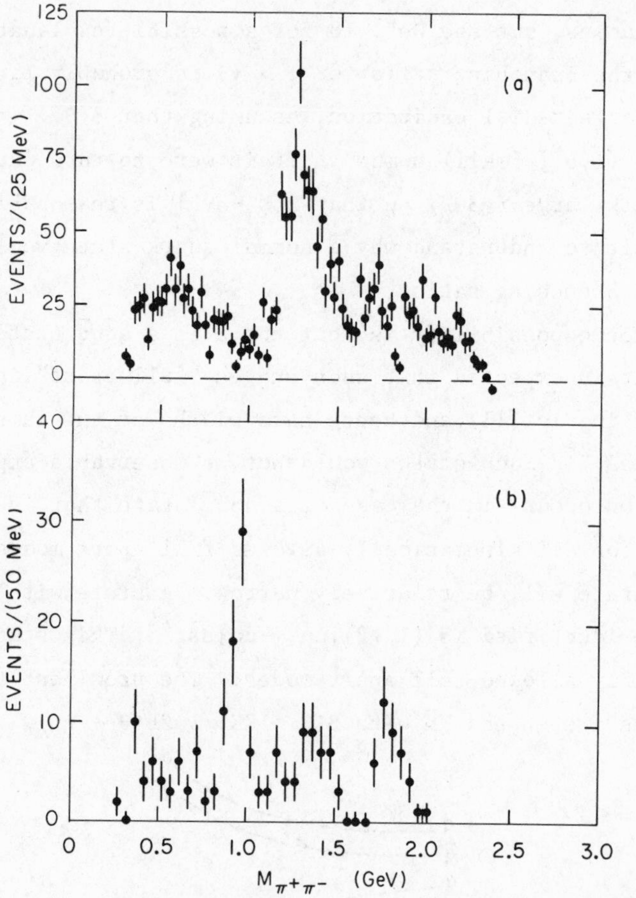

Fig. 12 $\pi^+\pi^-$ invariant mass distributions for (a) $\psi \to \omega\pi^+\pi^-$ and (b) $\psi \to \phi\pi^+\pi^-$ from the Mark II. Background subtractions for events under the ω and ϕ peaks have been made.

or ϕ). One expects glueballs to be produced only in the first process. The only final state X for which all three channels have been studied is $\pi\pi$. Figure 12(a) shows the $\pi^+\pi^-$ invariant mass distribution for $\psi \to \omega\pi^+\pi^-$ from the Mark II[27] after background subtraction for events under the peak. The distribution is dominated by the f(1270). An S*(980) peak is observed in Fig. 12(b) which shows a similar distribution for $\psi \to \phi\pi^+\pi^-$ from the Mark II.[27,28] In the process $\psi \to \gamma\pi\pi$ (see Figs. 8 and 9), one sees evidence for only an f(1270) peak. Not only is there no evidence for new resonance structure which might indicate a glueball, but there is also no evidence for the S*(980). Thus, the S*(980) is probably not a glueball.

V. $\phi\phi$ GLUEBALLS

The processes

$$\pi^- p \to K^+K^-K^+K^- n \tag{8}$$

$$\pi^- p \to K^+K^- \phi n \tag{9}$$

$$\pi^- p \to \phi\phi n \tag{10}$$

were studied by a BNL/CCNY collaboration at 22.6 GeV/c.[29] As discussed earlier, reactions (8) and (9) are OZI allowed while reaction (10) is OZI suppressed. Thus, one expects to see no enhancement of (10) over the background from (8) and (9). Experimentally, it is found that (10) is not suppressed, but instead is enhanced by a factor of approximately 1500 over the surrounding background from (8). One possible explanation for this enhancement is the existence of one or more glueballs which are produced in $\pi^- p$ interactions and which decay into $\phi\phi$. Figure 13 shows the $\phi\phi$ invariant mass distribution from the BNL/CCNY experiment.[9,29] (No background subtraction has been made for events under the ϕ peaks.) No significant resonance structure is observed, but the data sample will be increased by about a factor of 20 in the near future.

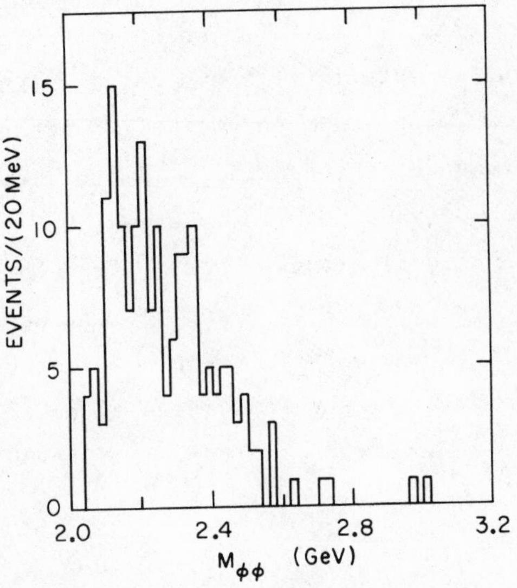

Fig. 13 $\phi\phi$ invariant mass distributions for $\pi^- p \to \phi\phi n$ at 22.6 GeV/c.

Inclusive $\phi\phi$ production has been studied in two other experiments,

$$\pi^- Be \to \phi\phi + X \tag{11}$$

at 175 and 100 GeV/c,[30] and

$$pp \to \phi\phi + X \tag{12}$$

at 400 GeV/c.[31] In (11), 48 ± 23 excess $\phi\phi$ events are observed in the 175 GeV/c data sample and 21 ± 8 excess $\phi\phi$ events in the 100 GeV/c data sample. Figure 14 shows the $\phi\phi$ invariant mass distributions at the two energies. The solid curves show the expected backgrounds from ϕK^+K^- and $K^+K^-K^+K^-$ events. The dashed curves show the total backgrounds including the estimated contributions from uncorrelated $\phi\phi$ production. There is a four standard deviation peak at about 2100 MeV in the 100 GeV/c data.

Only very preliminary data for (12) is now available but it is expected that more data will be taken soon. An excess of 9 ± 6 $\phi\phi$

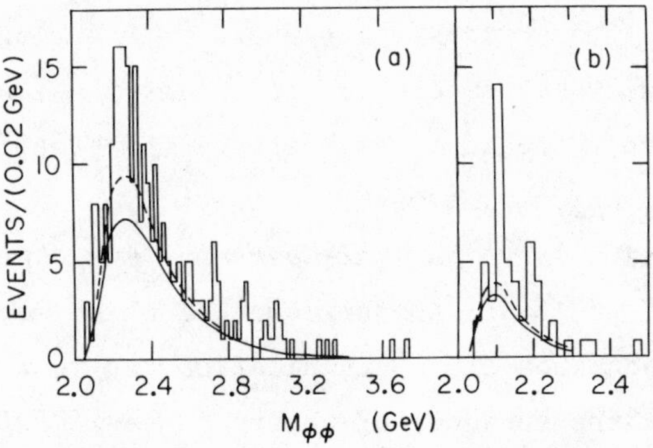

Fig. 14 $\phi\phi$ invariant mass distributions for $\pi^- Be \to \phi\phi + X$ at (a) 175 GeV/c and (b) 100 GeV/c. Curves are discussed in text.

events is observed in the current data sample. Interestingly enough, most of the excess events are in a single bin near 2100 MeV.

VI. CONCLUSIONS

Although there are no known problems or inconsistencies with a glueball interpretation of the $\iota(1440)$, the evidence in favor of this hypothesis is far from conclusive.

The nonobservation of $\iota \to \eta\pi\pi$ is somewhat of a worry since $\iota \to \delta\pi$ and the δ is believed to decay into both $K\bar{K}$ and $\eta\pi$.[22] However, the only published limits are fairly large.[10,15] It is important that a good measurement be made by the Crystal Ball. It is also important that a reliable determination of the relative $K\bar{K}$ and $\eta\pi$ branching ratios of the δ be made.

An effort should be made to find the partner of the $\zeta(1275)$ in hadronic interactions. Observation of such a state would rule out any possibility that the ι is a radial excitation. The partner of the ζ is expected to be somewhat higher in mass than the ι.[20] Thus, a study of $K\bar{K}\pi$ and $\eta\pi\pi$ production up to masses of 1.6 GeV is necessary.

The spin-parity determination of the E(1420) is based on a

single experiment. Additional studies of $K\bar{K}\pi$ production near 1.4 GeV in hadronic interactions might prove interesting. Some preliminary results of $K\bar{K}\pi$ production in $\bar{p}p$ interactions at 5 GeV/c are expected within the next few months.[32] A proposed study of π^-p, K^-p, and $\bar{p}p$ production of $K\bar{K}\pi$ between 6 and 8 GeV/c is expected to have enough data for a partial-wave analysis of the $K\bar{K}\pi$ system in the region of the E.[33] Another channel which would be useful to study is $K\bar{K}\pi$ production in $\bar{p}p$ annihilations at rest. There are indications that the enhancement observed near 1400 MeV in this channel is not the E(1420).[13,14] In fact, a spin analysis from one experiment[17] favors $J^P = 0^-$ over 1^+ for this state.

The present meager data on the $\theta(1640)$ seems to disfavor a glueball interpretation for the θ. It is crucial that a better theoretical understanding of the relative branching ratios for two-body decay modes be obtained as the naive arguments based on unitary symmetry may be totally wrong. In particular, it is necessary to understand whether the nonexistence of a $\pi\pi$ decay made of the θ really rules out a glueball interpretation.

Results from the Mark II on $\psi \to \gamma K^+K^-$ are expected soon. If the θ is a glueball, one expects $B(\theta \to K^+K^-) \approx 2B(\theta \to \eta\eta)$. The naive arguments may be more reliable in this case as the K and η masses are relatively close. If the θ is a $q\bar{q}q\bar{q}$ state, one expects $B(\theta \to K^+K^-) \approx B(\theta \to \eta\eta)$, and if the θ is just an ordinary $s\bar{s}$ state (which would also explain the lack of a $\pi\pi$ decay mode), then $B(\theta \to K^+K^-) \approx 1.5\ B(\theta \to \eta\eta)$.

Results on $\psi \to \gamma\rho\rho$ are also expected in the near future from both the Mark II and the Crystal Ball. The $\rho\rho$ decay of the θ is expected to be small for a $q\bar{q}q\bar{q}$ state, but not necessarily so for a glueball.

Conclusions on possible $\phi\phi$ glueballs will have to wait until the new data are available.

REFERENCES

1. H. Fritzsch and M. Gell-Mann, in Proceedings of the XVI International Conference on High Energy Physics, v. 2, edited by J. D. Jackson and A. Roberts (National Accelerator Laboratory, Batavia, Ill., 1973), p. 135; J. D. Bjorken, in Quantum Chromodynamics, edited by A. Mosher (Stanford, Ca., 1980), p. 219; J. F. Donoghue, in Experimental Meson Spectroscopy - 1980, edited by S. U. Chung and S. J. Lindenbaum (AIP, New York, 1981), p. 104; J. F. Donoghue, in High Energy Physics - 1980, edited by L. Durand and L. G. Pondrom (AIP, New York, 1981), p. 35; P. M. Fishbane, in Gauge Theories, Massive Neutrinos, and Proton Decay, edited by A. Perlmutter (Plenum Press, New York, 1981), p. 63; C. Quigg, Fermi National Accelerator Laboratory Report No. FERMILAB-Conf-81/78-THY, to be published in the Proceedings of the Les Houches Summer School in Theoretical Physics, Les Houches, France, August 3 to September 11, 1981; and references therein.
2. R. L. Jaffe and K. Johnson, Phys. Lett. 60B, 201 (1976).
3. K. Babu Joseph and M. N. Sreedharan Nair, Cochin University Preprint No. CUTP-81-1 (1981); T. Barnes, Z. Phys. C 10, 275 (1981); T. Barnes, F. E. Close, S. Monaghan, Rutherford Laboratory Preprint No. RL-81-090, 1981 (unpublished); J. F. Donoghue, K. Johnson, and B. A. Li, Phys. Lett. 99B, 416 (1981).
4. H. J. Lipkin, Argonne National Laboratory Report No. ANL-HEP-PR-81-35, 1981 (unpublished); P. M. Fishbane, D. Horn, G. Karl, and S. Meshkov, 1981 (unpublished); C. E. Carlson, J. J. Coyne, P. M. Fishbane, F. Gross, and S. Meshkov, Phys. Lett. 99B, 353 (1981); C. E. Carlson, J. J. Coyne, P. M. Fishbane, F. Gross, and S. Meshkov, 1981 (unpublished).
5. C. N. Yang, Phys. Rev. 77, 242 (1950).
6. T. Appelquist, A. De Rujula, H. D. Politzer, and S. L. Glashow, Phys. Rev. Lett. 34, 365 (1975); M. S. Chanowitz, Phys. Rev. D 12, 918 (1975); L. B. Okun and M. B. Voloshin, Institute of

Theoretical and Experimental Physics, Moscow, Report No. ITEP-95-1976, 1976 (unpublished); S. J. Brodsky, T. A. DeGrand, R. R. Horgan, and D. G. Coyne, Phys. Lett. 73B, 203 (1978); K. Koller and T. Walsh, Nucl. Phys. B140, 449 (1978).

7. G. S. Abrams et al., Phys. Rev. Lett. 44, 114 (1980); D. L. Scharre et al., Phys. Rev. D 23, 43 (1981).

8. A. Billoire, R. Lacaze, A. Morel, and H. Navelet, Phys. Lett. 80B, 381 (1979).

9. S. J. Lindenbaum, Nuovo Cimento 65A, 222 (1981).

10. D. L. Scharre et al., Phys. Lett. 97B, 329 (1980).

11. D. L. Scharre, in Proceedings of the 1981 International Symposium on Lepton and Photon Interactions at High Energies, edited by W. Pfeil (Physikalisches Institut, Universität, Bonn, West Germany, 1981), p. 163.

12. K. Ishikawa, Phys. Rev. Lett. 46, 978 (1981); K. Ishikawa, Phys. Lett. 101B, 344 (1981); M. Chanowitz, Phys. Rev. Lett. 46, 981 (1981); C. E. Carlson, J. J. Coyne, P. M. Fishbane, F. Gross, and S. Meshkov, Phys. Lett. 98B, 110 (1981); H. J. Lipkin, Phys. Lett. 106B, 114 (1981); T. Barnes, Nucl. Phys. B187, 42 (1981).

13. M. Chanowitz, Phys. Rev. Lett. 46, 981 (1981).

14. M. S. Chanowitz, Lawrence Berkeley Laboratory Report No. LBL-13398, to be published in the Proceedings of the 1981 Meeting of the Division of Particles and Fields of the American Physical Society, Santa Cruz, Ca., September 9-11, 1981.

15. D. L. Scharre, in Experimental Meson Spectroscopy - 1980, edited by S. U. Chung and S. J. Lindenbaum (AIP, New York, 1981), p. 329.

16. C. Dionisi et al., Nucl. Phys. B169, 1 (1980).

17. P. Baillon et al., Nuovo Cimento 50A, 393 (1967).

18. I. Cohen, N. Isgur, and H. J. Lipkin, Argonne National Laboratory Report No. ANL-HEP-PR-81-45, 1981 (unpublished);

C. E. Carlson, invited talk this conference.

19. N. P. Stanton et al., Phys. Rev. Lett. 42, 346 (1979).
20. I. Cohen and H. J. Lipkin, Nucl. Phys. B151, 16 (1979).
21. C. Edwards et al., Stanford Linear Accelerator Center Report No. SLAC-PUB-2822, 1981 (submitted to Phys. Rev. Lett.).
22. Particle Data Group, Rev. Mod. Phys. 52, S1 (1980).
23. P. K. Kabir and A.J.G. Hey, Phys. Rev. D 13, 3161 (1976).
24. C. Edwards et al., Stanford Linear Accelerator Center Report No. SLAC-PUB-2847, 1981 (submitted to Phys. Rev. D).
25. R. L. Jaffe, Phys. Rev. D 15, 267 (1977).
26. R. L. Jaffe, in Proceedings of the 1981 International Symposium on Lepton and Photon Interactions at High Energies, edited by W. Pfeil (Physikalisches Institut, Universität, Bonn, West Germany, 1981), p. 395.
27. G. Gidal, presented at the APS/AAPT Annual Joint Meeting, New York, N.Y., February 26-29, 1981.
28. G. Gidal et al., Phys. Lett. 107B, 153 (1981).
29. A. Etkin et al., Phys. Rev. Lett. 40, 422 (1978); A. Etkin et al., Phys. Rev. Lett. 41, 784 (1978).
30. C. Daum et al., Phys. Lett. 104B, 246 (1981).
31. D. R. Green, Fermi National Accelerator Laboratory Report No. FERMILAB-81/81-EXP, 1981 (unpublished).
32. S. U. Chung, private communication.
33. S. U. Chung, et al., Argonne National Laboratory Proposal No. AGS-771, 1981.

GLUEBALLS - SOME SELECTED THEORETICAL TOPICS

Carl E. Carlson*

William & Mary

Williamsburg, Virginia 23185

I. INTRODUCTION AND REMARKS ABOUT DISCOVERING GLUEBALLS

The topics selected for this talk are elementary considerations of how glueballs may be found, the identity of the candidate glueball states[1] found at SLAC, and admonitory remarks concerning mixing of glueballs and quarkonia. Discussion of the first topic follows in this introduction. The second topic (section II) is chosen because discovery of an actual glueball is the most important possible happening in glueball science. From this viewpoint, our discussion will be unwelcome in that it appears that one candidate, the ι (1440), can be accommodated as a radially excited pseudoscalar and cannot be simply accommodated as a glueball. The other candidate, θ(1640), has decay properties uncharacteristic of a glueball but characteristic of a state made from quarks. Much of the thinking about glueball properties is for states that are "pure" glueballs, but mixing with neutral, isoscalar quarkonia with the same J^{PC} is possible and some remarks about the extent and effects of such mixing are made in section III.

---

*Supported in part by the NSF

How can one find a glueball? They may look distressingly like a neutral, isoscalar $q\bar{q}$ meson. Several signatures may help.

(1) "Oddballs".[2] "Oddballs" or more formally "abnormal-C_n states"[3] are states having a J^{PC} that cannot be made as a $q\bar{q}$ state. There are two series of oddballs, plus one unserialed oddball, to wit,

$$J^{PC} = (\text{odd})^{+-} = 1^{+-},\ 3^{+-}, \ldots$$

$$J^{PC} = (\text{even})^{-+} = 0^{-+},\ 2^{-+}, \ldots$$

$$J^{PC} = 0^{--} \quad .$$

Discovering any of the above would be exciting. Further examination would be needed to verify that it was a glueball. The above quantum numbers can also be produced with $q^2\bar{q}^2$ or $q\bar{q}g$ states.[4]

(2) "Overpopulation."[5] Every quarkonium nonet of a given J^{PC} has two isospin-zero neutral members. Finding an extra isospin-zero neutral meson would immediately raise suspicions that a glueball was there. Again, $q^2\bar{q}^2$ or $q\bar{q}g$ states could give the same results.

(3) Decay democracy.[6] A glueball is a flavor singlet, so that barring special considerations and neglecting phase space, a glueball should decay equally into all flavor varieties of a given type of final state. ("Special considerations" can, among other things, mean mass dependences other than phase space. For example, a spin zero glueball decaying to $q\bar{q}$ will - like $\pi \to \ell\nu$ - prefer to go to heavier states because of helicity considerations.)[7] For example, if an $\eta\eta$ decay of a suspected glueball is found, then $K\bar{K}$ and $\pi\pi$ decay modes should be present also. This is in contrast to usual quarkonium decay where, for example, the f' made mainly of $s\bar{s}$ can decay dominantly into $K\bar{K}$.

II. THE GLUEBALL CANDIDATES

The $\iota(1440)$

The $\iota(1440)$ has been considered as a glueball candidate.[6] We shall make comments on three subtopics. First will come the question of whether the iota can be a radially excited 0^{-+} quarkonium state. Second, we shall see if interpreting the iota as a glueball leads to problems with the J/ψ radiative decay into iota. Finally, we shall consider hadronic production of the iota if it be a glueball.

Regarding the possibility that the iota is a radially excited $q\bar{q}$ state leads us to consider why the J/ψ should decay into an excited state more frequently than the corresponding ground state. The data read,

$$B(J/\psi \to \gamma\iota)\ B(\iota \to K\bar{K}\pi) \simeq 4.\times 10^{-3} \quad ,$$

$$B(J/\psi \to \gamma\eta') \simeq 2.5\times 10^{-3} \quad .$$

Apparently, there is a common expectation that radially excited states are harder to produce than ground states. This is not so, as pointed out by Cohen, Isgur, and Lipkin[7] and by Pene and Ono;[8] a qualitative exposition of this point follows. The radiative decay of the J/ψ proceeds via reactions like the one diagrammed in Fig. 1. The production of the final $q\bar{q}$ pair is (like the reverse

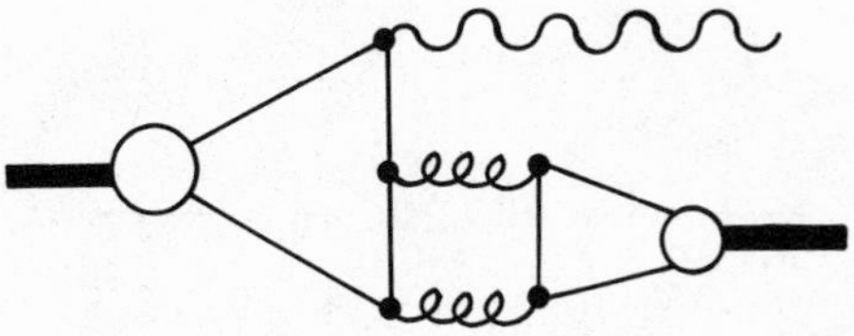

Fig. 1

annihilation into two gluons) a short range process proportional to $\psi(0)$, the wave function at the origin in coordinate space of the $q\bar{q}$ bound state. Hence, we shall take a short detour to consider this wave function.

One's intuition is often based on the coulomb potential. Here the wave function at the origin for successive S-state wave functions decreases sharply as principle quantum number increases. For a linear potential, on the other hand, $\psi(0)$ is independent of the principal quantum number. Quark confining potentials are well treated as some combination of linear and coulomb potentials. However, we have also to consider mixing between ground and excited states. This proceeds via annihilation diagrams like the one in Fig. 2 with two or more gluons in the intermediate states. The same processes are largely responsible for the η-η' mass splitting.

Start with the latter. The η and η' are both lowest lying $^1S_0$ states and (approximately) $\eta' \sim (u\bar{u} + d\bar{d} + s\bar{s})/\sqrt{3}$ and $\eta \sim (u\bar{u} + d\bar{d} - 2s\bar{s})/\sqrt{6}$. The annihilation diagrams are short distance dominated and are independent of quark flavors, and so self cancel for any SU(3) (flavor) nonsinglet.[9] Hence,

$$\langle \eta | H_{eff} | \eta \rangle = 0 \quad ,$$

but

$$\langle \eta' | H_{eff} | \eta' \rangle = A \, |\psi_{1S(0)}|^2 \quad .$$

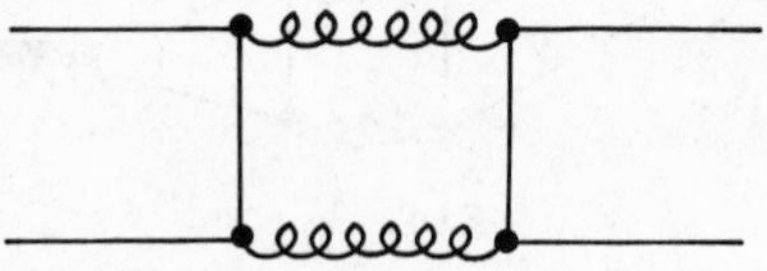

Fig. 2

Since the η' is heavier than the η, it must be that A is positive. This is enough for us to qualitatively understand the effect of 1S-2S mixing. Mixing between the η' and its radial excitation - η'^* - is proportional to

$$\langle\eta'|H_{eff}|\eta'^*\rangle = A\, \psi^*_{1S(0)}\, \psi_{2S(0)} \quad .$$

This matrix element (depending on one's phase choice for the two states) is positive. The wave functions of the physical η' and η'^* states are approximately

$$\psi_{\eta'} = \psi_{1S} + \frac{\langle\eta'^*|H_{eff}|\eta\rangle}{E_{1S} - E_{2S}}\, \psi_{2S}$$

and

$$\psi_{\eta'^*} = \psi_{2S} + \frac{\langle\eta'|H_{eff}|\eta\rangle}{E_{2S} - E_{1S}}\, \psi_{1S} \quad .$$

The differing signs of the energy denominators means destructive interference for $\psi_\eta'(0)$ and constructive interference for $\psi_\eta'^*(0)$. Details and numbers may be found in Cohen et al,[7] who suggest

$$\frac{\psi_{\eta'^*}(0)}{\psi_{\eta'}(0)} = 1.5 - 2 \quad .$$

The perturbative evaluation[10] of Fig. 1 and permutations have been done by Guberina and Kühn and by Devoto and Repko, who find

$$\Gamma(J/\psi\to\gamma\eta') = (3)e_c^{\,2}\, \alpha\, \alpha_s^{\,4}\, 2^{11}(\tfrac{2}{3})^3 \left(1 - \frac{m_{\eta'}^{\,2}}{m_{J/\psi}^{\,2}}\right)^3 \left(1 + \frac{m_{\eta'}^{\,2}}{m_{J/\psi}^{\,2}}\right) |I|^2$$

$$\times \frac{|\psi_{J/\psi}(0)|^2}{m_{J/\psi}^{\,3}} \times \frac{|\psi_{\eta'}(0)|^2}{m_{\eta'}^{\,2}} \quad ,$$

where (3) is a flavor factor for SU(3) flavor singlets and I is a known integral. One has approximately

$$\frac{\Gamma(J/\psi\to\gamma\eta'^*)}{\Gamma(J/\psi\to\gamma\eta)} \simeq \left(\frac{m_{\eta'}}{m_{\eta'^*}}\right)^2 \left|\frac{\psi_{\eta'^*}(0)}{\psi_{\eta'}(0)}\right|^2 \simeq 1\text{-}2 \quad .$$

Identifying the ι with the η'^* gives accord between the expected and observed radiative branching ratios.

The absolute calculation of $J/\psi \to \gamma\eta'$ is sensitive to the choice of α_s. Getting $\psi J/_{\psi}(0)$ from the J/ψ leptonic decay and $\psi_{\eta}'(0)$ from the known 5.3 keV width of $\eta' \to \gamma\gamma$ requires a not unreasonable $\alpha_s = 0.25$ to obtain the observed $\Gamma(J/\psi \to \gamma\eta')$.

We should note that if the ι is the η', then its $\gamma\gamma$ decay width is calculable,

$$\Gamma(\iota \to \gamma\gamma) = \frac{48\pi\alpha^2}{m_\iota^2} (e_{eff}) \, |\psi_{\eta'*}(0)|^2$$

$$= \frac{m_{\eta'}^2}{m_\iota^2} \left|\frac{\psi_{\eta'*}(0)}{\psi_{\eta'}(0)}\right|^2 \Gamma(\eta' \to \gamma\gamma)$$

$$\simeq 5 \text{ keV} \quad .$$

An experimental limit $\Gamma(\iota \to \gamma\gamma) < 8$ keV has already been quoted.[1]

The radiative decay of the J/ψ into a 0^{-+} glueball (G) can be similarly considered.[11] The Ggg vertex can be only

$$T \, \varepsilon_{\mu\nu\alpha\beta} \, k_1^\alpha \, k_2^\beta \quad ,$$

where μ and ν are polarization indices and k_1 and k_2 momenta of the gluons and T is a scalar function. We can estimate the radiative decays by making a simple and, we think, reasonable choice of T as a monopole form factor,

$$T = f/(k^2 - \Lambda^2) \quad ,$$

with $k = (k_1 - k_2)/2$ and $\Lambda = M_G/2$. The calculation along the lines of Ref. 10 gives[11]

$$\Gamma(J/\psi \to \gamma G) = \frac{1}{36\pi^3} \frac{\alpha_s^2}{\alpha} \frac{1}{m_{J/\psi}^2 - m_G^2} \Gamma(J/\psi \to e^+e^-) f^2 \, |I|^2 \quad .$$

The normalization constant f, playing a role analogous to $\psi(0)$ in the previous case, is not yet known. We can, however, attempt to determine it from the total decay width of the glueball.[12] If we are dealing with a low lying glueball, it must decay into quarks as in Fig. 3a or 3b, and if we give the quarks constituent masses (on the grounds that we are not considering final state confinement effects explicitly so should use the constituent mass to approximate a quark mass including its accompanying fields) then Fig. 3b, the "crowsfoot" is suppressed relative to 3a (even for spin-0 glueballs where a helicity conservation argument forbids 3a for massless on-shell quarks). Figure 3a leads to[12]

$$\Gamma(G \to \text{had}) = \frac{\alpha_s^2}{3\, m_G} \frac{f^2}{4\pi} \frac{m_q^2}{m_G^2} \times \ln^2 \frac{m_G}{m_q} \quad .$$

Now we may try to identify the ι as a glueball with $\Gamma(G \to \text{had}) = \Gamma(\iota \to \text{had}) \simeq 50$ MeV, leading to a large result,

$$\Gamma(J/\psi \to \gamma G) = 15 \text{ keV} \quad ,$$

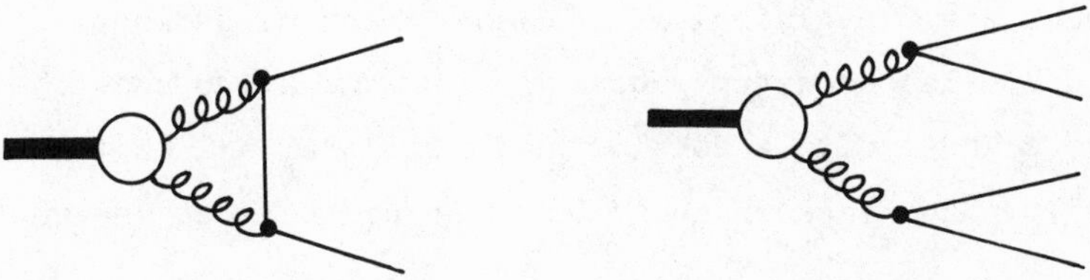

Fig. 3

compared to the experimental

$$\Gamma(J/\psi \to \gamma\iota) \simeq 0.25 \text{ keV} \quad .$$

The reader may note that our calculated $\Gamma(J/\psi \to \gamma G)$ is even larger than the expected total $\Gamma(J/\psi \to \gamma gg)$. If the ι is to be a glueball, its width should be narrower to prevent this tragedy.

We should also comment on producing glueballs in hadronic reactions. One would expect a significant cross section because there is so much glue - 50% as measured by momentum carried - in a hadron. We may compare the production probability of a glueball with the nominal properties of the iota to that of the E^o meson. The reactions proceed via gluon fusion or $s\bar{s}$ fusion. If the glueball is made from gluons about the same way the E^o is made from $s\bar{s}$, and if neither of the colliding hadrons has a valence s or $\bar{s}$ quarks, we can estimate,[13]

$$\frac{\sigma_G}{\sigma_E} = \frac{9}{64} \cdot \frac{1}{2} \cdot \left(\frac{\text{no. of gluons}}{\text{no. of s-quarks}}\right)^2 \quad .$$

The first factor is due to color and the second because we can exchange $s \leftrightarrow \bar{s}$. About 10% of a hadron (by momentum) is ocean quark and dividing this among the u,d,s and antiquarks gives about 1.7% strange quarks. This leads to

$$\frac{\sigma_G}{\sigma_E} \simeq 60 \quad .$$

Where, then, are the glueballs? An experiment of Bromberg _et al_[14] with 100 GeV/c lab momentum pions on a proton target sees the E^o in the $K\bar{K}\pi$ channel with a total cross section $\sigma_E \simeq 2$ μb (and also the D(1285) in the same channel) but does not seem to see the iota.

The θ(1640)

We will make only a few comments on the θ(1640) as we are interested in glueballs and it seems that one can now say on experimental grounds that the θ is not a glueball.

The θ was discovered[15] in radiative J/ψ decay, $J/\psi \to \gamma + \theta$ followed by $\theta \to \eta\eta$ followed in turn by each $\eta \to \gamma\gamma$. The width is about 220 MeV and the J^{PC} quantum numbers are 2^{++}.

A glueball should exhibit a certain democracy among its decays. If a 2^{++} glueball can decay into $\eta\eta$ it should decay 3 times as often into $\pi\pi$, counting charge states but not correcting for phase space. The $\pi\pi$ decay mode has been searched for and not found, with a limit[1]

$$B(J/\psi \to \gamma\theta) \cdot B(\theta \to \pi\pi) < 2 \times 10^{-4}$$

to be compared with

$$B(J/\psi \to \gamma\theta) \cdot B(\theta \to \eta\eta) = (4.9 \pm 1.4 \pm 1.0) \times 10^{-4} \quad .$$

What then is the θ? Two suggestions have been made. It may be a $q^2\bar{q}^2$ state,[16] specifically an $s\bar{s}(u\bar{u} + d\bar{d})$. This state would decay by falling apart and naturally explain why $\theta \to \pi\pi$ is small and predict significant $\theta \to K\bar{K}$. No explanation yet exists why such a state would not be extremely broad. Another suggestion[7] has the θ being a constructive interference effect among several $J^{PC} = 2^{++}$ $q\bar{q}$ states. The width then has to do with the spacing between states and is about right; also small $\theta \to \pi\pi$ and reasonable $\theta \to K\bar{K}$ seem to follow.

III. GLUONIUM-QUARKONIUM MIXING

A worry is that glueballs may mix with quark states with the same quantum number.

There exist some relevant calculations of the mixing of η and η' with 0^{-+} glueballs. One set of analyses[17] uses current algebra, including the anomaly in the singlet axial current,

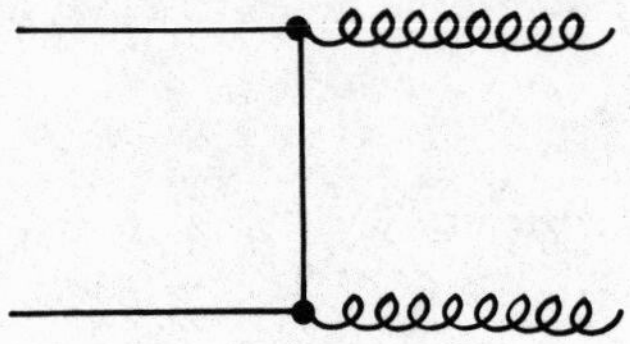

Fig. 4

$$\partial_\mu A^\mu(0) = \sqrt{\frac{3}{2}}\,\frac{\alpha_s}{4\pi}\,G_{\rho\sigma}{}^a\,\tilde{G}^{\rho\sigma,a} + \ldots. \quad ,$$

where $G_{\rho\sigma}{}^a$ is the gluon field strength tensor, together with data including the 2γ decays of pseudoscalar mesons and shows a need for a significant amount of glue in the η and η'.

Also a calculation has been done using the Bag model to directly calculate the mixing of the lowest $q\bar{q}$ and gg 0^{-+} states.[18] The mixing is calculated from Fig. 4, where the external lines are confined Bag states and the propagator has the same short distance singularities as the free propagator but satisfies the Bag boundary conditions. Using parameters obtained from studying light hadron spectroscopy [$\alpha_s = 2.2$, R = 1 fermi, $m_u = m_d = 0$, and $m_s = 300$ MeV] one finds for

$$|\eta\rangle = |q\bar{q}\rangle + c\,|gg\rangle \quad ,$$

$$|\eta'\rangle = |q\bar{q}\rangle + c'\,|gg\rangle \quad ,$$

that

$$|c|^2 \simeq 7\% \quad ,$$

$$|c'|^2 \simeq 20\% \quad .$$

ACKNOWLEDGEMENTS

Thanks are due to my collaborators Joe Coyne, Paul Fishbane, Franz Gross, Hans Hansson, Syd Meshkov, and Carsten Peterson, who deserve all of the credit and none of the blame for our work together. Thanks are also due N. Isgur for a lucid discussion of Ref. 7 and, of course, to the organizers of this very pleasant and educational conference.

REFERENCES

1. D.L. Scharer, "Glueballs - a Status Report", this conference (also SLAC-PUB-2880, Feb. 1982).
2. C.E. Carlson, J.J. Coyne, P.M. Fishbane, F. Gross, and S. Meshkov, Phys. Lett. 99B (1981) 353.
3. Review of Particle Properties, Rev. Mod. Phys. 52 (1980), S3-S5.
4. R. Jaffe, Phys. Rev. D15 (1977) 267.
5. J. Donoghue, "Expectations for Glueballs", Proceedings of 1981 APS Particles and Fields Divisional Meeting, Santa Cruz, CA.
6. H.J. Lipkin, Phys. Lett. 109B (1982) 326; P.M. Fishbane, D. Horn, G. Karl, and S. Meshkov, "Flavor Symmetry and Glueballs" NBS Preprint 1981.

6a. K. Ishikawa, Phys. Rev. D20 (1979) 731.

6b. K. Ishikawa, Phys. Rev. Lett. 46 (1981) 978; M. Chanowitz, Phys. Rev. Lett. 46 (1981) 981.

7. I. Cohen, N. Isgur, and H. Lipkin, "Conference Mixing, and Interference Phenomena in Radiative J/ψ Decays," Argonne Preprint ANL-HEP-PR-81-45, Oct. 1981.
8. S. Ono and O. Pesse, "Are ι(1440) and θ(1640) Glueballs or Quarkonia?," Orsay Preprint LEPTHE 81/21, Oct. 1981.
9. H. Fritzsch and J.D. Jackson, Phys. Lett. 66B (1977) 365; A. DeRujula, H. Georgi, and S. Glashow, Phys. Rev. D12 (1975) 147.
10. B. Guberina and J. Kuhn, Lett. Nuovo Cim. 32 (1982) 295; A. DeVoto and W. Repko, Phys. Lett. 106B (1981) 501.
11. P.M. Fishbane et al, to be published.
12. Ref. 2 and "Hadronic Decays of Low Lying Glueballs," NBS Preprint, 1981.
13. J.J. Coyne et al, Phys. Lett. 98B (1981) 110.
14. C. Bromberg et al, Phys. Rev. D22 (1980) 1513.
15. C. Edwards et al, Phys. Rev. Lett. 48 (1982) 458.

16. M. Chanowitz, Proceedings of the 1981 APS Particles and Fields Divisional Meeting, Santa Cruz, CA.

17. H. Goldberg, Phys. Rev. Lett. 44 (1980) 363; K. Milton, W. Palmer, and S. Pinsky, Phys. Rev. D22 (1980) 1124 and 1647.

18. C.E. Carlson and T.H. Hansson, "η-η'-Glueball Mixing," Nuclear Phys. B (to be published).

MONTE CARLO COMPUTATIONS OF THE HADRONIC MASS SPECTRUM

C. Rebbi

Brookhaven National Laboratory

Upton, New York 11973

ABSTRACT

This paper summarizes two talks presented at the Orbis Scientiae Meeting, 1982. Monte Carlo results on the mass gap (or glueball mass) and on the masses of the lightest quark-model hadrons are illustrated.

During the last few years Monte Carlo simulations have emerged as a very powerful technique to obtain numerical information on quantum gauge field theories.[1] They provide the necessary bridge between the perturbative results and the strong coupling results which the lattice formulation allows to derive. Monte Carlo computations have produced very substantial evidence for the validity of the lattice regularization, and for the correctness of quantum chromodynamics (QCD) as the theory of strong interactions. Moreover

---

The submitted manuscript has been authored under contract DE-AC02-76CH00016 with the U.S. Department of Energy. Accordingly, the U.S. Government retains a nonexclusive, royalty-free license to publish or reproduce the published form of this contribution, or allow others to do so, for U.S. Government purposes.

it has been possible to estimate the value of several quantities of great experimental and theoretical interest.

I wish to present here recent Monte Carlo results on the lightest masses in the spectrum of a quantized SU(2) or SU(3) gauge theory. I shall consider the states occurring in a pure gauge system as well as those present in a theory of coupled gauge and fermionic fields. In QCD the former states (frequently called glueballs) correspond to quantized colorless excitations of the gluonic field. The latter states would, of course, constitute the full hadronic spectrum in a complete theory. In a first approximation, however, they are the states of the quark model, i.e. states of a quark and an antiquark or of three quarks, bound together by the gluonic field. As we shall see, the computation has not proceeded yet beyond this approximation, but the results are already very satisfactory.

The common denominator to both analyses is the procedure by which the masses are evaluated from the rate of decay of suitable Green's functions. Let $\mathcal{O}(\vec{x},t)$ be an operator with nonvanishing matrix elements between the vacuum and the state under consideration. O is defined in the lattice-regularized theory and $\vec{x}$, t label lattice sites. By Monte Carlo simulations one calculates the quantum expectation value of the product of two such operators and forms the connected Green's function

$$G(\vec{x},t) = \langle \mathcal{O}(\vec{x},t)\, \mathcal{O}(0) \rangle - \langle \mathcal{O} \rangle^2 \quad . \tag{1}$$

Expressing G(x,t) as a sum over a complete set of intermediate states one obtains

$$G(x,t) = \sum_{n\neq 0} |\langle n|\mathcal{O}(0)|0\rangle|^2 \, e^{-E_n t + i\, \vec{p}_n \cdot \vec{x}} \quad , \tag{2}$$

where $|0\rangle$ stands for the vacuum state and E_n, $\vec{p}_n$ are energy and momentum of state $|n\rangle$ (recall that a Wick rotation to Euclidean space-time has been performed). It is convenient to project onto intermediate states of zero momentum by summing over all space lattice positions $\vec{x}$. In this way one arrives at

$$G_0(t) \equiv \sum_{\vec{x}} G(\vec{x},t) = \sum_{\substack{n\neq 0 \\ \vec{p}_n=0}} |\langle n|\mathcal{O}(0)|0\rangle|^2 e^{-m_n t} , \qquad (3)$$

where now m_n is the mass of the state. One can then estimate the mass of the lightest state from the rate of decay of $G_0(t)$ with t sufficiently large.

The problems met in the numerical determination of the Green's functions are however different in the pure gauge theory and in the theory with fermions and I shall proceed with a separate discussion of the two cases.

The simplest operator one can use to excite a pure gauge state out of the vacuum is the character of the parallel transporter around a plaquette, $U_\Box$, in the fundamental representation. It is convenient to sum over all orientations of the plaquette in order to project onto the states of lowest lattice angular momentum. Consequently, we define

$$\mathcal{O} = \sum_{\text{orientations}} \frac{1}{2}\,\mathrm{Tr}(U_\Box) \qquad . \qquad (4)$$

This operator reduces to the Lagrangian density $1/4\ \mathrm{Tr}(F_{\mu\nu}\ F^{\mu\nu})$ in the continuum limit.

The difficulty of the computation resides in the fact that the plaquette-plaquette correlation, i.e. $\langle\ \mathcal{O}(\vec{x},t)\ \mathcal{O}(0)\ \rangle$ connected, remains of very short range throughout the domain of couplings, where the simulation can produce meaningful results. This is a consequence of the rather large value of the mass-gap, when expressed in terms of the natural lattice scale Λ. When the separation between the plaquettes exceeds a few lattice units, the correlation falls below the level of statistical fluctuations inherent in the Monte Carlo procedure and cannot be reliably determined. In practice one is limited to a time separation $t \leq 3$ in Eq. (3).

In a SU(2) gauge theory the Monte Carlo simulation produces the results displayed in Fig. 1.[2] The graph shows the values of the effective mass-gap

$$m(t) = -\frac{1}{t} \ln \frac{G_0(t)}{G_0(0)} \tag{5}$$

as function of the coupling parameter $\beta = 4/g^2$, g being the unrenormalized coupling constant. The true mass-gap is given by

$$m(\beta) = \lim_{t\to\infty} m(t,\beta) \quad . \tag{6}$$

The lines in Fig. 1 illustrate the leading term in the strong coupling expansion for $m(\beta)$ (dashed line, all $m(t)$ are identical to leading order), the expected scaling behavior

$$m(\beta) = c \left[\frac{6\pi^2}{11}\beta\right]^{\frac{51}{121}} \exp\left[-\frac{3\pi^2}{11}\beta\right] \quad , \tag{7}$$

with three values of c representing a fit to the Monte Carlo results and estimated error (solid lines), and the constant asymptotic limits for $m(1)$, $m(2)$ and $m(3)$ (short dashed lines to the right), as derived from a perturbative weak-coupling expansion.

One notices that the Monte Carlo data agree well with the strong coupling expansion up to $\beta \approx 2$. After that, for a limited range of β, the values for $m(t,\beta)$ are consistent with a scaling behavior of the envelope $m(\beta) \equiv m(\infty,\beta)$. The interval over which the data appear to scale is rather small, because the effective masses $m(t,\beta)$ must eventually tend to the constant, weak-coupling limits. The overall pattern is reminiscent of the behavior of the effective string tension K_{eff}, as determined from finite size Wilson loops.[3,4] The range over which the effective string tension appears to scale

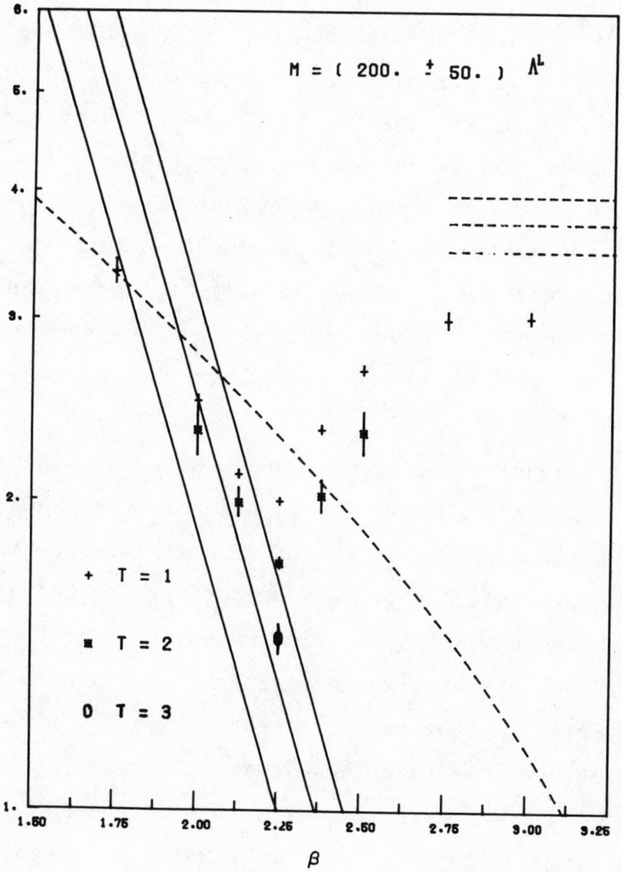

Fig. 1 Estimate of the SU(2) mass-gap from the plaquette-plaquette correlation function.

is however more extended, also because all K_{eff} tend to zero rather than to constant values in the weak coupling limit.

The computation of the mass gap can be ameliorated by selecting a more appropriate operator 0 to excite the glueball state from the vacuum. Indeed, if one could choose an operator such that $\langle n|\mathcal{O}|0\rangle$ is different from zero only for a specific state $|n\rangle$, then of course all m(t) in Eq. (5) would be identical and equal to m_n. Knowing the exact form of the operator with the above property would be tantamount to mastering the wave function of the lowest excited state, an unattainable feat for the moment. But in analogy with the variational technique used to find the lowest eigenstate of a Hamiltonian, one can allow for some free parameters in the definition of $\mathcal{O}$ and determine these so as to obtain the slowest rate of decay. This provides better upper bounds for the mass gap.

The variational procedure has been followed in Refs. (2), (5) and (6) for the SU(2) theory, and in Ref. (7) for the SU(3) theory. Linear combinations of characters of parallel transporters around various kinds of closed loops were used to excite the glueball state from the vacuum. Fig. 2 illustrates the results obtained in Ref. (2). The higher statistical errors deriving from the inclusion of several operators restrict the range of time separations to $t \leqslant 2$. Yet one sees that the minimization procedure allows to obtain, with $t \leqslant 2$, an estimate comparable to the one obtained with $t \leqslant 3$ and a single plaquette operator. Using the earlier determination of the string tension K, the value[2] found for the constant in Eq. (7) corresponds to

$$m_g = (2.4 \pm 0.6)\sqrt{K}$$

for the mass-gap, i.e.

$$m_g \simeq 1.2 \text{ GeV} \quad ,$$

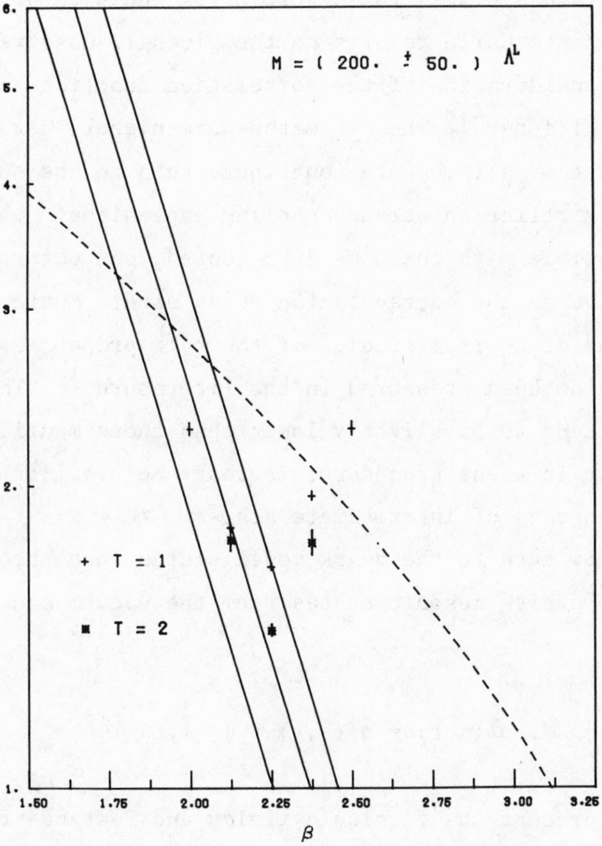

Fig. 2 Estimate of the mass gap using a variational technique.

assuming that the passage from SU(2) to SU(3) and the inclusion of fermions do not alter $m_g/\sqrt{K}$ by too much.

In Refs. (5) and (6) (where estimates for the masses of higher spin glueballs are also presented) the values $m_g = (2.9 \pm 0.4)\sqrt{K}$ and $(2.5 \pm 0.3)\sqrt{K}$ have been found. The authors of Ref. (7) give instead $m_g = (2.6 \pm 0.3)\sqrt{K}$ for a pure SU(3) gauge theory.

Earlier Monte Carlo results on the glueball mass were derived from direct consideration of the correlation function, with no sum over space positions.[4,8] The estimates were higher, not surprisingly since states of all momenta contribute then to the sum. Other estimates have relied on strong coupling expansions,[9] which can give results compatible with the ones just quoted, but unfortunately quite sensitive to the extrapolation being used. Monte Carlo determinations of m_g from studies of the bulk properties of the system have also been presented in the literature.[10] The values found for m_g tend to be slightly lower than those mentioned above; the difficulty in these procedures is that one has little control over the degeneracy of intermediate states.

Let me now turn to the quark-model states in a theory with fermions. To excite mesonic states from the vacuum one considers operators

$$\mathcal{O}(\vec{x},t) = \bar{\psi}(\vec{x},t)\,\Gamma\,\psi(\vec{x},t) \quad , \tag{8}$$

where ψ, $\bar{\psi}$ represent the fermionic fields and Γ stands for a suitable matrix acting over spin and flavor degrees of freedom. In the study of baryonic states one would instead consider expressions trilinear in the fermionic fields. The relevant Green's functions are given by

$$\begin{aligned} < \mathcal{O}(\vec{x},t)\, \mathcal{O}(0) > &= Z^{-1}\int DA \int D\psi D\bar{\psi}\; \mathcal{O}(\vec{x},t)\,\mathcal{O}(0)\,\times \\ &\times \exp\{-\,S_G(A) - \bar{\psi}\,(\not{D}(A) + m)\psi\} \quad . \end{aligned} \tag{9}$$

In this formula one integrates over all values of the gauge dynamical variables A, associated with the links of the lattice, and over the values of the anticommuting (Grassmann) variables ψ and $\bar{\psi}$, defined at the lattice sites. $S_G(A)$ stands for the pure gauge part of the action, whereas the fermionic action is given by the expression $\bar{\psi}\,(\not{D}(A) + m)\psi$, quadratic in the Fermi fields. I have used the continuum notation $\not{D}(A) + m$ to represent the lattice Dirac operator as well. I shall not discuss here the various ways to formulate the Dirac equation on the lattice, nor the difficulties one encounters. Let me only state that in the Monte Carlo applications both the method proposed by Wilson[11] and an extension of the method proposed by Susskind and collaborators[12] have been followed. Finally, Z in Eq. (9) stands for the vacuum to vacuum permanence amplitude, given by an integral analogous to the one shown in the formula, but without the O operators.

In the path integral formulation of theories with fermions one introduces Grassmann variables to represent sums over occupation numbers which can be 0 or 1. In a lattice of finite extent, the number of fermionic states which can be occupied or empty is finite. Thus one might think that the integration over fermionic degrees of freedom, representing a finite sum, should also be amenable to the Monte Carlo procedure. Unfortunately, if the quantum averages are recast as sums over occupation numbers the crucial positive-definiteness property of the measure is lost.[13]

Attempts to perform Monte Carlo simulations for systems with fermions have followed a different route. The Gaussian integral over fermionic variables can be done explicitly and one arrives at

$$< \mathcal{O}(\vec{x},t)\ \mathcal{O}(0) > =$$

$$= Z^{-1} \int DA < \mathcal{O}(\vec{x},t)\ \mathcal{O}(0) >_A\ e^{-S_G(A)}\ \mathrm{Det}\ (\not{D}(A) + m)\ , \qquad (10)$$

where $< \mathcal{O}(\vec{x},t)\ \mathcal{O}(0) >_A$ stands now for the Green's function of the operator O in the presence of a fixed background gauge field A. Assuming that Det ($\not{D}$ + m) is positive (which in QCD is guaranteed by the existence of a γ^5 transformation: $\gamma_5(\not{D} + m)\gamma_5 = - D + m = (\not{D} + m)^\dagger$), then one can evaluate the quantum averages as averages over gauge field configurations only with the new measure $\exp\{- S_{eff}(A)\}$, where

$$\exp\{- S_{eff}(A)\} \equiv \exp\{ - S_G(A) + \ln \text{Det} (\not{D}(A) + m)\} \ . \qquad (11)$$

This does not solve all problems, because the new, effection action $S_{eff}(A)$ is nonlocal and the computation of the variation ΔS_{eff} induced by a change ΔA becomes much more difficult. The evaluation of ΔS_{eff} is at the core of the Monte Carlo procedure and indeed the major efforts in the application of the method to fermions have gone precisely into finding efficient ways to calculate ΔS_{eff}.

Techniques to include the fermionic determinant in the measure are now available and have been tested on a variety of simplified models.[14,15] One can however conjecture that the major dynamical effects of the interaction between the gauge field and the fermions would already be manifest if the fixed-background Green's functions $< \mathcal{O}(\vec{x},t)\ \mathcal{O}(0) >_A$ were averaged with the pure gauge measure $\exp\{- S_G(A)\}$. In diagrammatic terms, leaving out the Det ($\not{D}$ + m) factor in Eq. (10) corresponds to considering only Feynman graphs without internal closed fermion loops (although, of course, non-perturbative effects are also included in a Monte Carlo computation). This goes beyond the so-called planar approximation which has often been advocated as a viable first order approximation for the theory of strong interactions.

Monte Carlo results for the massive Schwinger model, obtained with or without the inclusion of the Det ($\not{D}$ + m) factor, have been compared in Ref. 15. Borrowing from the theory of spin glasses, the term "quenched approximation" has been introduced to denote the neglect of the fermionic determinant in the measure. A crucial parameter is the expectation value $< \bar{\psi} \psi >$, which does not vanish as the mass m of the fermion goes to zero because of spontaneous chiral symmetry breaking. In the continuum limit $< \bar{\psi} \psi >_{m=0}$ is known from the exact solution of the theory. The Monte Carlo numerical analysis gave[15]

$$< \bar{\psi} \psi >_{m=0} \approx 0.84 < \bar{\psi} \psi >_{m=0,\ cont.}$$

when the effects of Det $(\not{D} + m)$ were included, and

$$< \bar{\psi} \psi >_{m=0} \approx 1.38 < \bar{\psi} \psi >_{m=0,\ cont.}$$

in the quenched approximation. Monte Carlo simulations making use of the quenched approximation have been performed for an SU(2) model of quarks and gluons in Refs. 16, 17 and 18; for the SU(3) theory in Ref. 19.

In Refs. 16 and 17 the fermionic action was defined adapting to the Euclidean lattice a formulation proposed by Susskind and collaborators in a Hamiltonian context.[13] The Lagrangian maintains then even in the lattice version a chiral invariance broken explicitly only by the mass term m_q for the quarks. In the limit $m_q = 0$ the action is chirally invariant (the conserved chiral charge is not a singlet with respect to flavor transformations, thus avoiding a conflict with theorems on the anomaly), but one expects the symmetry to be dynamically broken. Indeed, the Monte Carlo simulation gives evidence for a nonvanishing expectation value of $\bar{\psi} \psi$ (see Fig. 3), which moreover appears to scale with β in agreement with the renormalization group prediction (solid line). In Ref. 16 correlation functions of operators exciting pseudoscalar, vector and scalar mesons from the vacuum were evaluated. (Relaxation techniques have been used in all investigations[16-19] to compute the Green's functions in a fixed background field.) Fig. 4 illustrates the typical time dependence of a definite

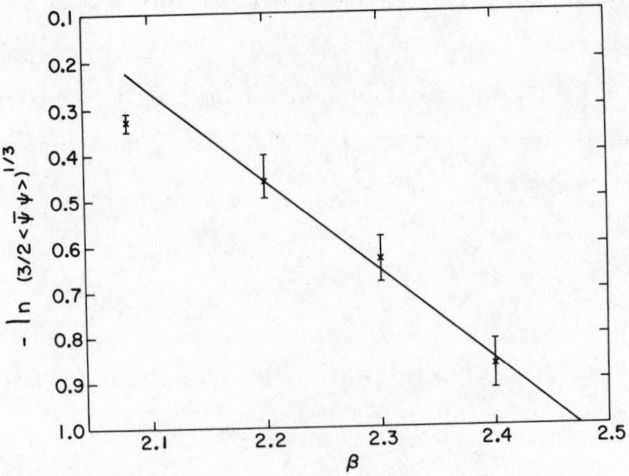

Fig. 3 Monte Carlo measurement of $< \psi \bar{\psi} >_{m_q=0}$ in an SU(2) model with fermions.

correlation (namely, the correlation of the pseudoscalar operator on a $8^3 \times 16$ lattice, with $m_q = 0.2$ in units of inverse lattice spacing and $\beta = 2.2$). The numerical computation shows that the mass squared of the lightest pseudoscalar object goes to zero together with the quark mass parameter. One can then use the experimental value of m_π and the independently determined value of the string tension[3,4] to fix m_q, which turns out approximately equal to 7 MeV (this must be interpreted as an average mass for the u and d quarks which are not differentiated in the analysis). The masses of the vector and scalar excitations, m_ρ and m_δ, exhibit much less dependence on the quark mass and tend to nonvanishing limits as $m_q = 0$. One finds $m_\rho = 800 \pm 80$ MeV and $m_\delta = 950 \pm 100$ MeV. Finally, the value measured for $< \bar{\psi}\psi >$ allows one to determine $f_\pi = 150 \pm 10$ MeV for the pion decay constant.

In Ref. 17 the parameters of the system (β and m_q) were modified so as to adapt the analysis to the states of charmonium. Using $m_{J/\psi} = 3.10$ GeV as input, values

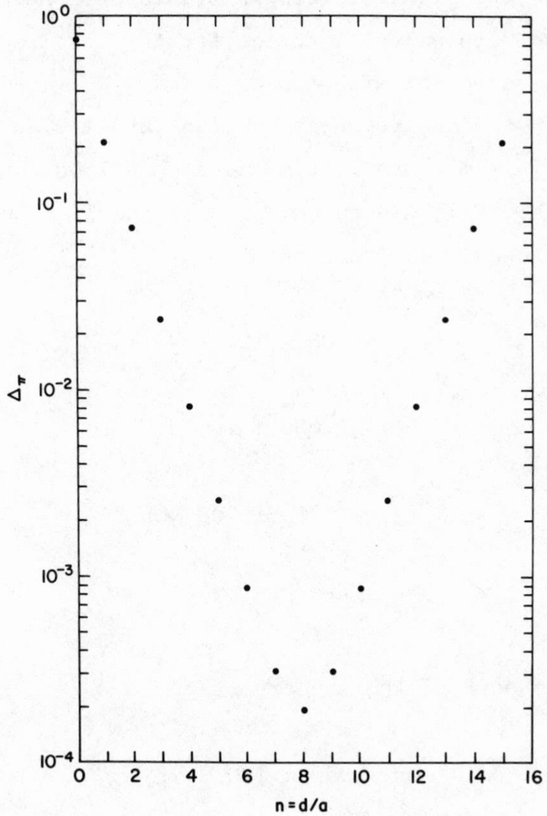

Fig. 4 Monte Carlo measurement of the correlation function for the pseudoscalar state.

$$m_{\eta_c} = (2.95 \pm 0.03)\text{GeV (exp } 2.979 \text{ GeV)},$$

$$m_{X_o} = (3.40 \pm 0.10)\text{GeV (exp } 3.414 \text{ GeV)},$$

$$m_{Pc} = (3.60 \pm 0.15)\text{GeV (exp } 3.507 \text{ GeV)},$$

$$m_{\eta'_c} = (3.65 \pm 0.15)\text{GeV, and}$$

$$m_P = (3.75 \pm 0.15)\text{GeV have been derived.}$$

In Ref. 18 the lightest mesonic states have been studied in a SU(2) model and with Wilson's action for the lattice fermions. The results are in agreement with those of Ref. 16.

In Ref. 19 the analysis was carried out with an SU(3) model, which is certainly more realistic and also allows considering baryonic states. Wilson's formulation of the lattice fermionic action was followed. The results:

$$m_\rho = 800 \pm 100 \text{ MeV},$$

$$m_\delta = 1000 \pm 100 \text{ MeV},$$

$$m_{A_1} = 1200 \pm 100 \text{ MeV},$$

$$f_\pi = 95 \pm 10 \text{ MeV},$$

for the mesonic part of the spectrum, and

$$m_p = 950 \pm 100 \text{ MeV},$$

$$m_\Delta = 1300 \pm 100 \text{ MeV},$$

for the baryonic sector, are in excellent agreement with the experimental data.

Altogether, the results on hadronic spectroscopy obtained with Monte Carlo simulations are highly satisfactory. Much work remains to be done. In particular, one needs to go beyond the quench approximation: research along this line is in progress. Moreover, the most rewarding, analytical understanding of the properties of the theory is still evading us. But what has been achieved up to now is very gratifying and reinforces our conviction of the correctness of QCD as the theory of strong interactions.

REFERENCES

1. See for instance:
M. Creutz, lectures given at the 1981 Ettore Majorana School in Erice; Brookhaven preprint 1981.
C. Rebbi, lectures given at the 1981 GIFT School in San Feliú de Guixols and at the ICTP, Trieste, ICTP preprint 1981.
2. B. Berg, A. Billoire and C. Rebbi, CERN preprint 1982, to be published in Annals of Physics.
3. M. Creutz, Phys. Rev. Letts. 45, (1980) 313.
4. G. Bhanot and C. Rebbi, Nucl. Phys. B180[FS2], (1981) 469.
5. M. Falcioni, E. Marinari, M.L. Paciello, G. Parisi, F. Rapuano, B. Taglienti and Zhang Yi-Cheng, Rome Univ. preprint 1981.
6. K. Ishikawa, M. Teper and G. Schierholz, Hamburg Univ. preprint 1981.
7. B. Berg and A. Billoire, CERN preprint 1982.
8. B. Berg, Phys. Lett. 97B (198) 401.
9. G. Muenster, Nucl. Phys. B190[FS3] (1981) 454;
G. Bhanot, Phys. Lett. 101B (1981) 95.
10. J. Engels, F. Karsch, H. Satz and I. Montvay, Phys. Lett. 102B (1981) 332;
R. Brower, M. Nauenberg and T. Schalk, Phys. Rev. D24 (1981) 548;
R. Brower, M. Creutz and M. Nauenberg, Univ. Cal./Santa Cruz, and BNL preprint, 1981.
11. K. Wilson, Phys. Rev. D10, (1974) 2445, and in New Phenomena in Subnuclear Physics, ed. A. Zichichi (Plenum, New York, 1977).
12. T. Banks, S. Raby, L. Susskind, J. Kogus, D.R.T. Jones, P.N. Scharbach and D.K. Sinclair, Phys. Rev. D15 (1977) 1111;
L. Susskind, Phys. Rev. D16, (1977) 3031.
13. The positive-definiteness of the measure may be maintained in two dimensional models by an appropriate rearrangement of the sum over fermionic degrees of freedom. Then a direct Monte Carlo simulation becomes possible. See:

J.E. Hirsch, D.J. Scalapino and R.L. Sugar, Phys. Rev. Lett. 47 (1981) 1628.

14. F. Fucito, E. Marinari, G. Parisi and C. Rebbi, Nucl. Phys. B180 (1981) 369;
F. Fucito and E. Marinari, Nucl. Phys. B190 (1981) 266;
D.J. Scalapino and R.L. Sugar, Phys. Rev. Lett. 46 (1981) 519;
D. Weingarten and D. Petcher, Phys. Lett. 99B (1981) 333;
H. Hamber, Phys. Rev. D24 (1981) 951;
A. Duncan and M. Furman, Columbia Univ. Report No. CU-TP-194 (to be published).

15. E. Marinari, G. Parisi and C. Rebbi, Nucl. Phys. B190 (1981) 734.

16. E. Marinari, G. Parisi and C. Rebbi, Phys. Lett. 47 (1981) 1795.

17. H. Hamber, E. Marinari, G. Parisi and C. Rebbi, Phys. Letts. 108B (1982) 314.

18. D. Weingarten, Phys. Letts. 190B (1982) 57.

19. H. Hamber and G. Parisi, Phys. Rev. Lett. 47 (1981) 1792.

LEARNING FROM MASS CALCULATIONS IN QUARKLESS QCD*

Richard C. Brower

University of California

Santa Cruz, California 94064

ABSTRACT

Attempts to calculate mass ratios in SU(2) quarkless QCD by Monte Carlo simulations have led to increased understanding of the dynamics of lattice gauge theory. A brief review of the role of precocious scaling, Z_2 flux condensation and finite lattice effects is presented.

The progress so far in calculating mass ratios by Monte Carlo simulations of quantum chromodynamics (QCD) has been encouraging but not spectacular.[1] Even in quarkless two-color (SU(2)) QCD, only rough estimates for four mass ratios have been made (string tension[2] $/\Lambda_{\overline{ms}} = \sqrt{\kappa}/\Lambda_{\overline{ms}} \simeq 3.9 \pm .6$, glueball mass[3] $/\sqrt{\kappa} \simeq 2.0$, deconfinement temperature[4] $/\sqrt{\kappa} \simeq .5 \pm .1$, η-mass parameter[5] $/\sqrt{\kappa} \simeq .11 \pm .02$). What, you might ask, is all the excitement about?

The answer is that we are rapidly learning about the dynamics of (lattice) QCD from these preliminary attempts to calculate mass

*Supported by a grant from the National Science Foundation.

ratios. Even with very modest use of computer time, in comparison with that devoted to other nonlinear field problems encountered in astrophysics or fluid dynamics, we can see some very general features of gauge theories. This insight can help us to devise more sophisticated Monte Carlo algorithms for the really prodigious effort to predict accurate (to 1% or better) mass ratios in the near future. Moreover, as physicists, we wish not only to compute, but to gain insight into the dynamics of gauge theories, insight which may be vital in applying gauge theories to unification and/or cosmology.

The bulk of the computations to date have been performed in pure (quarkless) SU(2) gauge theory with the Wilson action

$$S_W(\beta) = \sum_P \frac{\beta}{2} \mathrm{Tr}(U_P) \quad , \tag{1}$$

where the sum extends over all the plaquettes (or 1 x 1 squares) on a finite periodic lattice. Creutz[2] considered the first mass ratio: the string tension ($\sqrt{\kappa}$, Wilson loop $\simeq \exp(-\kappa$ Area)) relative to the lattice Λ_o parameter. The calculation involves a computation of the inverse correlation length for the string tension, $\xi_{st}^{-1} = a\sqrt{\kappa}$. One observes a rather abrupt departure from the strong coupling behavior with logarithmic growth ($\xi_{st} \sim (\log(\frac{1}{\beta}))^{-\frac{1}{2}}$) to a regime with very rapidly increasing correlation length. This so called "string tension crossover" at $\beta \approx 2.2$ (or $\alpha_{strong} = g_o^2/4\pi \simeq 1/2.2\pi$) is thought to be the onset of the weak coupling scaling behavior, referred to as asymptotic freedom. On the basis of this bold (or is it foolish) presumption of precocious scaling, the string tension can be extracted by matching to the universal exponential divergence of correlation lengths

$$\xi_{st} = \frac{1}{a\sqrt{\kappa}} \simeq \frac{\Lambda_o}{\sqrt{\kappa}} \left(\frac{6\pi^2}{11} \beta\right)^{-51/121} e^{3\pi\beta/11} \quad , \tag{2}$$

as dictated by the Gell-Mann Low equation to two loops. Immediately, one must ask whether this is indeed the correct interpretation for the string tension crossover. For instance, there is a sudden "crossover" phenomenon (actually 1st order transition) in Z_2 gauge theory at $\beta \sim 1$, but this is due to flux condensation and marks the end of confinement, not the onset of any scaling property.

Next a similar crossover was observed in the specific heat at about the same value of coupling[6] ($\beta \simeq 2.2$), and many workers assumed a similar dynamical mechanism must be operating for this bulk quantity. Brower, Nauenberg and Schalk[7] calculated the lattice size dependence of the internal energy (see Fig. 1)

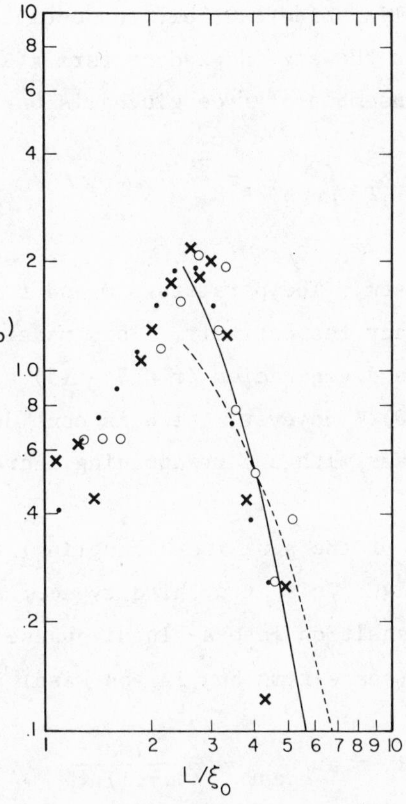

Fig. 1 Glueball gas fits to the Monte Carlo data of Ref. 7 on lattices of size L=4 (·), L=5 (×) and L=6 (o): dotted line is r=5, $m/\sqrt{\kappa}$ = 1.3 and solid line is r=15, $m/\sqrt{\kappa}$ = 1.9.

$$\varepsilon(\xi/L) \equiv L^4 \left(\langle Tr(U_P)\rangle_L - \langle Tr(U_P)\rangle_\infty\right) \quad , \tag{3}$$

and showed that it scaled. Indeed by assuming a modified beta function

$$a \frac{d}{da} g_o = - \delta g_o - \gamma g_o^3 \quad , \tag{4}$$

the scaling fits favored $|\delta| < \frac{1}{45}$ and $\gamma \simeq .041$ in good agreement with the asymptotic freedom prediction ($\delta = 0$, $\gamma = \frac{11}{45\pi^2} \simeq .046$). This analysis, combined with a similar analysis by Creutz[8] for the string tension, gives us direct evidence for the precocious scaling hypothesis for both string tension and specific heat, just to the weak coupling side of their respective crossovers.

A model[9] for the size dependent term $\varepsilon(L/\xi)$ has been devised in terms of an ensemble of free glueballs on a finite lattice

$$\varepsilon(x) = \frac{r}{11} \frac{x^3}{16} \int_o^\infty d\lambda \, e^{-\frac{x}{2\lambda}} \left\{ \left(\sum_{n=-\infty}^{\infty} e^{-\frac{n^2 x \lambda}{2}} \right)^4 - 1 \right\} \quad , \tag{5}$$

where $x = L/\xi = aLm$. The parameters m and r are the glueball mass and spin degeneracy respectively. This model fits well for $L/\xi > 3$ and suggest large degeneracies ($r \simeq 5 - 15$) and rather small masses ($m/\sqrt{\kappa} \simeq 1.2 - 2.0$). However, there is considerable ambiguity by increasing the mass with a corresponding increase in degeneracy (see Fig. 1).

To the left of the peak (weak coupling) the glueball picture fails dramatically. This is a third type of crossover due to the deconfinement transition into a gluonic phase. An estimate of the critical radius (space-time bag if you wish) is

$$R_c = aL/_{peak} = \frac{1}{(.5\pm.1)\sqrt{\kappa}} \quad , \tag{6}$$

or an equivalent critical temperature $T_c \simeq (.5\pm.1)\sqrt{\kappa}$. This agrees with the proper definition of critical temperature carried out on an asymmetric lattice by others.[4] A more precise understanding of this transition from a glueball to a gluon gas would be helpful to remove the mass-degeneracy ambiguity in our fit. Recently very nice calculations[10] to measure the glueball mass from the plaquette-plaquette correlations length give a value $m/\sqrt{\kappa} \approx 2.35 \pm .30$, which is quite different from the finite lattice scaling result. Possibly the glueball gas model is seriously incomplete in representing the dynamics of finite lattice scaling. Again investigations of asymmetric lattice could be very helpful in resolving this issue.

A quantitative understanding of the behavior of the free energy (and thus of the specific heat) has been obtained[11] through a study of the Z_2 degrees of freedom (monopoles and their strings) present in the theory. The specific heat peak coincided with the condensation point of the monopole-string "gas"; or, looking at it from strong coupling, with the point where the exponential decrease in the density of these objects set in. Furthermore, by introducing a chemical potential, λ, for the monopoles,

$$S = \frac{\beta}{2} \sum_P \mathrm{Tr}(U_P) + \lambda \sum_c \sigma_c \quad , \tag{7}$$

where on each cube c

$$\sigma_c \equiv \prod_{P \varepsilon \partial c} \mathrm{sign}\, \mathrm{Tr}(U_P) \quad , \tag{8}$$

we would change the peak into a first order transition for $\lambda \gtrsim .45$. The location of the transition then moved toward smaller β for increasing λ, reaching $\beta \simeq .97$ at $\lambda = \infty$. Finally, by integrating out the nonabelian degrees of freedom, we were able to derive an effective theory, dual to a Z_2-Higgs model, which well predicts the phase structure (see Fig. 2). Results along similar lines have been obtained by Halliday and Schwimmer,[12] and Mack and Pietarinen.[13]

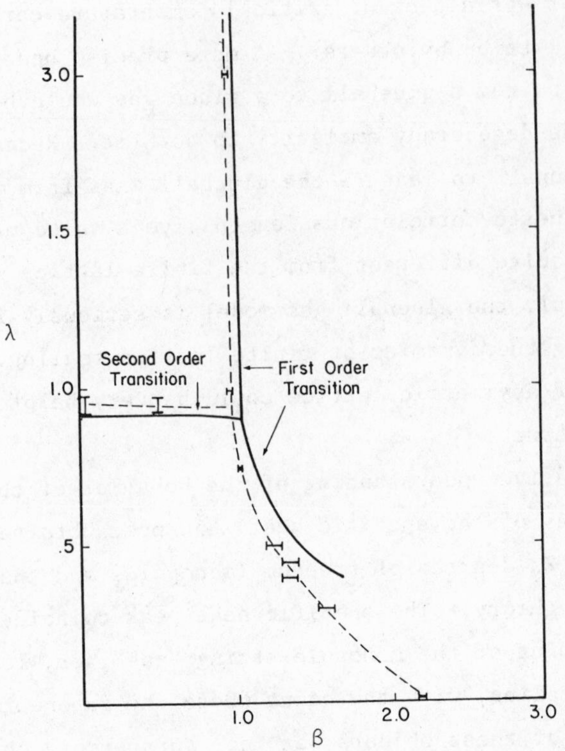

Fig. 2 A comparison of the predicted (solid line) and Monte Carlo simulations for the β-λ phase plane for SU(2) gauge theory with a variable (λ) monopole activity.

Thus we concluded that the onset of scaling (where the lattice string tension $a\sqrt{\kappa}$ begins its exponential dive to zero) is in effect Z_2-deconfinement at the lowest scale flux, measured by single lattice spacing Wilson loops, $W(1,1) = \langle \frac{1}{2}\,\mathrm{Tr}(U_P)\rangle$. This is intuitively plausible, since the onset of asymptotic freedom viewed from strong coupling should in fact be the replacement of the confining linear potential with a deconfined Coulombic regime at short distances. Unlike an abelian theory, where flux is strictly additive, the nonabelian problem in principle allows flux on larger scales (measured by $I \times I$ Wilson loops, $W(I,I)$) to survive its demise on the lowest I = 1 scale.

Thus with the success[11] of the Z_2-Higgs model in characterizing the bulk crossover, it is important to see what effect λ has on the string tension. We can argue that the string tension in the modified model ($\lambda \neq 0$) must agree precisely with the standard Wilson model ($\lambda = 0$) on the weak coupling side of the crossover, both scaling in accordance with asymptotic freedom. The reason for this is that our modified action is zero to all orders in weak coupling and thus the relationship between the lattice Λ_o and the scaling violation parameter $\Lambda_{\overline{ms}}$ is unaffected[14] by λ. Physically λ only effects the single lattice spacing monopoles, which are exponentially suppressed as β increases.

In the Wilson model, the exponential scaling regime cuts off abruptly at $\beta \sim 2.2$ after which the string tension increases only logarithmically, following the strong coupling expansion. As we go to positive values of λ, we expect that the strong coupling string tension will uniformly decrease; since the presence of a nonzero λ acts as a constraint that tends to inhibit disorder and any increase in order will undoubtedly decrease the area law coefficient. Therefore, the break in the exponential rise of the string tension ought to take place at larger β than the string tension crossover for Wilson action. However, as noted above, the peak in the specific heat moves to smaller β as we increase λ. We therefore must conclude that the crossover in the string tension does not track with the specific heat peak. In fact, in at least this case it moves in the opposite direction from the bulk transition. This is the main observation which we would like to qualitatively understand.

First, let us review the standard ideas regarding the causes of asymptotically free scaling in the weak coupling region, so that we can better understand its breakdown. This scaling is a result of the universal behavior exhibited by a theory near a critical point, in this case at $\beta = \infty$. Any theory can be equivalently described by an effective theory on a larger scale, obtained by some type of renormalization group transformation such as block spinning, until this scale (renormalized lattice spacing) is equal the correlation length. According to standard lore, after repeated block spinning a theory forgets its nonrelevant couplings and approaches a universal theory which is then invariant under further transformations up to a change of scale. Now, as we move closer to the critical point the correlation length goes to infinity. Therefore, any fixed starting theory has to undergo more block spins to reach the effective theory at the correlation length scale and hence this effective theory becomes closer to the universal (fixed point) theory.[15] In this fashion, we obtain universal behavior in some region around the critical point. In our case, the correlation length is given by the inverse square root of the string tension; its falloff then agrees with the correlation length getting large. Furthermore, the way in which this occurs (analog of critical exponents) can be determined by perturbation theory, which in our case leads to asymptotic freedom and its concomitant exponential law for the correlation length (see eq. (2)). To summarize then, we expect on general grounds that regardless of our initial action, the string tension for sufficiently weak coupling will be governed by the scaling laws of asymptotically free perturbation theory.

Now, we can examine the significance of deviations from universal behavior. As we come in from weak coupling, the correlation length shrinks rapidly; eventually, the theory does not have enough time (in terms of block-spin steps) to approach the universal theory. The particular details of the lattice action start to come into play and scaling no longer holds. Using this picture, we can appreciate the apparent "specialness" of the Wilson action. Here, the string tension scales all the way to $\beta \sim 2.2$, where the correlation length is only $1.6 \sim 2$ (lattice spacings). Thus, the Wilson action scales according to asymptotic freedom for about as far as it possibly can; empirically, it seems to be very close to the universal theory.

In a recent paper,[16] a more quantitative discussion of the delay of the onset of asymptotic freedom for the string tension is given for a constrained model with $1/2 \text{ Tr } U_P \geq \cos\alpha$ for $\pi \geq \alpha \geq 0$. This contraint causes both Z_2 flux and Z_2 monopoles to be squeezed out simultaneously. A sympathetic reader can see there that the effect of the constraint (say to a positive plaquette model $\text{Tr}(U_P) > 0$ or $\alpha = 0$) does indeed delay the onset of asymptotic freedom to larger $\beta \approx 2\ 1/2$. We believe that the correlative length ξ_{st} in the positive plaquette model must get to about 3 or 4 lattice spacing before the excised flux on the unit scale ($\text{Tr}(U_P)>0$) is healed by fluctuations on larger scales.

Creutz and Bhanot[17] have explored another modified Wilson action by adding an adjoint trace

$$S_A = \sum_P \frac{\beta}{2} \text{Tr}(U_P) + \sum_P \frac{\beta_A}{3} (\text{Tr}(U_P))^2 \tag{9}$$

with a phase plane similar to Fig. 2 - a sharpening of the crossover for positive β_A into a first order transition (beginning at $(\beta, \beta_A) \approx (1.6, .9)$) due presumably to the Z_2 flux dynamics discussed for the λ-model (Eq. 7). Recently, Bhanot and Dashen[18] have investigated the string tension for a range of nonzero β_A. They are alarmed that the effects are large if $|\beta_A| > .5$, and conclude that since any β_A is by universality an equally "good" action, no reliable calculation of the string tension can be made at intermediate coupling ($2 < \beta + 2\beta_A < 3$).

Our attitude is just the opposite. They note that very stable values for moderate adjoint admixtures, $\beta_A \leq 1/4$. If you notice that $\beta_{eff} = \beta + 2\beta_A$, really $2\beta_A \sim O(1)$ is a large admixture. At $2\beta_A = -1$, the bare coupling $g_o^2 \to \infty$, and as $\beta_A/\beta \to \infty$, no string tension can even be defined! In the pure adjoint model, there is a symmetry under the Z_2 center ($U_\mu \to -U_\mu$) so that all Wilson loops are exactly zero. In strong coupling $a^2\kappa \sim \log(\frac{1}{\beta}) \to \infty$ is $\beta_A/\beta \to \infty$.

However, the fact that the string tension is very insensitive to shifts in β_A around $\beta_A = 0$ is consistent with our contention that the Wilson action is close to the optimal fixed point action using only single plaquette terms. Clearly, more investigation should be done on the renormalization group to see what local or single plaquette terms should be added to give the best precocious scaling. Wilson[15] has reported that larger scale loops on several plaquettes are needed to give precocious Lorentz invariance.

Much work remains to be done to straighten out the general picture presented here. In pure SU(2) lattice QCD, there are several important correlation lengths (string tension, glueball, deconfinement, finite lattice scaling lengths) and their interrelation is still not clear. The introduction of modified actions is a powerful tool to separately study different dynamical effects. Also the investigation of the fixed point action[15] in the multivariable space is worth pursuing to search for ways to quantify and further supress the finite lattice spacing distortions.

One difficulty with the present investigation of gluonic dynamics is the experimental inaccessibility of all the mass parameters that are most easily calculated. A little reflection indicates that $\Lambda_{\overline{ms}}$, the string tension, the glueball masses, the η-mass parameter and the deconfinement temperature all present severe problems in relating them to precise experimental measurements. To overcome this, a superb probe of gluonic dynamics, being pursued at Santa Cruz, is the introduction of heavy quarks which allows the calculation of the charmonium, upsilonium and toponium levels in the quenched (or valence) approximation. Even the effects of light quarks can be estimated by using modified actions. This should give abundant predictions of lattice QCD to confront with the growing wealth of data.

It is a pleasure to acknowledge the fruitful collaboration

with D. Kessler and H. Levine in which many of these ideas germinated.

REFERENCES

1. R. Brower, Discrete Quantum Chromodynamics, 1981 Les Houches Lectures, UCSC preprint TH-147-81 and references therein for a review of the current status.
2. M. Creutz, Phys. Rev. Lett. 45, 313 (1980); Phys. Rev. D21, 2308 (1980).
3. G. Bhanot, and C. Rebbi, Nucl. Phys. B180, 469 (1981) and Ref. 9 and 10 below.
4. L. McLerran and B. Svetitsky, Phys. Lett. 98B, 195 (1981); J. Engles, F. Karsch, I. Montvay, and H. Satz, Bielefeld, Phys. Lett. 101B, 89 (1981); ibid Phys. Lett. 102B, 332 (1981).
5. P. DiVecchia, K. Fabrius, G. Rossi and G. Veneziano, Nucl. Phys. B192, 392 (1981).
6. B. Lautrup and M. Nauenberg, Phys. Rev. Lett. 45, 410 (1980).
7. R. Brower, M. Nauenberg and T. Schalk, Phys. Rev. D24, 548 (1981).
8. M. Creutz, Phys. Rev. D23, 1815 (1981).
9. R. Brower, M. Creutz and M. Nauenberg, BNL-30964 (1982).
10. K.M. Mutter and K. Schilling, TH-3246 CERN (1982).
11. R.C. Brower, D.A. Kessler and H. Levine, Phys. Rev. Lett. 47, 621 (1981); Nucl. Phys. B205, 77 (1982).
12. I. Halliday and A. Schwimmer, Phys. Lett. 101B, 327 (1981), 102B, 337 (1981).
13. G. Mack and E. Petarinen, "Monopoles, Vortices and Confinement", DESY 81/067 (1981).
14. This should be contrasted with the adjoint modification in Eq. (9) discussed below.
15. This idea has been developed in K. Wilson, Cornell report 1979 (unpublished) and unpublished lectures.

16. R. Brower, D. Kessler and H. Levine, Schlumberger-Doll Print-82-0312 (1982).

17. G. Bhanot and M. Creutz, BNL-29640 (1982).

18. G. Bhanot and R. Dashen, preprint Princeton IAP, Print-82-0251 (1982).

BEYOND QCD: WHY AND HOW

Giuliano Preparata

Istituto di Fisica, Universita di Bari and

Istituto Nazionale di Fisica Nuclear, Sezione di Bari,

Italy

ABSTRACT

Arguments based on recent experimental information are presented to stress the necessity of going beyond the present formulation of Quantum Chromo Dynamics (QCD). A new theory, Anisotropic Chromo-Dynamics (ACD), based on the hadrodynamical pillars: quarks, color and local symmetry, is discussed and its first successful steps in describing hadrons are outlined.

The aim of this talk its two-fold: first to argue in favor of going beyond the theoretical paradigm of the day: QCD, and then to present a concrete proposal of how can one proceed to go beyond QCD.

In order to clear the way from any ambiguity and misunderstanding, I would like to reiterate with all clarity that I believe that the basic theoretical notions that underly QCD:

(i) Quarks,

(ii) Color,

(iii) a gauge-principle;

are destined to remain with us. They do in fact represent an important step forward in our understanding of subnuclear physics. Such notions have been willed to us by two decades of immense efforts, both experimental and theoretical, and have shown their validity in a countless number of physical situations.

Thus my criticism of QCD will not question the above mentioned pillars, upon which rests all our understanding of subnuclear phenomena, but rather the natural but logically unwarranted step that has led almost everybody to conclude that QCD must be the theory of hadrons. For if the latter dictum is accepted, then experimental difficulties with the present understanding of QCD would imply that there is something wrong with our basic hadrodynamical notions, and this would certainly leave us in a hopeless mire.

In the following I shall give arguments, based on the presently available experimental information, that we should abandon the generally accepted notion of QCD, based both on perturbative calculations (Perturbative QCD) and on lattice calculations, whose physical meaning and relevance, at least at the present stage, are far from being well established. This shall constitute the "pars destruens" of my discourse.

The "pars construens" shall be focussed on the description of a new theory of hadrons which, making use of the three basic pillars of hadrodynamics, not only shall avoid but will also explain away the difficulties encountered by the "accepted" QCD framework.

1. BEYOND QCD: WHY?

In this section I shall produce arguments, based on present experimental knowledge, which strongly suggest that we should try to go beyond the generally accepted QCD paradigm. I shall concentrate on two points:

(A) Gluons;

(B) Asymptotic Freedom.

1A. Gluons

One of the qualifying features of QCD is the existence of 8 colored gluons.

These gluons have in QCD a dual role: to provide for the color-confining force (scalar and longitudinal gluons) and to give rise to new, independent degrees of freedom of hadronic matter (transverse gluons). Thus if we are to make any sense of QCD as a physically relevant theory, the transverse degrees of freedom of the color-fields should give rise to a number of characteristic physical effects that, I believe, could not have escaped our observation. Let's briefly review them:

(i) Glueballs

While we have good and unequivocal experimental evidence for several hundred states of the $q\bar{q}$, qqq-type, clear evidence for glueball states has so far eluded the most sophisticated experimental attempts to observe them. Several reasons have been advocated for the elusiveness of glueballs, most notably lack of a clear signature, mixing to the $q\bar{q}$ states, etc. But it should be recalled that simple MIT-bag calculations (which yield acceptable spectra for $q\bar{q}$ and qqq ground states) make us expect 0^{++}, 2^{++} gluon-gluon states with masses smaller or equal to 1 GeV. Now it would not appear entirely reasonable that such states had escaped the experimental search, being located in a well studied region of the hadronic spectrum. The recently reported states $\iota(1440)$ and $\theta(1640)$ by the Crystal Ball group at SPEAR,[1] have been convincingly argued not to be glueball states.[2] In any event it is certainly fair to say that we have come to a point where the elusiveness of glueball states begins to appear as a serious embarrassment for the generally accepted form of QCD-theory.

(ii) "Hermaphrodite states"

Such are the states of the type $q\bar{q}g$, qqqg,.... where g has the characteristics of a transverse gluon. The existence of such states would considerably enrich the particle spectrum even in the mass range below 2 GeV. Again this is what one would expect from simple MIT-bag calculations. But from experimental knowledge, which in the case of baryon resonances with $m \lesssim 2$ GeV is remarkably detailed, no room seems available for hermaphrodite states.[3] Furthermore, predictions for narrow states of the $b\bar{b}g$-type have been contradicted by recent experiments at CESR.[4] The outlook for gluonic states either pure (glueballs) or accompanied by quarks (hermaphrodite states) seems at present particularly dim.

(iii) "Glue jets"

The change in the pattern of hadronic final states in high energy e^+e^- collisions, first observed at PETRA about three years ago,[5] has been universally taken as strong evidence for the active presence of the (transverse) gluon degree of freedom. Even though the hint for the radiation of a hard gluon appears at first rather strong, a detailed comparison of the experimental information with theoretical expectations reveals grave difficulties. In fact, were we allowed to consider "partons" only, the situation would appear quite comfortable for the hard gluon interpretation. However, when we give a closer look to the hadron fragmentation properties of the "three-jet events" observed in such experiments we find the strange result that all jets look alike. This cannot be easily understood. For, when we try to form a picture of the gluon fragmentation properties along the lines, embodied in the Field-Feynman model, that have been rather successful for quarks, we fall immediately into

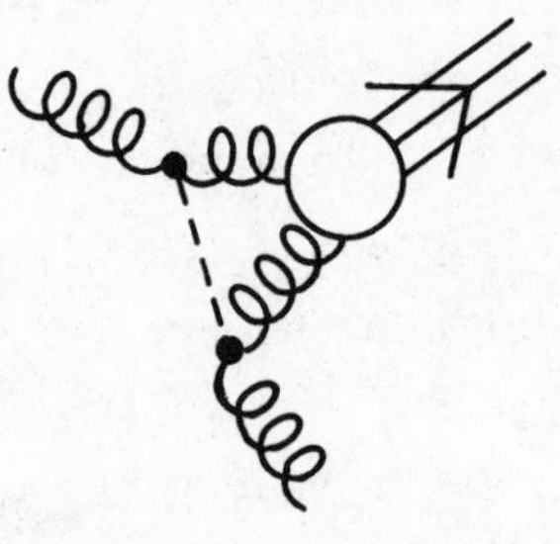

Fig. 1 The main fragmentation process for the gluon.

ℓℓℓ = transverse gluon

----- = longitudinal gluon

the difficulty that

(a) in its color field the gluon prefers (9:4) to create a gluon-pair rather than a quark pair:

(b) the gluon pair gives rise to "glueballs". (See Fig. 1).

As glueballs (if they exist at all) must be quite heavier than low-lying $q\bar{q}$-states (pseudoscalar and vector mesons), secondary hadrons (π,K,...) would be produced much more copiously in gluon than in quark-fragmentation. Thus it appears inevitable to expect gluon jets to have a considerably softer energy distribution and higher multiplicity than quark-jets. As we have recalled, experimentally this is not borne out. No believable way out has so far been proposed from what appears as a very serious, possibly fatal, difficulty. Should we fail within the QCD framework to gain any understanding on this puzzling behavior of gluon-jets, we would be faced with the somewhat ironical situation that a strong evidence <u>for</u>, on closer look, becomes strong evidence <u>against</u> the (transverse) gluonic degrees of freedom of hadronic matter.

Admittedly the status of gluons is still unclear, but it seems to me that in the few points discussed above the QCD enthusiasts may find more than one reason to worry.

1B. Asymptotic Freedom

As is well known the first indication (judged very strong by many people) that QCD might be the theory of strong interactions has come from the discovery that in perturbation theory a nonabelian gauge theory, such as QCD, is asymptotically free.[10] For this would then imply that the remarkable scaling properties observed in deep inelastic phenomena could thus find an elegant theoretical explanation.

Furthermore, if one gives for granted the highly nontrivial circumstance that the properties of a perturbative theory go over to the (unproved) confined situation without any change, Asymptotic Freedom (AF) predicts a characteristic pattern of scaling violations which in the last few years have been claimed to be shown by the

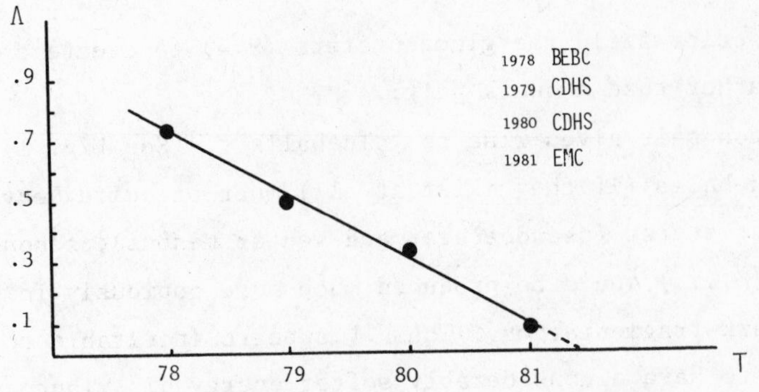

Fig. 2 A rough sketch of the "generally accepted" values of Λ [See Eq. (2.1)] as a function of time.

data. However, the size of such violations is controlled (modulo unimportant theoretical subtleties) by the all-important Λ-parameter which determines the "running coupling constant" $\alpha_s(Q^2)$:

$$\alpha_s(Q^2) \sim \frac{1}{[\log Q^2/\Lambda^2]} \quad . \tag{2.1}$$

And over the years success has been claimed with values of Λ which have shown an ominous trend to decrease. Fig. 2 gives a rough description of how the generally accepted Λ-values have changed as a function of time, Λ being now almost consistent with zero (no asymptotic scaling violations). The interpretation of Fig. 2 might become easier if we recall that the maximum values of Q^2 (current's momentum transfer) have consistently increased over the last few years. The aspect which I find most striking in the high Q^2 experiments now completed, is that for $Q^2 \gtrsim 10$ GeV^2 Bjorken scaling is exhibited almost unadulterated by the data. In order to fit the experimental points with AF one must consider a value of Λ so low ($\Lambda \simeq 100$ MeV) that elaborate analysis of nonleading effects (the so called "higher twists") is necessary before one can disentangle the minute AF-corrections.

Be as it may, it has also been shown that all of scaling violations in deep-inelastic scattering can be very economically described by subasymptotic contributions of $O(\frac{1}{\sqrt{Q^2}})$,[7] which within the MQM-framework were predicted long before their observation.

Suppose, however, that AF holds and Λ is as low as 100 MeV, then the possibly naive question that one might ask is: who keeps the quarks inside the bag? For with such a low value of Λ we find that the color forces become nonperturbative at distances of the order of 2 Fermis, well beyond the radius of a reasonable bag.

Again, the situation is far from clear, but here are a few more subjects for meditation offered to the QCD enthusiasts:

(i) The "explanation" of the $\Delta I = \frac{1}{2}$ rule in nonleptonic decays offered by AF at short distance[8] with $\Lambda \simeq 100$ MeV completely evaporates.[9]

(ii) There occurs in deep-inelastic muon-nucleon scattering the bizarre fact[10] that in order to describe the structure functions one needs $\Lambda \simeq 100$ MeV, while the p_T-distribution of planar events requires $\Lambda \simeq 500$ MeV. Note that the data used in the two analyses are the same.

(iii) The "big successes" of perturbative QCD are only qualitative. However, I would like to ask a few questions:

(a) onia: where is the Coulomb potential expected at short distances?[11]

(b) high p_T-physics: where are jets?[12]

(c) lepton-pair hadroproduction: what happened to the K-factor?

(d) exclusive physics, form factors, etc.: how can we make it to agree with experiments, especially when spin is involved?

Let me end this Section with one question (Q) and one remark (R).

Q: Aren't we being too wishful thinking in claiming that the data strongly suggest that QCD is the theory of hadrons?

R: Perturbative QCD cannot provide a basis for "explaining"

the phenomena to which it is currently applied, because it lacks in a most serious fashion any provision for confinement. Only after the confinement phenomenon has been correctly and realistically taken into account can we assess the meaning and the limits of applicability of perturbative QCD.

2. BEYOND QCD: HOW?

It is a remarkable fact that a number of puzzling aspects of hadrodynamics, which QCD finds so difficult to account for, are natural and well understood in two-dimensional (one space - one time) gauge theories. For instance

(i) Confinement is a natural property of one-dimensional space due to the nature of the Coulomb potential which in one-space dimension at large distances behaves as $V_{1+1}(r) = \mu^2 r$ instead of the familiar $V_{3+1}(r) = \frac{e^2}{r}$.

(ii) Freedom at short distances is also natural in 1+1 dimensions. As is well known, 2-dimensional theories are very well behaved in the ultraviolet region (technically, they are "superrenormalizable"), so that radiative corrections do not spoil free-field short distance behavior.

(iii) The gauge fields in two-dimensions do not carry any independent dynamical degree of freedom; transverse dimensions in fact do not exist. Thus "physical" gluons disappear from the theory.

Should we take up this clue? And if so, how?

Anisotropic Chromo-Dynamics (ACD) has been proposed in the attempt to conjugate the basic pillars of hadrodynamics: quarks, color and local $SU(3)_c$ symmetry, with the peculiar characteristics of two-dimensional gauge dynamics, just mentioned. In the rest of the talk I shall be concerned with a brief discussion of its ideas and its present achievements. The interested reader is invited to consult the existing literature.[13-14]

2A. The Theory[13]

The basic idea of ACD is to construct a gauge dynamics for the color field which is isomorphic to a two-dimensional gauge-theory. In order to achieve this one enlarges the base-space of the theory from the Minkowskian manifold M_4 to a seven-dimensional space-time structure [$\bar{S}_3$ is a 3-dimensional pseudosphere]

$$M_4 \times \bar{S}_3 \quad ,$$

to be called Anisotropic Space-Time (AST). The points of AST are the elementary physical events E, which are represented by a pair of 4-vectors: $E = (x,n)$, where $x_\mu \varepsilon M_4$ and $n_\mu \varepsilon \bar{S}_3$ with $n_\mu n^\mu = -1$. The principle of relativity is schematically represented by the following diagram:

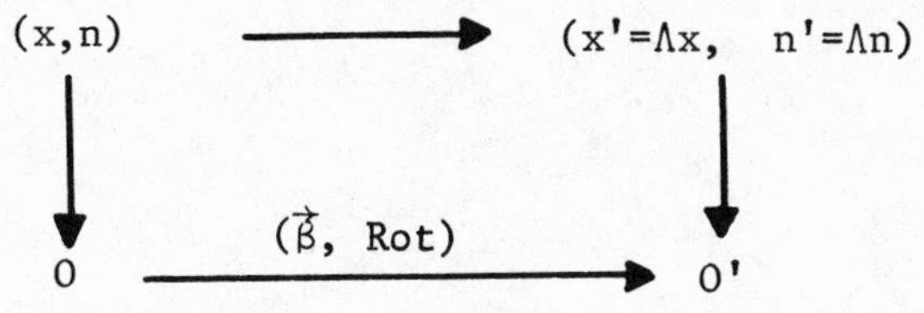

that indicates that if the inertial observer 0 gives the events E the coordinates (x,n), the inertial observer 0' assigns the same event E the coordinates (x',n'), with both vectors x' and n' obtained from x and n by the <u>same</u> homogeneous Lorentz transformation Λ, connecting 0 and 0'.

The theory is then defined by the action:

$$S = S_F + S_{AYM} + S_{INT} \quad , \tag{2.1}$$

where [$d\mu(n)$ is the invariant measure of $\bar{S}_3$]

$$S_F = \int d\mu(n) \int d\mu(n') \int d^4x \; \bar{q}(x,n) \; [i\not{\partial}-m] \; q(x,n') \tag{2.2}$$

is the quark-Dirac type action;

$$S_{AYM} = \int d\mu(n) \int d^4x\, L_{AYM}(x,n) \qquad (2.3)$$

is the gauge-field action (AYM stands for Anisotropic Yang-Mills), whose Lagrangian density is

$$L_{AYM}(x,n) = -\frac{1}{4} f^a_{\mu\nu}(x,n)\, F^{\mu\nu}_a(x,n) \quad (a=1,\ldots,8), \qquad (2.4)$$

with the usual field "intensity"-tensor

$$F^a_{\mu\nu}(x,n) = \partial_\mu A^a_\nu(x,n) - \partial_\nu A^a_\mu(x,n) + g\, f^{abc} A^b_\mu(x,n) A^c_\nu(x,n), \qquad (2.5)$$

and the field "magnitude" -tensor given by

$$f^a_{\mu\nu}(x,n) = \varepsilon^{\alpha\beta}_{\mu\nu}(n)\, F^a_{\alpha\beta}(x,n) \quad , \qquad (2.6)$$

the "anisotropy"-tensor being given by

$$\varepsilon^{\alpha\beta}_{\mu\nu}(n) = -2\, [\delta^\alpha_\mu n_\nu n^\beta + \delta^\beta_\nu n_\mu n^\alpha] \quad . \qquad (2.7)$$

Finally the interaction term S_{INT} has the form:

$$S_{INT} = g \int d\mu(n) \int d^4x\, \bar{q}(x,n)\, A\!\!\!/^a(x,n) \frac{\lambda^a}{2} q(x,n) \; . \qquad (2.7)$$

It is straightforward to check[13] that at the classical level S[Eq. (2.1)] describes a confined theory of quarks, whose dynamical structure is isomorphic to that of a two dimensional theory. This is made possible by the "anisotropy direction" n_μ, which allows us to construct a theory whose spatial dynamics evolves along the space-direction of n_μ only. Fig. 3 attempts to give an intuitive picture of the dynamical possibilities afforded by our extension of the usual Minkowskian space-time to AST.

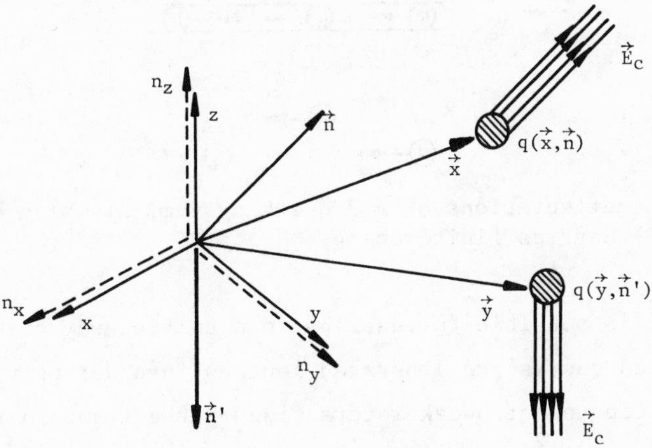

Fig. 3 $q(\vec{x},\vec{n})$ creates a color charge at $(\vec{x},\vec{n})$, while $q(\vec{y},\vec{n}')$ creates a charge at $(\vec{y},\vec{n}')$. The color electric $\vec{E}_c$ field configurations produced by the charges according to the action (2.1) are described by their lines of force.

We see that creating for instance a charge at $(\vec{x},\vec{n})$ produces a color electric field whose lines of force are completely focused along the direction $\vec{n}$. This is characteristic of two-dimensional gauge dynamics, where the electric lines of force have no other direction to follow but the only spatial direction. As a result the work to be done to displace a charge to infinity along the direction of the electric flux is infinite, due to the constancy of the intensity of the electric field.

2B. The Quantum Structure[13]

As remarked above any gauge theory constructed according to (2.1) is confined at the classical level. The crucial question to ask now is whether this is also true when quantum fluctuations are properly taken into account. By analyzing this problem[13] we find the remarkable result that a bifurcation occurs between vector (parity-conserving) and chiral theories (maximally parity-violating): the former preserving the classical confining structure, the latter losing confinement and becoming usual, isotropic Yang-Mills gauge theories.

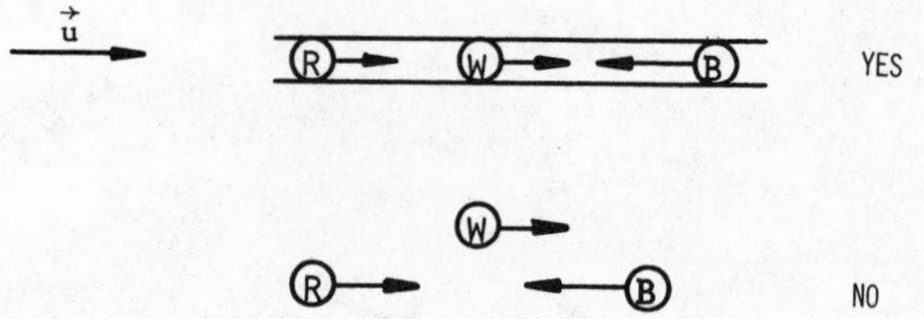

Fig. 4 Two configurations of a 3 quark system, of which only one (YES) carries finite energy.

Thus it is possible to describe in a unified way both confined, colored quarks and liberated leptons interacting via a GSW parity-violating electroweak interaction of the standard type. As a further bonus $\sin^2\theta_W$ becomes calculable and, to lowest order in electroweak interactions, turns out to be equal to 0.25.

Reverting now to the hadronic theory, quark confinement implies:

(i) only color singlets can have finite energies and therefore by physically realizable;

(ii) the classical trajectories of zero triality states must have the following structure [p=1,...,N]:

$$\vec{x}_p(t) = \vec{x}_\perp(t) + x_p(t)\vec{u}(t) \quad , \tag{2.9}$$

where $\vec{x}_\perp(t).\vec{u}(t) = 0$. In Fig. 4 we report two configurations for a baryonic system, of which one is forbidden. Thus in this theory hadrons form linear structures.

(iii) The number of independent degrees of freedom (excluding C.M. motion) for a N-quark system is instead of 3(N-1),

$$n_{DF} = N+1 \quad . \tag{2.10}$$

For a meson system $(q\bar{q})$ there is no difference with the quark-model, but for baryons (2.10) gives 4 degrees of freedom, while the naive quark model has 6. This prediction of our approach can be tested by determining whether the baryon spectrum has fewer states than expected on the basis of the quark model.

2C. The Perturbative Structure[14]

Another very interesting feature of ACD is that it leads naturally to a long sought property of hadronic interactions: its perturbative structure. Several theoretical notions have in the past been tried out to account for the surprising validity of the quark model, the Zweig rule, the (relative) smallness of hadronic widths etc., most notably: planarity in the dual model, and $\frac{1}{N}$ expansions in QCD. This is how ACD brings out such a structure:

(i) we write the Hamiltonian as:

$$H = H_{KIN} + H_f \,, \tag{2.11}$$

where the kinetic part of the hamiltonian is the usual Dirac hamiltonianand the field part H_f is given by $[n_\mu \equiv (o,\vec{u})]$ the current-current expression[13]

$$H_f = \int d\mu(\vec{u}) \int d^3x \;\; \Delta(\vec{x},\vec{u}) \; \bar{J}^a_\mu(\frac{\vec{x}}{2},\vec{u}) \; \bar{J}^\mu_a(-\frac{\vec{x}}{2},\vec{u}) \; . \tag{2.12}$$

(ii) we quantize the quark fields canonically on a surface t = const and decompose it in creation and annihilation operators for quarks (b^+,b) and antiquarks (d^+,d).

(iii) we decompose the color-currents $J^a_\mu(\vec{x},\vec{u})$ appearing in (2.12) in their no-pair creation part (b^+b,d^+d) J^{np}_μ, and pair creation part (bd,b^+d^+) J^p_μ, and write

$$H_f = H_{np} + H_p \quad , \tag{2.13}$$

where H_{np} contains J^{np}_μ only and H_p the rest.

(iv) we further decompose the Hamiltonian (2.11) as

$$H = H_o + H' \quad , \tag{2.14}$$

where

$$H_o = H_{KIN} + H_{np} , \tag{2.15}$$

and

$$H' = H_p . \tag{2.16}$$

(v) we treat H' as a perturbation.

Why are the steps that bring us from (2.11) to (2.16) meaningful? The reason is that when we compute the Hamiltonian H_f, it is only the H_{np} piece that exhibits the behavior $\sim\mu^2 r$, where r is the relative distance between color charges. H_p is perfectly well behaved when $r \to \infty$ and is explicitly proportional to the "string tension" μ^2. Suppose now that μ^2 is very small; then we can treat H_p as a perturbation while H_{np} can never be so treated. It is sufficient to take r large enough. Thus in the limit of small μ^2 a perturbative structure is seen to arise naturally. An analysis of the spectrum yields $\mu^2 \simeq .3\ GeV^2$,[13] which on the hadronic standard is a fairly small number. Thus we obtain a perturbative structure in the number of quark pairs, which is readily seen to be strongly related to the experimentally observed features of hadrons.

2D. The $q\bar{q}$-spectrum[14]

I will not elaborate on this first step toward calculating strong interactions which is treated in J.L. Basdevant's[15] contribution. It suffices to say that with 5 parameters, the string tension μ^2 and the 4 quark masses $m_u=m_d$, m_s, m_c and m_b we calculate the meson spectrum and get agreement with experiments within 100MeV, perfectly consistent with the size of the corrections that we expect from the neglected part of the Hamiltonian, H'.[14]

2E. PCAC and Chirality[16]

In solving H_o in the $q\bar{q}$-sector, for the lowest-lying states S=0 and S=1 we can calculate their masses M as a function of the quark masses m_q. The result of such calculations is reported in Fig. 5. We obtain the remarkable fact that for $\frac{m_q}{\mu} \lesssim .16$ the

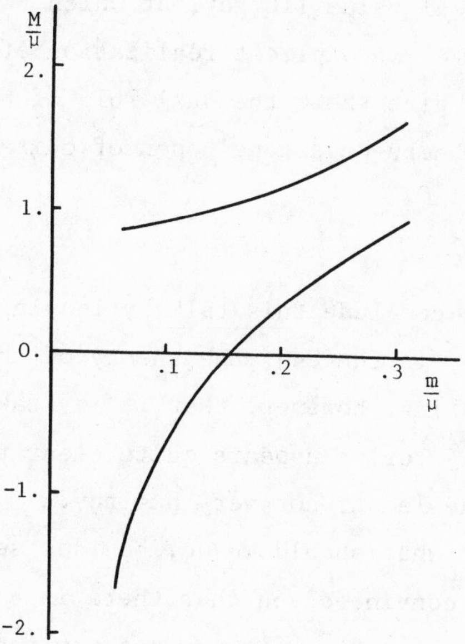

Fig. 5 The lowest lying states of a $q\bar{q}$-system [S=0,1] as a function of the quark mass, in units of μ, the string tension.

pseudoscalar meson (S=0) acquires a negative mass, while the vector state (S=1) mass remains positive. The occurrence of a negative mass triggers an instability, $q\bar{q}$-pairs condense and we obtain a new ground state, through a mechanism akin to BCS superconductivity. Such a peculiar phenomenon appears as a genuine consequence of our "two-dimensional" dynamics coupled to a "four-dimenionsal" spin, that engenders a magnetic interaction which is singular when $m_q \to 0$. Thus, when $m_q \to 0$ we have

(i) a BCS condensation;

(ii) a "constituent" quark mass m_c;

(iii) the pseudoscalar meson (π) becomes the Nambu-Goldstone boson of chiral symmetry.

It turns out that we can calculate m_q and we find[16] the value

$$m_q = 17 \text{ MeV} , \tag{2.17}$$

well below the critical value 110 MeV, at which BCS condensation occurs. Thus ACD gives an explicit realization of the long sought mechanism that would give the π the dual role of a $q\bar{q}$-bound state and of the (almost) Nambu-Goldstone boson of chiral symmetry.

3. CONCLUSION

I would like to conclude this talk by leaving the possibility open that QCD may be, in the end, the theory of the hadronic world, with the nagging feeling, however, that it may take a very long time before we know. For it appears quite clear that its present status is, to say the least, not very healthy.

In the meantime what should we do, besides seeking for the real QCD? I hope I convinced you that there is a definite advantage in keeping physics open, and in trying out new realizations of the fundamental pillars upon which rest all our understanding of hadronic interactions. In this direction, ACD does seem to be able to conquer the hadronic world with surprising ease and at an intellectual level which is not obviously inferior to QCD.

REFERENCES

1. See Aschman's contribution to Rencontre de Moriond (1982).
2. Carlson's report at the "Orbis Scientiae" Conference in Coral Gables (January 1982) contains the most articulate discussion of the subject that I know of.
3. Consult for instance F. Close's report at the EPS Lisbon Conference (1981) (to be published).
4. See J. Lee Franzini's Contribution to Recontre de Moriond (1982).
5. The first reports of the PETRA discovery were given on the occasion of the 1979 Fermilab Conference [see the Contributions by H. Newman, Ch. Berger, G. Wolf, and S. Orito, in Proceedings of the International Symposium on Lepton and Photon Interactions at High Energies, T.B.W. Kirk and

H.D.I. Abarbanel ed, Batavia (1979)] and were hailed as the discovery of the "gluon".

6. H.D. Politzer, Phys. Rev. Letters 26, 1346 (1973). D.J. Gross and F. Wilczek, Phys. Rev. Letters 26, 1343 (1981).
7. P. Castorina, G. Nardulli and G. Preparata, Phys. Rev. Letters 47, 468 (1981).
8. M.K. Gaillard and B.W. Lee, Phys. Rev. Letters 33, 108 (1974). G. Altarelli and L. Maiani, Phys. Letters 52B, 351 (1974).
9. G. Nardulli and G. Preparata, Phys. Letters 104B, 399 (1981).
10. See P. Renton's report at the EPS Lisbon Conference (1981), to be published.
11. See in this context Buchmüller's contribution to Recontre de Moriond (1982).
12. I am referring to the interesting results reported by the NA5 Collaboration at the CERN-SPS. See C. de Marzo et al., A study of deep inelastic hadron-hadron collision with a large acceptance calorimeter trigger (submitted to Phys. Letters).
13. G. Preparata, Phys. Lett. 102B, 327 (1981); ibid 108B, 187 (1982). G. Preparata, Nuovo Cimento A, 66, 205 (1981).
14. J.L. Basdevant and G. Preparata, The Structure of Strong Interactions in Anisotropic Chromodynamics I, Nuovo Cimento A, 67, 19, (1982). J.L. Basdevant, P. Colangelo and G. Preparata, The Structure of Strong Interactions in ACD II: The Meson Spectrum. Preprint BA-GT-81/23.
15. Cf. J.L. Basdevant's contribution to Recontre de Moriond (1982).
16. P. Castorina, P. Cea, P. Colangelo, G. Nardulli and G. Preparata, Chirality and its breaking in a new theory of hadrons, Preprint BA-GT-81/22.

LARGE MAGNETIC MOMENT EFFECTS AT HIGH ENERGIES AND COMPOSITE PARTICLE MODELS

A. O. Barut

The University of Colorado

Boulder, CO 80309

I. INTRODUCTION

Spin effects are generally thought to be inessential complications in particle physics. The discovery of large spin-dependent terms in recent pp and np experiments with polarized particles, which have been discussed extensively in the previous Proceedings of this conference,[1] shows that this is not the case. I shall show that spin effects not only play a major role in scattering at high energies, but also dominate the bound-state problems at short distances. Consequently any model or theory of hadrons, or any subconstituent model using spin-$\frac{1}{2}$-components (with magnetic moments) must take into account the following purely electromagnetic effect: The (electro)magnetic interactions become very strong at short distances and give rise to the formation of massive, narrow resonances between light spin-$\frac{1}{2}$ particles which subsequently decay very weakly by tunnelling. In the following we explain this phenomenon and extend it to the case of the neutrino, assumed to have a small anomalous magnetic moment.

If physicists have a new effect, they usually consider it to be the only one for the processes under consideration in order to determine its full scope and extent. Thus, assuming that the magnetic forces are the only cause of strong and weak interactions, we have attempted to construct a particle model based only on the two stable constituents; electron e, and neutrino ν_e. It turns out, as we shall see, that this is in principle possible, and the multiplet structures and symmetries of particle states can be regained from this simple starting point.

Because the theory is based on magnetic interactions, it is also a dynamical theory. The only parameter, besides the fine structure constant $1/\alpha = 137$ and the masses m_e and $m_\nu \approx 0$ of electron and neutrino, is the anomalous magnetic moment of the neutrino which we determine from scattering experiments and, independently, from the decay of $(e\text{-}\bar{\nu}_e)$-resonances. There are many tests of the theory: theoretically by the calculation of absolute values of the masses of hadrons and heavy leptons, e.g. M_ν, M_π, $M_{neutron}$, ..., their magnetic moments and lifetimes, some of which can already be estimated pretty well; experimentally, by some energy and angular distributions in scattering which can distinguish this model from others.

Now we discuss some of these questions in detail.[2]

II. THE DYNAMICS OF RELATIVISTIC SPIN INTERACTIONS

In quantum electrodynamics, the energy operator $\hat{H}$ describing the interaction of two spin $\frac{1}{2}$ fields ψ_1 and ψ_2, defined by

$$\int H\, d\vec{x} = \int d\vec{x}d\vec{y}\; \psi_1^+(x)\psi_2(y)\; \hat{H}\; \psi_1(x)\psi_2(y) \quad , \tag{1}$$

is given by

$$\hat{H} = (\overset{1}{\alpha}\cdot\vec{p}_1+\overset{1}{\beta}m_1)+(\overset{2}{\alpha}\cdot\vec{p}_2+\overset{2}{\beta}m_2)+V(\vec{r}_1-\vec{r}_2) \quad . \tag{2}$$

For minimal coupling the relativistic potential V is the well-known Breit-interaction

$$V(\vec{r}) = \frac{e_1 e_2}{4\pi} \frac{1-\overset{1}{\alpha}\cdot\overset{2}{\alpha}}{r} \quad . \tag{3}$$

The operator $\hat{H}$ acts on the 2-body functions Φ defined by

$$\Phi(\vec{x},\vec{y},t) = \Psi_1(\vec{x},t)\Psi_2(\vec{y},t-|\vec{x}-\vec{y}|) \ , \tag{4}$$

i.e. product of two fields separated by a retarded time. This formalism is relativistic and exact. We shall look for stationary solutions of $\hat{H}$ which constitute special solutions of the coupled field equations.

The Breit-interaction induces the well-known spin-orbit and spin-spin interactions. In addition, if the constituents have anomalous magnetic moments, we may introduce a new Pauli-coupling in the Lagrangian, i.e. a $\overline{\psi}\sigma_{\mu\nu}\psi F^{\mu\nu}$, and derive a number of new terms in the potential V(r) of the type

$$e_1 a_2 \ \overset{1}{\alpha}\cdot\frac{\beta\overset{2}{\sigma}\times\vec{r}}{r^3} \ , \ a_1 a_2 \ \frac{\sigma_1\cdot\sigma_2}{r^3} \ , \ldots \quad .$$

The full lengthy expressions are given elsewhere.[3]

The main point is that one must solve these relativistic potentials in a nonperturbative way in order to see the new effect. To illustrate this, consider the radial equation of a Dirac-particle e_1 in the field of a fixed charged magnetic dipole $\vec{\mu}_2 = \mu_2 \vec{S}_2$:

$$\begin{aligned}\left[\left(E-\frac{e_1 e_2}{r}\right)^2 - m^2\right]\Psi = \Big[&-\frac{d^2}{dr^2} + \frac{\ell(\ell+1)}{r^2} - \frac{4e_1\mu_2}{r^3}\vec{S}_2\cdot\vec{L} \\ &+ \frac{2e_1^2\mu_2^2}{r^4} + e_1\mu_2 \ S_{12} + ie_1 e_2 \vec{\alpha}_1\cdot\vec{E}\Big]\Psi \ ,\end{aligned} \tag{5}$$

where S_{12} is the well-known tensor force proportional to $1/r^3$. Even simpler is the equation of a spinless relativistic charged particle in the field of a fixed magnetic dipole:

$$\left(p^2+2m\,\frac{e_1e_2}{r} - 2e_1\mu_2\,\frac{\sigma_2\cdot\vec{L}}{r^3} + 2\,\frac{e_1^2\mu_2^2}{r^4}\right)u = (E^2-m^2)u \quad . \tag{6}$$

In the solutions of all these equations, the standard procedure has been (in atomic, nuclear and particle physics, e.g. charmonium models with spin terms) to solve the Coulomb (or scalar) part of the equation exactly, to drop the very singular $1/r^4$-part, and to treat the $1/r^3$-term as perturbation. One drops the $1/r^4$-term ("it is too small!"), because it cannot be treated as a perturbation at $r = 0$; but then with a $(-1/r^3)$-term the Hamiltonian is not essentially-self-adjoint! However if we plot the potential of eq. (6) we find a striking structure at short distances and the equation can be treated nonperturbatively. It is essential for the self-adjointness of the problem not to drop the $1/r^4$-term. One finds, in addition to the bound levels of the Coulomb (or scalar) part, a new region at short distances, of a deep narrow magnetic potential well, where very narrow very massive resonances can exist which can never be reached in any order of perturbation theory. The resonances very weakly decay through the large potential barrier. The orders of magnitudes of the two potential wells in Fig. 1 for the e^+e^--system are $r \sim 1/\alpha m$, $E \sim -\frac{1}{4}\,m\alpha^2$ for the positronium, and $r \sim \alpha/m \cong 2.8$ f , $E \sim m/\alpha \cong 70$ MeV for the magnetic resonances (called superpositronium). The resonance masses increase with increasing quantum numbers n and ℓ.

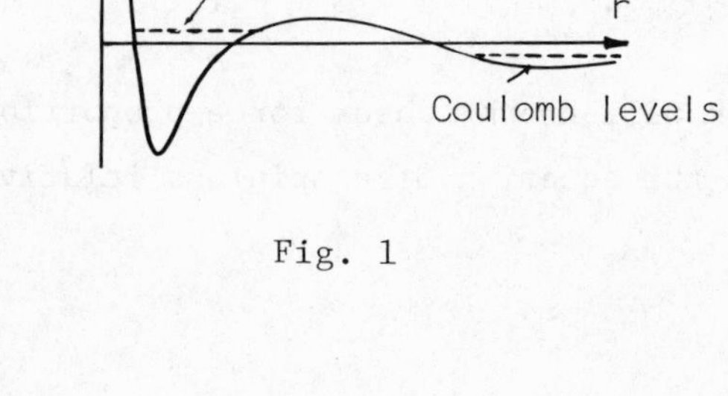

Fig. 1

Actually this resonance-phenomenon is well-known in physics, namely in the trapping of cosmic ray particles in the dipole field of the earth (Størmer's problem). One finds in the classical relativistic dynamics that these charge-dipole resonances occur at distances $r \sim \frac{g_E}{\ell_z} \alpha/m$ with a resonance energy $E \sim \frac{\ell_z}{g_E} m/\alpha$, where ℓ_z is the z-component of the angular momentum and g_E the g-factor of the earth, in units of $\hbar$. It is remarkable that we have an elementary particle version of these resonances, and these formulas for $g \sim 2$, $\ell_z \sim 1$, go over to the quantum formulas.

III. BUILDING-UP OF HADRONS AND LEPTONS FROM MAGNETIC RESONANCES

We now use the principle of magnetic resonances to build the states of heavy leptons and hadrons, and construct a model incorporating both the "strong" and "weak" interactions into electromagnetism.

First we discuss the construction of particle states and quantum numbers kinematically. The idea is what one may call the "chemical principle": electrons and neutrinos form semistable resonances which in turn act as building blocks, together with e and ν again, for new less stable resonances, and so on. We do not introduce, a new "quark" for every new species of hadrons.

The most stable hadron state is the proton which has the quantum numbers of the state $(e^+e^+e^-)$. Because the proton is so far absolutely stable, it is more economical to take the stable particles: proton, electron and neutrino as the starting point. (This is the stable particle PEN-Model). However, for some purposes it is convenient to think of proton as $(e^+e^+e^-)$, although the existence of such a stable state has not yet been shown dynamically.

Next we construct ν_μ and μ as

$$\nu_\mu = (\nu_e \nu_e \bar{\nu}_e) \quad , \quad \mu = (e\bar{\nu}_e \nu_\mu) \quad .$$

The first one is the analog of $(e^+e^+e^-)$ and may be also very stable.

(We do not yet know experimentally the precise mass and lifetime of ν_μ). The second represents μ from its decay products, and some preliminary magnetic resonance models do indeed calculate the lifetime[4] of μ. Continuing this process we obtain the lepton sequence ℓ: (e, ν_e), (μ,ν_μ), (τ,ν_τ), ... which are less and less stable.

All baryons eventually decay into a proton and pairs of leptons. We first build the neutron as $n = (pe\bar{\nu}_e)$, which by the way is the original neutron model of Pauli when he postulated the neutrino, and is an obvious one because the neutron decays into proton, electron and an antineutrino. It is interesting that this bound-state model of the neutron was not pursued further and was replaced by the phenomenological Fermi model, because "one did not know a deep enough potential well to bind e and $\bar{\nu}$ to the proton and to compensate the large magnetic moment of e". Because we now have a deep enough potential well, we must reconsider this model. We then proceed as in the case of lepton building, $\lambda_s = (p\mu\,\bar{\nu}_e)$, $\lambda_c = (p\nu_\mu\bar{\nu}_e)$.... and obtain a sequence b: (p,n), (λ_c,λ_s),... . All meson states are of the form

$$M = \ell \otimes \bar{\ell}' + \ell' \otimes \bar{\ell}$$

and all baryon states of the form

$$B = b \otimes \ell \otimes \bar{\ell} \quad .$$

It has been shown[5] that the standard SO(3) or SO(4)-classification of mesons and baryons can be obtained in this way with the following b and ℓ assignments:

Table 1. Correspondence between b, ℓ, $\bar{\ell}'$ and "quarks"

Flavor / color	b Green	ℓ Blúe	$\bar{\ell}'$ Red	<Q>
u	p	ν_e	e^+	2/3
d	n	e^-	$\bar{\nu}_e$	-1/3
s	λ_s	μ^-	$\bar{\nu}_\mu$	-1/3
c	λ_c	ν_μ	μ^+	2/3

In this form our b, ℓ and ℓ' are in one-to-one correspondence with the integrally charged Han-Nambu "quarks", enlarged from 3 to 9, which corresponds precisely to the socalled "color" degree of freedom. Note however that here "color" is not a new degree of freedom (we do not increase the number of "quarks"), but corresponds to a reordering of the basic constituents. In the construction of baryons according to eq. (8) one takes one constituent from each column of Table I. The average of the charges in each row of Table I is equal to the quark charges so that the rows correspond to the "quarks" (u,d,s,c). Thus all group-theoretic results are equally valid in the present theory.

This theory (unlike QCD for example) incorporates in principle not only the formation but also the decay of all mesons and baryons as narrow massive resonances, not stable bound states, since only e and ν (and perhaps proton) are stable and can be called "the elements". If they are truly indestructable the chain of model building stops there.

IV. SOME FURTHER CONSEQUENCES AND TESTS

1) The calculation of hadron masses from first principles

is surely one of the best tests of any theory of strong interactions. This is being pursued at the moment in QCD and other theories with a number of input parameters.[6] In the magnetic resonance model, the only parameter besides those of QED (m_e,α) is the neutrino magnetic moment a_ν, as we have noted above. For the (e^+e^--system) there are no new parameters at all. A relativistic WKB approximation to the 2-body resonances gives a mass spectrum for the neutral meson resonances (π^o,ζ,ω,..ψ,γ..): $M_n = A \frac{m}{\alpha} n^k$, where A is a constant of the order of 1, and k is between 2 and 3. This is in qualitative agreement with the meson spectrum to which spin-spin splittings must be added. Exact relativistic calculations will be reported elsewhere.

2) An upper limit for the ν-magnetic moment derived from the neutral-current process $e + \nu \rightarrow e + \nu$ is $a_\nu \cong 10^{-10} \mu_o$. This value also gives correct lifetimes for charged pions and muons assumed to be magnetic resonances of the type ($e\bar{\nu}_e$) and ($e\nu_e\bar{\nu}_\mu$) decaying via barrier penetration. Hence the Fermi-coupling constant can be related to the small magnetic moment of the neutrino.[4] The angular distribution of $e + \nu \rightarrow e + \nu$ would be a good test to distinguish this theory from the standard model.[7]

3) The neutron $n = (pe\bar{\nu}_e)$ is heavier than the proton, the n-p mass difference can be estimated as the magnetic energy of ($e\bar{\nu}_e$) in the field of the proton.

4) The deep magnetic potential well containing the light particles (e, ν_e) is a definite realization of the "bag" model which seems to be more appropriate here than in the quark model. This model is an approximation of the relativistic many-body dynamics. However, we do not have the problem of confinement (the constituents eventually leak out through the barrier), nor the spin-statistics problem.

There are many more remarkable features of the magnetic interactions which give a clear physical picture to strong and weak interactions. These have been discussed in the reviews and elsewhere.[2] We note only with reference to the large spin effects in polarized pp and np-experiments, that the effect due to the spin-spin term for pp has almost equal but the opposite sign as the np, because the neutron has a negative magnetic moment. This is indeed experimentally confirmed.[8]

Finally it has been shown[9] that the approximate internal symmetries of hadron processes, isospin and SU(3)-invariance (flavor)can all be reduced to the finite group of permutations, of the constituents (e, ν,...) in hadron collisions. There is no need to introduce Lie groups for these internal symmetries, hence no reason for gauging the flavor-groups.

REFERENCES

1. A. Yokosawa, in Orbis Scientiae, Studies in Natural Sciences, Vol. 12, (1977), Plenum Press, K. Abe, in Orbis Scientiae, Studies in Natural Sciences, Vol. 12 (1977), Plenum Press, A.D. Krisch, in Orbis Scientiae, Studies in Natural Sciences, Vol. 16 (1979), Plenum Press.
2. For some recent reviews see A.O. Barut, Surveys in High Energy Physics, 1, 113-140 (1981); in American Institute of Physics Conference Proceedings, Vol. 71 (edit. T. Seligman), p. 73-107 (1981); in Quantum-electrodynamics of Strong Fields (edit. W. Greiner), Plenum Press, 1982.
3. A.O. Barut and Bo-wei Xu, Derivation of Pauli Spin-orbit and Spin-Spin Potentials from Field Theory, preprint 1980 (to be published).
4. A.O. Barut and G.L. Strobel, Composite Models of Pion and Muon and a Calculation of the Fermi Coupling Constant from their Decay (to be published).
5. A.O. Barut and S.A. Basri, Connection between the Stable Particle Model and the Integrally Charged Quark Model (to be published).

6. See the contribution of C. Rebbi and D. Gross, G. Preparata, these proceedings.
7. A.O. Barut, Z.Z. Aydin and I.H. Duru, An Upper Limit on the Magnetic Moment of the Neutrino (to be published).
8. D.G. Crabb et al, Phys. Rev. Lett. 43, 983 (1979).
9. A.O. Barut, Physica (in press).

THE NEXT LAYER OF THE ONION?

COMPOSITE MODELS OF QUARKS AND LEPTONS

O.W. Greenberg

University of Maryland

College Park, Maryland 20742

ABSTRACT

We give motivations to consider composite models of quarks, leptons, and other particles now considered elementary. The increase in the number of quarks and leptons associated with the discovery of the higher generations, and the large number of arbitrary parameters in the present description of elementary particle physics are reasons to consider composite models. There is no evidence for extended structure of quarks and leptons. Present data provides the upper limit $mr \sim m/M \lesssim 10^{-6}$ for the electron and muon, where m is the lepton mass, r its size, and M is the mass scale associated with binding. Chiral symmetry is a possible way to keep this number small. Some composite models are sketched.

I want to introduce this session on composite models of quarks and leptons by giving the motivation for considering a new level of compositeness in the architecture of matter.

Let me remind you of the situation in elementary particle physics in the 1960's. Data for hadronic spectroscopy had accumulated, and a number of sets of baryonic and mesonic resonances had been found. I will concentrate on the eight positive-parity baryons which were known at that time, namely p, n, Λ^0, $\Sigma^{+,0,-}$, and $\Xi^{0,-}$. Attempts were made to give a unified description of these eight positive parity baryons. Among these attempts were global symmetry, which treated the eight baryons on an equal footing, and gave the mass formula

$$\frac{N+\Xi}{2} = \frac{\Lambda+3\Sigma}{4} .$$

In this mass formula I let the symbol of the baryon stand for its mass. This mass formula was in contradiction with the observed masses of these particles. There was also the eightfold way or SU(3) model, which led to the mass formula

$$\frac{N+\Xi}{2} = \frac{3\Lambda+\Sigma}{4} .$$

This mass formula was in rather good agreement with experiment. Both of these theories of the eight positive parity baryons can be considered as analogs of the present efforts to give grand unified theories of the quarks and leptons. We now believe that one of these models, global symmetry, was incorrect; and the other, the eightfold way or SU(3), was correct. However, the true understanding of the eightfold way was provided by the quark model, which gave an interpretation of the eightfold way based on a deeper level of compositeness. It seems plausible that we are now in the situation of the early 1960's, and that some of the presently considered grand unified theories will turn out to be correct, others incorrect, but -and this is the important point here- that the true understanding of the nature of matter in the small will depend upon discovery of a further level of compositeness. It is clear that the present grand unified theories, being local gauge theories, are more sophisticated than the global symmetries of the 1960's; nonetheless, the analogy holds, and the true understanding may come from a deeper level of structure.

Now I want to give a quick review of the physics of the small, based on four known levels of structure; and I also want to introduce the possible fifth level of structure which is the subject of this morning's session. In each case I want to identify the composite objects, the constituents, the binding forces which bind the constituents into the composites, and in several of the cases, the reinterpretation of the binding forces at a given level of structure as residual effects of more fundamental binding mechanisms at a deeper level of structure. This review of physics of the small is given in Table I.

Table I

Physics of the Small

	Constituents	Binding forces	Reinterpretation of binding forces
1. Molecules	atoms	ionic, covalent, van der Waals	residuals of Coulomb
2. Atoms	nuclei, electrons	Coulomb	fundamental
3. Nuclei	protons, neutrons	Yukawa	residual of QCD-string breaking
4. Hadrons	quarks, antiquarks	QCD confinement	?
5. Quarks, leptons	fermions fermions + bosons (supersymmetry?)	abelian -- nonabelian --	U(1) U(1)-dual SU(n) SO(n) spin(n)

At each level, clues were discovered by studying the given level, but fundamental progress was made by going to a deeper level of structure. Philosophic, esthetic and theoretical arguments alone, however useful, do not suffice. Concrete experimental evidence is needed. Considerations of economy, for example, counting the number of fundamental fermions, have some significance in our search for an understanding of a deeper level of compositeness. Nonetheless, counting the number of states by itself is not a decisive criterion. The prescientific view of the nature of matter, which comes to us from antiquity, was that all matter was made of four elements: earth, air, fire and water. The first scientific view of the nature of matter, that matter is constructed of molecules, had, as the fundamental constituents, vastly more objects than the four elements of the Greek philosophers; nonetheless, one would have to say that the theory based on molecules is closer to the truth than the model of antiquity. Another example in which deeper understanding produced more rather than fewer constituents is the transition from the period when all matter was considered to be composed of protons and electrons to the eight-fold way. Even the quark model, successful as it is, has more than two constituents.

In addition to a large number of fundamental objects, and the larger symmetry groups considered in grand unified models, there is also a large number of parameters, including fermion masses and mixing parameters, parameters associated with the Higgs potential, the gauge coupling, etc. All of these facts make it very appealing to look for a more fundamental description and a deeper level of compositeness.

The signals for compositeness at the hadronic level were the large number of states, and the observation of hadronic form factors not associated with radiative effects at the hadronic level. This latter signal for compositeness is clearer to us now than it was then: at the time many attempts were made to understand the

Table II

The Three Generations of Quarks and Leptons

I	II	III	
u_i	c_i	$(t_i?)$	i=1,2,3 is
d_i	s_i	b_i	the color index
ν_e	ν_μ	ν_τ	
e^-	μ^-	τ^-	

hadronic form factors as associated with radiative effects of the mesonic clouds in a theory of the strong interactions. It is not always easy to separate radiative effects at a given level from the effects of compositeness at a deeper level.

At present, we have evidence for the multiplicity of quarks and leptons that is a likely signal of compositeness. The three generations of quarks and leptons are given in Table II.

In addition to this large number of quarks and leptons, there are also the photon, $W^\pm$, Z^0, X, Y, Higgs bosons, and so forth. It is worth pointing out that the presently discovered spectrum of quarks and leptons is very peculiar. In Table III, I give the masses of the three generations of quarks and leptons.

Table III

Peculiar Spectrum of Quarks and Leptons

(Masses in MeV)

u = 4.3	C = 1600	t > 18,000
d = 7.5	S = 150	b = 4,500
$\nu_e < 6 \times 10^{-5}$	$\nu_\mu < 1$	$\nu_\tau < 250$
e = 0.51	μ = 106	τ = 1,850

In contrast to the signal of the multiplicity of quarks and leptons, there is no evidence at all for form factors of quarks and leptons. On the contrary, at present there are strong experimental limits on the size of quarks and leptons, particularly on the size of the electron and muon. Gordon Shaw will present an analysis of the present limits on structure for quarks and leptons. The electron and muon, have size $r \lesssim 10^{-16}$ cm., which corresponds to a mass scale $M = r^{-1} \gtrsim 200$ GeV. I will take this limit as a clue regarding the nature of the bound states which constitute the quarks and leptons.

In the present session, we are emphasizing the search for compositeness at the largest possible distance or, equivalently, at the smallest possible mass scale. We want to do realistic physics which will have visible experimental consequences in a finite time. If the scale of compositeness occurs only at the grand unified mass $\sim 10^{14}$ GeV or higher or the Planck mass of the order of 10^{19} GeV, then the compositeness will be invisible for the forseeable future, and there will be no direct evidence for it. We are particularly interested in the possibility that composite structure of quarks and leptons will be associated with mass scales which can fill in the gap between the W mass ~ 80 GeV/c^2 and the grand unified mass of order 2×10^{14} GeV/c^2. It seems unlikely to us that no new physics will occur over this enormous range of masses.

Now I will review the clues for compositeness, at the same time giving a survey of some of the models of composite quarks and leptons which have been proposed. The first clue is the spectrum of quarks and leptons and other particles. The spectrum includes the counting of states, the assignment of quantum numbers, the peculiar pattern of masses of the quarks and leptons, and finally the repetition of generations. In surveying models, one can tabulate several of their properties as shown in Table IV. A second clue is the small value of the product of the mass of the particle times its characteristic size. These values are

Table IV

Properties of Some Composite Models

Model	Color, flavor at constituent level	Connected with grand unified theory	Fermions(F) Bosons(B)
Reinterpretation of the standard model	Yes	No	F,B
Color-flavor	Yes	Yes	F,B
SO(10) or Spin(10)	No	Yes	F
Harari-Shupe	No flavor no color(color)	No	F

tabulated in Table V, together with some other parameters of interest for the hydrogen atom, the proton and the electron. The value of mr = m/M, where m is the mass of the particle and M is the characteristic mass of the binding scale, is very small for the electron. This is an unprecedented situation in elementary particle physics; such small values of this dimensionless quantity have not appeared before. One can ask whether this small number is connected to other small mass ratios of relevance in elementary particle physics. It seems likely that the smallness of this parameter is not accidental, but rather is the consequence of an almost exact

Table V

Parameters for Hydrogen, the Proton, and the Electron

	r(cm)	M(MeV)	$\rho(\frac{g}{cm^3})$	$\frac{Mc}{\hbar} r$
H	0.5×10^{-8}	940	1	2×10^5
p	2×10^{-14}	940	10^{17}	1
e	$<10^{-16}$	0.5	$>10^{21}$	$>2 \times 10^{-6}$

symmetry. The most likely symmetry which would produce this result is chiral symmetry. In the chiral symmetry limit, the fermions have mass zero. Various mass generation mechanisms, to be discussed by Stuart Raby at his talk later this morning, provide ways of generating small masses by breaking the chiral symmetry. In the absence of chiral symmetry, the likely effective mass for the composite states would be of the order of the mass associated with the binding scale, which as we discussed above, is greater than 400 GeV. Clearly, something like chiral symmetry is essential to avoid such large masses. (Another possible symmetry which might avoid large masses is supersymmetry.)

The model which can be considered the minimal composite model is a reinterpretation of the standard electroweak model using the same Lagrangian as that model but using a confining phase of the model rather than the usual Higgs phase. In this confinement phase, the SU(2) gauge interaction becomes strong and produces permanently confined left-handed fermions which are composites of the left-hand elementary Lagrangian fermions, together with the Higgs field. The right-handed fermions in this model are elementary. This model has been discussed by Dimopoulos, Raby and Susskind, who introduced the word "complementarity" for the possibility that a model with the Higgs field in the fundamental representation could have a continuous variation between the Higgs phase and the confinement phase. The model is also discussed by 't Hooft and was developed in detail by Abbott and Farhi. The model has the same number of phenomenological parameters in the low energy effective Lagrangian as the standard model. This model realizes, in concrete composite models, suggestions put forth by Bjorken, and by Hung and Sakurai, concerning the possibility to obtain gauge theory results without using the standard model.

The color-flavor models are typified by the model of Pati and Salam in which the first two generations of quarks and leptons are unified into a $4 \times 4 = 16$ multiplet of particles with the horizontal direction corresponding to color and lepton number as the fourth color and the vertical direction corresponding to the up and down flavors of the first two generations. It is natural to take this model apart and to consider that the flavor degree of freedom might be carried by one constituent and the color and lepton number degree of freedom by another constituent. Indeed, Pati and Salam made this suggestion in the same article in which they discussed their grand unified model. Since the weak interactions, which act on the flavor degree of freedom, have chiral character, it is natural to assume that the flavor carrying constituent is a fermion. A simple model can then be constructed by taking the other constituent carrying color and lepton number to be a scalar. That gives the model of the kind indicated below:

$$\begin{matrix} u_i & & F_u C_i & \\ d_i & & F_d C_i & \\ \nu_e & \sim & F_u C_4 & i = 1,2,3 \\ e & & F_d C_4 & \end{matrix}$$

Models of this kind were considered by Pati and Salam, as mentioned above, as well as by Pati, Salam and Strathdee, Greenberg, Matumoto, Ne'eman, Greenberg and Sucher, Fritzsch and Mandelbaum, Visnjic-Triantafillou, Veltman and Derman. Other particles can also be constructed as composites, for example the W bosons have the structure $W_\mu^+ \sim \bar{F}_d \gamma_\mu F_u$, the color gluons can be constructed as $G_\mu^a \sim i\, C^+ \lambda^a\, \partial_\mu C$, the Higgs mesons can be constructed as $\bar{F}F$, etc. The electric charge assignments for this model can be chosen with a one parameter degree of freedom, as indicated in Table VI. All of these models allow quark-line graphs of which we will give one typical illustration, as ways of representing the interaction of the composites. A variant of this model is one in which a third constituent is introduced to carry the generation degree of freedom. In this case, the quarks and leptons are three-body

Table VI

Electric Charges in the Color-Flavor Model

	Q/e	Color	Lepton No.
F_u	q	1	0
F_d	q^{-1}	1	0
C_i	$\frac{2}{3} - q$	3	0
C_4	$-q$	1	1

constituents, and one can choose all three of the constituents to be fermions. A model of this kind was first considered by Pati and Salam. Here the structure of the quarks and leptons is

$$\begin{matrix} F_u \\ \\ F_d \end{matrix} \times \begin{matrix} C_i \\ \\ C_4 \end{matrix} \times \begin{matrix} G_e \\ G_\mu \\ G_\tau \end{matrix}$$

In these models, flavor and color exist at the constituent level.

Another type of model is typified by a model motivated by the grand unified theory SO(10). Here, following Mansouri, the spinor representations, the 16 and $\overline{16}$ of SO(10) can be constructed with five fermions f_α and the corresponding antiparticles $\bar{f}_\alpha$. The construction makes the quarks and leptons as five body bound states of these fermions. In each bound state, either f_α or $\bar{f}_\alpha$ occurs

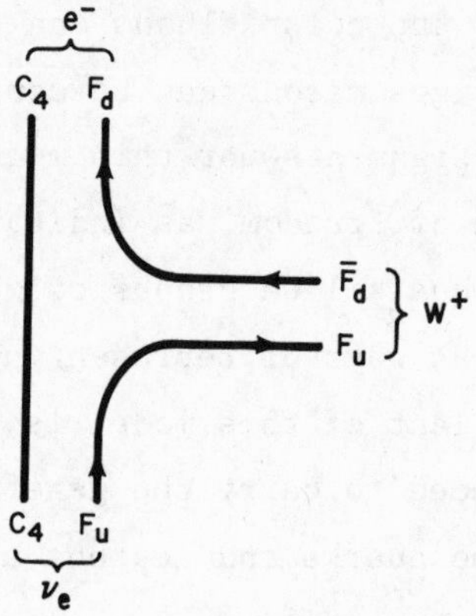

Fig. 1 Typical Weak Interaction Vertex

Table VII

States in an SO(10) Model

$f_1\ f_2\ f_3\ f_4\ f_5$	1
$\bar{f}_1\ f_2\ f_3\ f_4\ f_5, \ldots$	5
$\bar{f}_1\ \bar{f}_2\ f_3\ f_4\ f_5, \ldots$	10
$\bar{f}_1\ \bar{f}_2\ \bar{f}_3\ f_4\ f_5, \ldots$	10
$\bar{f}_1\ \bar{f}_2\ \bar{f}_3\ \bar{f}_4\ f_5, \ldots$	5
$\bar{f}_1\ \bar{f}_2\ \bar{f}_3\ \bar{f}_4\ \bar{f}_5$	1
	32

once for each α. We then generate the 32 states, given in Table VII. Since SO(10) is a rank five group, there are five (abelian) quantum numbers associated with SO(10). One can assume that the constituents are representations of $[U(1)]^5$. Choosing the five quantum numbers as given in Table VIII provides the proper quantum numbers for the composite states which appear as the quarks and leptons of the 16 and $\overline{16}$. The quantum numbers for the constituents are given in Table VIII. In this table, the electric charge is given by the Gell-Mann-Nishijima formula: $Q = I_3^W + \frac{1}{2} Y^W$.

Table VIII

Quantum Numbers for the Constituents

Constituent	I_3^W	Y^W	Q	I_3^C	Y^C	B-L
f_1	0	1/3	1/6	-1/4	-1/6	1/3
f_2	0	1/3	1/6	1/4	-1/6	1/3
f_3	0	1/3	1/6	0	1/3	1/3
f_4	1/4	-1/2	0	0	0	0
f_5	-1/4	-1/2	-1/2	0	0	0

The composites in the 16 spinor representation of SO(10) now have the quantum numbers given in Table IX.

Table IX

Quantum Numbers For The Composite States Which Form the 16 of SO(10)

Composite state		I_3^W	Y^W	Q	I_3^C	Y^C	Particle
$\bar{f}_1 f_2 f_3$					1/2	1/3	
$f_1 \bar{f}_2 f_3$	$\bar{f}_4 f_5$	-1/2	1/3	-1/3	-1/2	1/3	d_L
$f_1 f_2 \bar{f}_3$					0	-2/3	
$\bar{f}_1 f_2 f_3$					1/2	1/3	
$f_1 \bar{f}_2 f_3$	$f_4 \bar{f}_5$	1/2	1/3	2/3	-1/2	1/3	u_L
$f_1 f_2 \bar{f}_3$					0	-2/3	
$\bar{f}_1 \bar{f}_2 \bar{f}_3$	$f_4 \bar{f}_5$	1/2	-1	0	0	0	ν_L
$\bar{f}_1 \bar{f}_2 \bar{f}_3$	$\bar{f}_4 f_5$	-1/2	-1	-1	0	0	e_L^-
$\bar{f}_1 f_2 f_3$					1/2	1/3	
$f_1 \bar{f}_2 f_3$	$f_4 f_5$	0	-2/3	-1/3	-1/2	1/3	d_R
$f_1 f_2 \bar{f}_3$					0	-2/3	
$\bar{f}_1 f_2 f_3$					1/2	1/3	
$f_1 \bar{f}_2 f_3$	$\bar{f}_4 \bar{f}_5$	0	4/3	2/3	-1/2	1/3	u_R
$f_1 f_2 \bar{f}_3$					0	-2/3	
$\bar{f}_1 \bar{f}_2 \bar{f}_3$	$\bar{f}_4 \bar{f}_5$	0	0	0	0	0	ν_R
$\bar{f}_1 \bar{f}_2 \bar{f}_3$	$f_4 f_5$	0	-2	-1	0	0	e_R^-

These five-body bound states can have spin -1/2; however, even without orbital excitation, they can have higher spins, up to spin 5/2. As discussed above, chiral symmetry can keep the spin -1/2 composites light. The Weinberg-Witten theorem forces higher spin composites to be very massive, of order of the binding mass scale, which is equal to the inverse of the size, 1/r. The Weinberg-Witten theorem is the statement that massless particles with spin greater than 1/2 cannot carry a charge associated with a conserved Lorentz covariant vector. If there is a Lorentz covariant energy momentum tensor, the massless particles must have spin < 1. Provisionally, we can take this theorem as a justification for ignoring higher spin composites; based on the assumption that such composites will have very large mass. The detailed dynamical basis of this theorem is not yet understood; this is an important theoretical question.

I want to emphasize that one cannot get generations from space excitations because the mass scale associated with space excitations is of the order of M, the binding mass scale, therefore much, much too large. Among the mechanisms which can give generations are additional $f_i\bar{f}_i$ pairs, additional scalars added to the fundamental configuration, and additional constituents, for example if there is an f_4' and an f_5' and their associated antiparticles, then the combinations $f_1f_2f_3$ together with either f_4f_5, $f_4'f_5$, f_4f_5', or $f_4'f_5'$ can give four generations.

Finally I come to the rishon model of Harari and independently of Schupe. The quantum numbers of this model are given in Table X.

Table X

Fundamental Particles in the Rishon Model

	Q/e	(Color)	(Hyper Color)
T	1/3	(3)	(3)
V	0	$(\bar{3})$	(3)

The color and hypercolor assignments in parentheses were not present in the original version of this model. Here the states are as given in Table XI.

Table XI

Quarks and Leptons in the Rishon Model

	TTT		e^+
TTV,	TVT,	VTT	u_i
VVT,	VTV,	TVV	$\bar{d}^i$
	VVV		ν_e

Initially the authors suggested that the three permutations, TTV, TVT, and VTT, of the orders of the T's and V's in the state, should be associated with the color degree of freedom. This possibility fails for two reasons: first, these states are not linearly independent if the T's and V's have either Bose or Fermi commutation relations. Secondly, as J.C. Pati has pointed out, there is no binding mechanism with respect to which these combinations of rishons are neutral. Also, there is no way to bind these combinations and not other combinations which are not wanted in the model, such as $TT\bar{T}$, $T\bar{T}$, etc. Thus the attempt to derive color in the rishon model fails. The model has been revised by Harari and Seiberg, to introduce color and also hypercolor as degrees of freedom carried by the rishons as indicated in parentheses in Table X. This revised model avoids the objections just mentioned, but may have other difficulties. In the revised model, as in the original model, the proton can decay by rearrangement of rishons into antirishons and this decay rate might be too large if the binding scale is of the order of a TeV.

Gary Feinberg has pointed out an amusing consequence of the rishon model and the SO(10) model concerning baryon asymmetry. In the rishon model, hydrogen has an equal number of rishons and antirishons. Thus baryon asymmetry in the rishon model is not the question of why there are more rishons than antirishons; there are not, the number should be equal. Rather, the problem is why the rishons aggregate primarily in u-quarks and the antirishons aggregate primarily in d-quark and electrons.

We need phenomenological studies of signatures for compositeness and of experimental distinction between different models. Some work in this direction has already been done and will be reviewed by Gordon Shaw; more work needs to be done. We should be particularly alert for small effects of compositeness which are visible at low energies; as well as effects which occur at high energies, of which I will mention only one, namely that the rearrangement of the constituents can lead to large lepton and quark multiplicities.

Composite models allow a different kind of unification of the interactions and of the types of matter than occur in grand unified theories.

The outstanding theoretical challenge for composite models of quarks and leptons is to calculate the quark-lepton mass spectrum. To do this we must develop a new intuition about zero-mass composite states, which occur in the presence of unbroken chiral symmetry. We also need to go beyond estimates of effective interactions of composites based on dimensional analysis. In general, we need to derive the properties of the field theory of the composites from the field theory of the constituents. The dynamics of this is not yet understood and is an important subject for future theoretical investigation.

This work was supported in part by a grant from the National Science Foundation and by a Faculty Research Grant from the University of Maryland. It is a pleasure to acknowledge discussions with Shmuel Nussinov, Jogesh Pati, and Joseph Sucher.

REFERENCE

1. Early references and some later ones are collected in O. W. Greenberg, Resource Letter Q-1: Quarks, University of Maryland preprint PP 82-096 (1981). Recent references can be traced from M. Peskin, Compositeness of Quarks and Leptons, Cornell University preprint CLNS 81/516 (1981).

COMPOSITE FERMIONS: CONSTRAINTS AND TESTS

Gordon L. Shaw and Dennis J. Silverman

University of California

Irvine, California 92717

(Presented by Gordon L. Shaw)

ABSTRACT

We review the substantial constraints and tests of models of composite leptons and quarks set by experiment and theory. New limits on the anomalous magnetic moment of the tau and of the quarks are presented.

INTRODUCTION

There have been <u>many</u> recent papers in which leptons and quarks are considered to be composites of more fundamental constituents. Perhaps the main motivation for this large effort, and the only present indication for compositeness, is the large number of "elementary" fermions (three SO_{10} $\underline{16}$'s). In this paper, we review a number of possible indications or tests for compositeness. Some of these place very strong constraints on such composite models, e.g., the phenomenal agreement between experiment and present theory

---

Technical Report No. 82-21

(QED plus hadronic) for both the electron[1] and muon[2] anomalous magnetic moments F_2: a part in 10^{10} for F_2^e and a part in 10^8 for F_2^μ. In addition there are theoretical constraints such as those of 't Hooft[3] which are discussed in detail elsewhere in this Session.[4] The topic of composite fermions has thus developed to a stage such that all serious model builders should take all these constraints into account in a substantive manner and/or have some novel testable predictions.

Most efforts have naturally been directed toward the symmetry aspects of composite models. The very difficult aspect of the necessary dynamics has been largely neglected. The above mentioned constraints on F_2 indicate that if the e and μ are indeed composite, a "new" type of dynamics with $mR \ll 1$ (where m and R are the composite mass and size) is involved as compared to, e.g., that of hadrons composed of quarks with $mR > 1$. We review some recent relativistic calculations[5] for a "point-like" composite fermion in which a composite mass of approximately one tenth of the sum of the constituent masses was achieved. This work illustrates the difficulties in obtaining a dynamical calculation with heavy constituents of a composite e or μ. Perhaps models with light constituents with approximate chiral symmetry are a better approach. However, in this case we have no full dynamical calculation.[6]

Although nowhere in the same ballpark, useful limits have been set[7] on F_2 for the τ and for the quarks from PETRA data on e^+e^- annihilation. These limits as well as a new low-energy test[7] are presented. To illustrate the usefulness of these tests, we discuss a unified SO_{14} model[8] in which the SO_{10} electron and muon families are elementary and the tau family is composite.

POSSIBLE INDICATIONS FOR COMPOSITENESS: TESTS AND CONSTRAINTS

Here we list a number of possible indications for composite fermions and briefly comment on associated tests and constraints.

1. <u>Families</u>: The observed left-handed fermions (along with

the expected top quark) fit into three SO_{10} $\underline{16}$ irreducible representations (irreps), the electron family f_e (e, ν_e, u, d), the muon family f_μ (μ, ν_μ, s, c) and the tau family f_τ (τ, ν_τ, b, t). Many composite models predict lots of new fermions. For example, the SO_{14} model[8] presented below has a composite SO_{10} $\underline{16}$ and two $\underline{144}$'s not far above the tau family.

2. Excitations: Here, one would look for excited leptons, ℓ^*, and quarks, e.g., an ℓ^* which could decay rapidly into $\ell + \gamma$. This has been discussed in detail[9] and limits are obtained[10] from measurements of $e^+e^- \to \mu^+\mu^-\gamma$, $\mu^+\mu^-\gamma\gamma$ and $\gamma\gamma$ final states. (This latter process $\gamma\gamma$ would be modified by the exchange of a virtual e^* in the t-channel). These limits place bounds on the effective coupling λ of $\ell^*\ell\gamma$ and the mass of ℓ^*. For example,[10] for $\lambda_{e^*e\gamma} < .5$, $m_{e^*} > 40$ GeV.

3. Actual Constituents: If there is a new confining force among the constituents, and it is associated with a nonabelian unbroken gauge symmetry, then the constituents are probably confined as in the case of quarks in unbroken[11] QCD.

4. Moments and Transitions: As noted above, the incredible agreement between theory and experiment (a part in 10^{10}) for F_2^e and (a part in 10^8) for F_2^μ imposes very strong constraints for composite models for the e and μ. Nonrelativistic models could be ruled out.[12] However, relativistically, these constraints can be satisfied.[13] The electromagnetic vertex of a spin 1/2 fermion with charge +e and mass m has the usual form

$$M = i\,e\,\bar{u}(\gamma_\mu F_1 + (i\,\sigma_{\mu\nu}/2m)q^\nu F_2)u \quad , \tag{1}$$

where $F_2(0)$ is the anomalous magnetic moment in units of e/2m. Now, F_2 = "usual" corrections (electroweak + hadronic) + "new" composite corrections (${}^cF_2$). It was shown[13] in a variety of derivations that in general the corrections

$$^cF_2(0) \sim O(m/m_c) \quad , \tag{2}$$

where m_c is the "constituent" mass. Note that $F_2(q^2) \simeq F_2(0)$ for $|q^2| \ll m_c^2$. Furthermore for special cases,[13]

$$^cF_2(0) \sim O(mm_f/m_s^2) \quad , \tag{3}$$

for $m_f \ll m_s$, where f is a spin 1/2 constituent and s is a spin 0 or spin 1 particle. While clearly, it is in principle possible to implement these constraints, no one has presented such a dynamical calculation for the e or μ for which we must have[1,2]

$$^cF_2^e < 10^{-10} \quad ,$$

$$^cF_2^\mu < 10^{-8} \quad . \tag{4}$$

As reviewed below, much weaker yet useful limits have been determined[7] for F_2 of the tau and of the quarks. Additional strong constraints come from transition moments.[14] For example, $\mu \to e + \gamma$ has a branching ratio $< 2 \times 10^{-10}$, which tends to rule out models in which the μ is a radial excitation of a composite e.

5. Complicated "Effective Forces": This is a subjective feature, but perhaps a useful one. In retrospect, the enormously complicated nucleon-nucleon potentials determined phenomenologically in the 1950's with up to 40 parameters are clearly an indication that the nucleon and exchanged mesons are not all "elementary" particles. Conversely, the "beauty and simplicity" of the gauge theories argue strongly against the photon being composite, i.e., it is hard to imagine that QED is just an "effective" low-energy phenomenology.

6. Form Factors: Here we expect q^2 dependence of the order q^2/m_c^2 where again m_c is the constituent mass. For example,

$F_1(q^2)$ in (1) might be

$$F_1(q^2) = 1 - O(q^2/m_c^2) \quad . \tag{5}$$

Of course, the QED and hadronic contributions must be corrected for. This is now possible for the QED and QCD corrections to F_1 for the quarks and no additional composite structure is needed (see below).

7. High Spin: The discovery of a spin $\frac{3}{2}$ quark or lepton would be a likely candidate for a composite fermion. Outside of supersymmetric theories, a spin $\frac{3}{2}$ field would not be renormalizable, unlike the present electroweak and QCD gauge theories with spin 1/2 fermions. Thus such higher spin objects would generally be considered to be composite. Alternately, a composite model should have no low-lying spin $\frac{3}{2}$ states since these are contrary to experiment.

8. 't Hooft's Conditions[3]: Implementations of these constraints are discussed in detail later in this session.[4] In general they can be quite restrictive. However, 't Hooft's condition on absence of anomalies can be ensured by choosing a symmetry group such as SO_N in which all the irreps are anomaly free. His second "decoupling" condition which restricts the spectrum of light composite fermions now seems to be somewhat controversial.[15]

DYNAMICS

We review here the relativistic calculation[5] (via a strong, short-ranged force) of a point-like composite fermion having about one-tenth the total mass of the constituents and zero anomalous magnetic moment. The successes include obtaining a Dirac moment which was a factor of five larger than the constituent moment, (in contradiction to a nonrelativistic picture,[12] but of course, in agreement with the relativistic description), the absence of any other bound states in any spin state, and a g_A/g_V roughly unity. The difficulties include: a) the parameters had to be finely tuned and b) for smaller bound state masses, $F_2(0) = 0$ could still be "tuned" but g_A/g_V became small.

The relativistic equation[16] used is based on the constituent fermion field equation

$$(\not{p}-m_f)\psi(p) = \frac{g_f}{(2\pi)^4}\int d^4q\,\not{A}(p-q)\psi(q) \quad , \tag{6}$$

where g_f is the coupling of the fermion to the gauge field potential A_μ, see Fig. 1. We study the bound state wave function $\Psi_{B,\lambda}(\underline{s})$ by taking the matrix element of (6) between the bound state $|\underline{B},\lambda\rangle$ and the constituent scalar $|\underline{s}\rangle$:

$$\Psi_{\underline{B},\lambda}(\underline{s}) \equiv \langle \underline{s}|\psi(B-s)|\underline{B},\lambda\rangle \quad . \tag{7}$$

Insert a complete set of intermediate states $|n\rangle\langle n|$ between $\not{A}$ and ψ and keep only the heavy (on-shell) scalar particle itself, $|s'\rangle$. Then

$$(\not{B}-\not{s}-m_f)\Psi_{\underline{B},\lambda}(\underline{s}) = \frac{g_s g_f}{(2\pi)^3}\int \frac{ds'}{2\omega_{s'}}\,\frac{(\not{s}+\not{s}')}{(s-s')^2-m_g^2}\,F(s-s')\Psi_{\underline{B},\lambda}(\underline{s}') \quad , \tag{8}$$

with $\omega_s^2 = s^2 + m_s^2$ and m_g the mass of the exchanged vector particle. A form factor

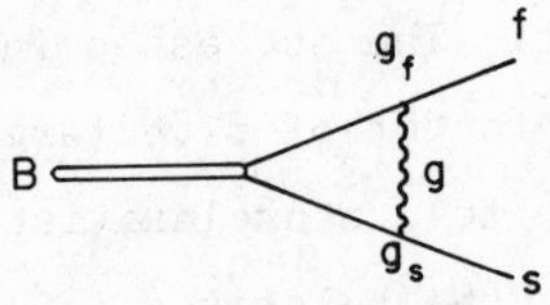

Fig. 1 Bound state composite fermion B with constituent fermion f and scalar s which couple with strength g_f and g_s to the massive gauge vector boson g.

$$F(s-s') = \frac{\Lambda^2}{\Lambda^2-(s-s')^2} \tag{9}$$

was included to make the resultant equations Fredholm. Such a form factor arises naturally if the scalar is itself a composite of two fermions. We then go to the rest frame of the bound state, $B = (m_B, 0)$ and perform a partial-wave analysis, reducing the equations to two coupled integral equations in s for the upper and lower components. These equations for $j^P = 1/2^+$ are solved numerically (by matrix diagonalization) and the coupling $g_s g_f$ adjusted to give the various m_B. Since[5] m_B is determined by this formalism to be $|m_s - m_f| < m_B < m_s - m_f$, m_f is chosen approximately equal to m_s. Fig. 2 shows the results for $G_M = 1 + F_2$, G_M/m_B and

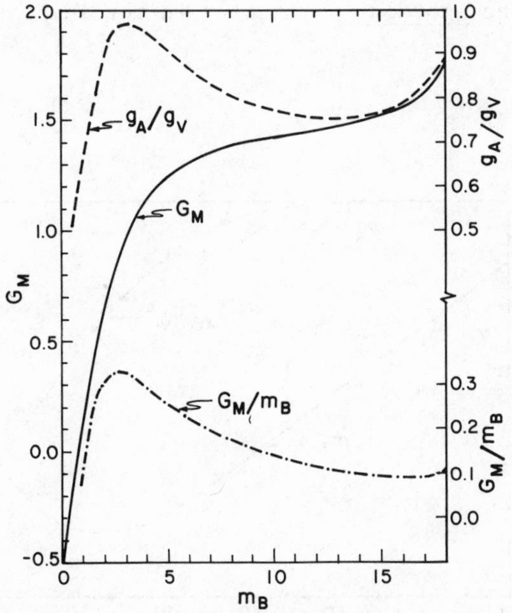

Fig. 2 Plots of $G_M = 1 + F_2$, G_M/m_B and g_A/g_V for the bound state fermion as a function of its mass m_B (i.e., vary $g_f g_s$ to obtain a given m_B) for $m_g = 30$, $\Lambda = m_g$, $m_s = 9.0$ and $m_f = 8.99$.

and g_A/g_V as a function of m_B for fixed m_g, Λ, m_s and m_f. Note that g_A/g_V and G_M/m_B attain maximum values when $G_M = 1$ ($F_2 = 0$). We see in Fig. 2 that G_M/m_B reaches a value at the peak for the deeply bound state which is considerably larger than that of individual constituents, in contrast to the nonrelativistic picture.[12] Figure 3 shows the plot of the bound state mass at which $G_M = 1$ as a function of m_g along with the corresponding values of g_a/g_V. We see that if we require $g_A/g_V \simeq 1$ as well as $F_2 \simeq 0$ then we are restricted to $m_B/(m_s + m_f) \gtrsim .1$. No other bound states are found for this or other partial waves at these values of $g_s g_f$.

SO_{14} UNIFIED COMPOSITE MODEL

Although the results presented in the previous section are substantial for a model which neglects multiparticle states, it clearly shows the enormous difficulty in obtaining a full dynamical

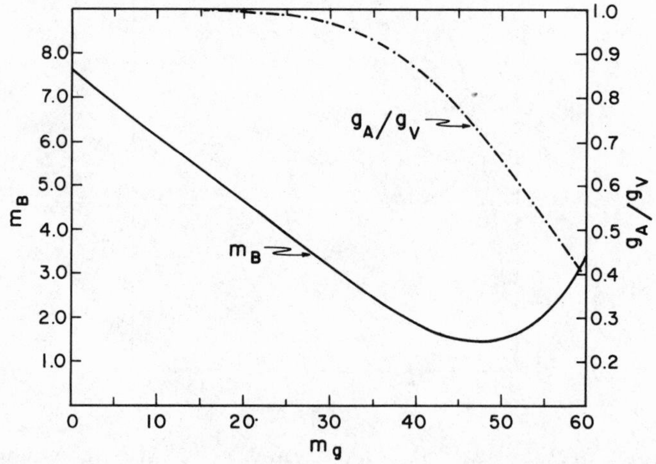

Fig. 3 Plot of m_B at which $F_2 = 0$ as a function of m_g (with $\Lambda = m_g$), with m_s and m_f as in Fig. 1. The corresponding values of g_A/g_V are also given.

calculation for a composite electron and muon as constrained by Eq. (4). With this in mind, we briefly review an SO_{14} unified model[8] in which both the electron and muon SO_{10} families are elementary but the tau family is composite. Another composite SO_{10} $\underline{16}$ and two $\underline{144}$'s are predicted. The model is a straightforward extension of a unified model[17] based on the exceptional group E_6 in which f_e and f_μ are elementary, and f_τ is composite.

A maximal subgroup fo SO_{14} is[18]

$$SO_{14} \supset SO_{10} \times SU_2^{c'} \times SU_2^{fam.} \quad . \tag{10}$$

The SO_{10} is the usual unified group containing $SU_2^w \times U_1^w \times SU_3^c$ in which the left-handed f_e, f_μ and f_τ families are each in the $\underline{16}$ irreducible representation. We take[8] the first $SU_2^{c'}$ in (10) as a new unbroken gauge symmetry which will provide the binding and confining mechanism for particles with color prime or c'. The second $SU_2^{fam.}$ will be broken at some energy scale, but the symmetry will provide a useful family classification. We assume an elementary $\underline{64}$ of fermions, and one $\underline{91}$ of scalars which branch to $SO_{10} \times SU_2^{c'} \times SU_2^{fam}$ as[18]

$$\underline{64} = (\underline{16}, \underline{1}, \underline{2}) + (\overline{\underline{16}}, \underline{2}, \underline{1}) \tag{11}$$

and

$$\underline{91} = (\underline{45}, \underline{1}, \underline{1}) + (\underline{10}, \underline{2}, \underline{2}) + (\underline{1}, \underline{1}, \underline{3}) + (\underline{1}, \underline{3}, \underline{1}) \quad . \tag{12}$$

Thus the elementary $\underline{64}$ contains two c' singlet SO_{10} $\underline{16}$'s to which we assign the left-handed electron and muon families. The right-handed $\overline{\underline{16}}$'s in (11) would not be observed since they are c' doublets and are thus confined in any reaction. However, c' singlet composite fermions can be formed from this $\overline{\underline{16}}$ together with the scalar $\underline{10}$'s in (12). Now

$$\overline{\underline{16}} \times \underline{10} = \underline{16} + \underline{144} \quad . \tag{13}$$

Thus $(\overline{\underline{16}}, \underline{2}, \underline{1}) \times (\underline{10}, \underline{2}, \underline{2})$ gives two c' singlet composite $\underline{16}$'s and two $\underline{144}$'s, accommodating the τ family as one of the composite $\underline{16}$'s as well as many new composite fermions. (Note that these composite fermions are not excited f_e or f_μ families). The $\underline{144}$ contains[17,18] fermions with fairly exotic properties: color $\underline{6}$ and $\underline{8}$ quarks and leptons with electric charge 2. The masses of these particles should not be far above that of the top quark. Another prediction of this model is that there are enough composite fermions to temporarily drive all the couplings α_i (color, electroweak and color prime) strong. As in the analogous E_6 case (see Fig. 1 of Ref. 17), we imagine that the composite fermions are tightly bound enough so that they contribute to the one-loop approximation to the running coupling constant equations over, e.g., the range of ~ 100 GeV to $\lesssim$ 1 TeV and quickly drive the couplings into a strong, nonperturbative region. At larger q^2, at which the bound state fermions are no longer point-like, the $\alpha_i(q^2)$ all decrease again to an asymptotically free regime.

LIMITS ON F_2 FOR τ AND QUARKS FROM e^+e^- ANNIHILATION

Recently, limits on F_2 for the τ and for the quarks were obtained[7] from PETRA data[19,20] on e^+e^- annihilation. In addition, lower energy experiments on the angular distribution of τ production and decay that could give an independent measurement of F_2^τ were suggested.[7] (Future experiments to determine F_2 at energies near the Z^o peak were also discussed).[7] We briefly summarize the results.

The differential cross section for $e^+e^- \to \tau^+\tau^-$ or $q\bar{q}$ (mass m and charge Q) is

$$\frac{d\sigma}{d\cos\theta} = Q^2 \frac{\pi\alpha^2}{2q^2} \beta \, [(2-\beta^2\sin^2\theta) \, F_1^2 + 4F_1F_2 + 2F_2^2$$

$$+ \beta^2\sin^2\theta \, (1 + \frac{\beta^2 q^2}{4m^2}) \, F_2^2] \quad . \tag{14}$$

For $\beta \to 1$ the total cross section is

$$\sigma_{\beta\to1} = Q^2 \frac{4\pi\alpha^2}{3q^2} \, [F_1^2 + 3F_1F_2 + (2 + \frac{q^2}{8m^2}) \, F_2^2] \quad . \tag{15}$$

The preliminary results at PETRA[19] at q^2 up to $(37 \text{ GeV})^2$ show that τ production agrees with the pointlike result $4\pi\alpha^2/3q^2$ to 10% at two standard deviations (2 S.D. = 95% CL) or 5% at 1 S.D. Using Eq. (15) with $F_1 = 1$ and $q^2 = 1350 \text{ GeV}^2$ gives a 2 S.D. bound $F_2^\tau \leqslant .023$ or at 1 S.D. $F_2^\tau \leqslant .014$.

We may also set limits on the anomalous magnetic moments of quarks at high q^2 using the accuracy of the agreement of the PETRA value[4] of R_{had} with that expected from pointlike quarks[7] as compared to that given by Eq. (2) for quarks with anomalous magnetic moments at high q^2. The smallest absolute normalization errors are those of PLUTO[20] and TASSO[20] which are 5% at 1 S.D. We keep the principal terms in Eq. (15) to compare with the data at $q^2 = (37 \text{ GeV})^2$ using constituent quark masses $m_u = m_d = .3$ GeV, $m_s = .45$ GeV, $m_c = 1.8$ GeV, and $m_b = 4.5$ GeV. With a 5% limit on agreement with R = 11/3 we get the bounds

$$R = \frac{11}{3} \, (1 \pm .05) = \frac{4}{3} \, (1 + 1900 \, (F_2^u)^2) + \frac{1}{3} \, (1 + 1900 \, (F_2^d)^2)$$

$$+ \frac{4}{3} \, (1 + 3 \, F_2^c + 53 \, (F_2^c)^2) + \frac{1}{3} \, (1 + 850 \, (F_2^s)^2) \tag{16}$$

$$+ \frac{1}{3} \, (1 + 3 \, F_2^b + 8.5 \, (F_2^b)^2) \, .$$

Subtracting off 11/3 and in turn separately assigning the $\pm$ .18

error to each term gives the bounds at 1 S.D. and $q^2 = (37\ \text{GeV})^2$;

$$F_2^u \leqslant .008,\ F_2^d \leqslant .017,\ F_2^s \leqslant .025,\ F_2^c \leqslant .030,\ F_2^b \leqslant .13 \quad . \qquad (17)$$

Use of the current algebra or bag model masses of m_u, $m_d \sim 10$ MeV would reduce the limits on F_2^u and F_2^d by a factor of 30 to $F_2^u \leqslant .0003$ and $F_2^d \leqslant .0006$.

Finally, we review here a method of measuring or bounding the τ anomalous moment to an accuracy of the order of 1% at relatively low $E_{cm} \leqslant 10$ GeV ($q^2 = E_{cm}^2$) by a high statistics measurement of the angular distribution of τ decay products.[7] At lower E_{cm} the τ direction is not precisely know because undetected neutrinos are emitted and the heavy τ mass can lead to a momentum for a charged decay particle of up to $m_\tau/2$ transverse to the τ direction compared to a longitudinal momentum of less than $E_{cm}/2$. However, since the nature of the two and three body decays of the τ are known, namely $\tau^- \to \nu_\tau \rho^-$ (22%), $\tau^- \to \nu_\tau \pi^-$ (8%), $\tau^- \to \nu_\tau \bar{\nu}_\mu \mu^-$ (18%), and $\tau^- \to \nu_\tau \bar{\nu}_e e^-$ (17%), we can integrate over the unobserved neutrino directions and predict the angular distribution for the <u>charged particle</u> including the effect of the F_2^τ form factor (for $F_2 << F_1 = 1$):

$$\frac{d\sigma}{d\cos\theta_c} \propto [1 + \frac{6}{3-\beta^2} F_2 + b(E_{cm}) P_2(\cos\theta_c)] \quad . \qquad (18)$$

The coefficient $b(E_{cm})$ determines how well we can separate the orthogonal angular distributions and set a limit on F_2. By its decrease from the value for the τ itself before decay, $b^\tau = \beta^2/(3-\beta^2)$, we find how much the decay has washed out the angular variation. We find that b^ρ is the largest and $b^\pi \simeq b^\mu = b^e$. Values of the b's are presented in Table I for various total E_{cm}. Using a simplified two bin method for estimation (bins with $P_2(\cos\theta_c)$ positive or negative) in isolating an angular distribution using statistical errors only we find the following limit can be set on F_2:

TABLE I. Coefficients $b(E_{cm})$ for Eq. (18) and statistical limits on F_2^τ obtainable at various E_{cm} from a simplified two bin analysis of the τ angular distribution using Eq. (19) with $(L_{31}YA) = 1$.

E_{cm}(GeV)	$b^{e,\mu,\pi}$	b^ρ	$F_{2\tau}^{(JOINT)}$
5	.056	.081	.014
6	.11	.18	.0074
8	.20	.30	.0053
10	.27	.37	.0049
12	.32	.41	.0051
20	.41	.47	.0068
30	.45	.47	.0096

$$F_2 < \frac{1.2 \times 10^{-4} E_{cm}(1 + 2m_\tau^2/E_{cm}^2)}{b(E_{cm})\sqrt{B.R.}\ L_{31}YA} , \qquad (19)$$

where L_{31} is the luminosity in units of $10^{31}/cm^2$-sec., Y is effective years running time and A is the acceptance. The limits that can be obtained at various energies from the e, μ and π lumped together with the total B.R. of 43% (since they have the same b's) and for the ρ are nearly equal. In Table I we show the weighted limit that can be obtained by combining both of these limits. The table shows that a limit $F_2^\tau \leqslant .02$ is obtainable statistically and the favored energy range is 7 GeV $\leqslant E_{cm} \leqslant$ 20GeV.

This work was supported in part by the National Science Foundation.

REFERENCES

1. T. Kinoshita and W.B. Lindquist, Phys. Rev. Lett. 47, 1573 (1981).
2. J. Bailey et al., Nuc. Phys. B150, 1 (1979).
3. G. 't Hooft, Cargese Summer Institute Lectures 1979.
4. See the paper by E. Eichten in this Session on Composite Fermions.
5. M. Bander, T.-W. Chiu, G.L. Shaw and D. Silverman, Phys. Rev. Lett. 47, 549 (1981).
6. See the approach of O.W. Greenberg in this Session on Composite Fermions.
7. D.J. Silverman and G.L. Shaw, UCI Report 82-11.
8. G.L. Shaw and F. Daghighian, UCI Report 82-19.
9. See, e.g., M.L. Perl et al., Phys. Rev. Lett. 35, 1489 (1975); S. Kovesi-Domokos and G. Domokos, Phys. Rev. D24, 2866 (1981).
10. B. Adeva et al., MIT Report 123, 1982.
11. For a scenario with spontaneously broken QCD, see, R. Slansky, T. Goldman and G.L. Shaw, Phys. Rev. Lett. 47, 887 (1981).
12. M. Gluck, Phys. Lett. 87B, 274 (1979); H.J. Lipkin, Phys. Lett. 89B, 358 (1980).
13. G.L. Shaw, D. Silverman and R. Slansky, Phys. Lett. 94B, 57 (1980); S.J. Brodsky and S.D. Drell, Phys. Rev. D22, 2236 (1980).
14. M.A. Beg, Int. Conf. on High Energy Physics, Lisbon (1981); H.J. Lipkin, Phys. Lett. 103B, 440 (1981).
15. J. Preskill and S. Weinberg, Phys. Rev. D24, 1059 (1981).
16. O.W. Greenberg and R. Genolio, Phys. Rev. 150, 1070 (1966); F. Gross, Phys. Rev. 186, 1448 (1969); K. Johnson, Phys. Rev. D4, 1101 (1972); A. Raychaudhui, Phys. Rev. D18, 4658 (1978).
17. G.L. Shaw and R. Slansky, Phys. Rev. D32, 1760 (1980).
18. R. Slansky, Phys. Rep. 79, 1 (1981).
19. R. Felst, 1981 Int. Symposium on Lepton Photon Interactions at High Energies, Bonn.

20. J.G. Branson, 1981 Int. Symposium on Lepton and Photon Interactions at High Energies, Bonn, and DESY Report 82-010.

LIGHT COMPOSITE FERMIONS - AN OVERVIEW

Stuart Raby

Los Alamos National Laboratory

Los Alamos, New Mexico 87545

1. RULES

I know that Estia Eichten (in these proceedings), has already discussed one set of rules which describe massless composite fermions and the possible breaking of chiral symmetries. I would just like to spend a few minutes now discussing the status of some rules which have been proposed and also to emphasize a few points. In particular, I will discuss (a) 't Hooft's anomaly conditions[1] (and left-right symmetric theories), (b) 't Hooft's decoupling conditions (or persistent mass hypothesis of Preskill and Weinberg[2]), and (c) "tumbling"[3] vs. no "tumbling."

(a) 't Hooft's anomaly conditions are by now well established.[4] However, the anomaly conditions are not sufficient to uniquely determine the spectrum of massless composites. Additional dynamical assumptions (RULES) are necessary in order to obtain an unambiguous spectrum. For example, the anomaly conditions are only applicable when the strong interaction dynamics (i.e., the binding forces) preserve the global chiral symmetries of the theory. If they are broken however, then there is no need for massless composite fermions even if the consistency conditions have solutions. The breaking of the chiral symmetries is a dynamical question. This brings us to the subject of left-right symmetric theories such as QCD. In QCD there is an $SU(M)_L \otimes SU(M)_R$ chiral symmetry, where M is the number of left and right-handed quark flavors. Solutions can be found to the anomaly conditions for particular values of M. However, we believe that the QCD forces create fermion condensates that break $SU(M)_L \otimes SU(M)_R$ to $SU(M)_{vector}$ giving all M flavors a dynamical mass. The resulting symmetry $SU(M)_{vector}$ does not require any massless composite fermions and there are none.

What evidence do we have that left-right symmetric theories will generically not require massless composite fermions.

1) Coleman and Witten[5] have shown that in the large N limit of an SU(N) gauge theory with fermions in the fundamental representation the chiral symmetries are spontaneously broken.

2) It has been shown that the chiral symmetries are spontaneously broken in an SU(N) lattice gauge theory with fermions in the fundamental representation either by using a selfconsistent strong coupling expansion[6] or in Monte Carlo calculations which neglect internal fermion loops.[7,8]

   Thus it is evident that SU(N) gauge theories with M flavors of left-right symmetric fermions in the fundamental representation do not lead to massless composite fermions.

3) The situation for an O(N) gauge group with fermions in the N dimensional representation is not as clear. For example, Banks and Kaplunovsky[9] have found evidence in an O(N) lattice gauge theory that the chiral symmetries are unbroken and that massless composite fermions (solutions to 't Hooft's anomaly conditions) exist.*

---

*After this talk was given, we received a paper by J. Kogut et al.[10] who discuss chiral symmetry breaking in lattice gauge theory. They find that for N = 3, the chiral symmetries are in fact spontaneously broken.

To summarize, it's a dynamical question whether chiral symmetries remain unbroken in a theory. This question can be answered in some cases with present techniques. It is crucial information which can eliminate certain classes of theories with regards to their use in constructing models of massless composite fermions.

(b) 't Hooft suggested that the anomaly conditions should be supplemented by the decoupling condition based on the Appelquist-Carrazone theorem. He argued that the Appelquist-Carrazone theorem would imply that if a solution to the consistency conditions is found in which the constituent can obtain a small mass, then the composite fermion would also obtain a small mass (since it must eventually decouple) and one should throw out such solutions on the grounds that they are "unnatural", i.e., if a particle is "naturally" massless, it should remain massless under small changes in parameters. However, it has been realized that the composite fermion may remain massless, even though its constituent has a small mass, and still be consistent with the Appelquist-Carrazone theorem, i.e., if there is a phase transition when $m_{constituent} \sim \Lambda_{binding}$. ($\Lambda_{binding}$ = binding scale.) The persistent mass hypothesis[2] makes it a dynamical principle that such phase transitions don't occur and that $m_{composite}$ is proportional to $m_{constituent}$. Recently Dimopoulos and Preskill[11] and Bars and Yankielowicz[12] have made it very plausible that massless composites can be made of massive constituents. Their analysis relies on the dynamics of subgroup-alignment and I refer you to their papers for the details. To summarize, it is clear that the decoupling condition as phrased in its strictest form is <u>not</u> a direct consequence of the Appelquiest-Carrazone theorem and is probably <u>not</u> valid as a general rule.

(c) The phenomenon of tumbling (i.e., a gauge group breaking itself by forming fermion condensates in a nontrivial representation) is the consequence of a rule suggested by Dimopoulos, Susskind, and myself.[3] In some cases it has no effect on the spectrum of massless composites whether or not the system actually tumbles. This is a result of a principle of complementarity[13] which I have no time to discuss now. In these cases tumbling manifests itself on the structure of intermediate scales. There do exist cases however in which tumbling is signalled by the presence of phase boundaries in the system which distinguish the broken and unbroken chiral phases. (These are necessarily noncomplementary examples.)

Consider the gauge group SU(6) with one left-handed multiplet of fermions in the 20-dimensional representation. The fermion field is given by the expression

$$\psi_{[ijk]} \qquad i,j,k = 1,\ldots,6 \tag{1}$$

and [ijk] denotes antisymmetrization. It is easy to see for this system that it is not possible to break chirality without simultaneously breaking either Lorentz invariance or SU(6). Consider the Lorentz scalar, gauge invariant operator

$$\psi_{[ijk]}\psi_{[rst]}\varepsilon^{ijkrst} \equiv 0 \quad . \tag{2}$$

It vanishes by Fermi statistics. The next possible channel in which a condensate might form is then given by the Lorentz scalar

$$\psi_{[ijk]}\psi_{[rst]}\varepsilon^{jkrst\ell} \equiv \phi_i^\ell \quad . \tag{3}$$

This is actually the Maximally Attractive Channel as defined in Ref. 3. If ϕ_i^ℓ condenses it will break SU(6). One possible breaking pattern is

$$SU(6) \rightarrow SU(3)\otimes SU(3)\otimes U(1) \quad . \tag{4}$$

In the symmetric phase there are no finite energy fermion states, whereas in the broken phase there are SU(3)⊗SU(3) singlet fermions carrying long range U(1) forces. There also exist U(1) monopoles in this phase. Whether or not the system tumbles is thus a dynamical question which deserves further study.

Recently Eichten and Feinberg[14] have argued that tumbling does not occur in four dimensions. They find, in a perturbative approximation, that the gauge contributions to the vacuum energy dominate over the Fermion contributions and tend to preserve the symmetric phase. A. D'Adda et al.,[15] on the other hand, have studied the non-Abelian generalization of CP^n models in 2 dimensions. These models contain an $SU(\ell)$ local gauge symmetry in addition to a global $U(n)$ invariance. They find, in a $1/n$ expansion, that the local $SU(\ell)$ gauge group breaks down due to the formation of the fermion bilinear condensate

$$\phi_{(\alpha\beta)} = \langle\bar{\psi}_{R\alpha}\psi_{L\beta}\rangle \neq 0 \quad . \tag{5}$$

$\alpha,\beta = 1,\ldots,\ell$. Clearly, much work must still be done to understand the tumbling phenomenon.

2. MASS GENERATION MECHANISMS

We have all these massless composite fermions flying around and we would like to give them some small masses $m \ll \Lambda_{binding}$. Recall the reason why they are massless is due to an exact chiracl symmetry G_{flavor} which as a result of 't Hooft's anomaly conditions protects these fermions from getting mass.* Thus, in order to obtain mass, we must break G_{flavor} and the degree to which G_{fl} is broken determines the ratio $m/\Lambda_{binding}$. I am aware of four mechanisms for this breaking.

1) Weak gauging I[16]

In this scenario G_{fl} is assumed to spontaneously break to a

---

*For a discussion of this point see E. Eichten, in these proceedings.

subgroup H_{f1} as a result of the strong interaction dynamics. Consider an irreducible representation under G_{f1} of massless composite fermions--solutions to the anomaly conditions for G_{f1}. This becomes a reducible representation under the subgroup H_{f1}. Not all of these states are protected by H_{f1} and in fact it is easy to see that some obtain mass either directly from the many fermion condensate that broke G_{f1} to H_{f1} or via instantons and the condensate.

$$D(G_{f1}) = \sum_{\oplus i} D_i(H_{f1})$$

Now imagine weakly gauging a subgroup G_w of G_{f1} so that G_w is not contained in H_{f1}, but explicitly breaks H_{f1}. Then the fermions which are protected by H_{f1} will obtain mass radiatively via weak exchanges. This is represented by the wiggly lines in the figure below, i.e., the weak exchanges take massless states into massive ones. These corrections will typically be of order

$$m \sim \alpha\Lambda_{binding} \ , \ \alpha^2\Lambda_{binding} \ , \tag{6}$$

where α is the weak fine structure constant.

2) Weak gauging II[17]

In this scenario G_{f1} does not spontaneously break. However,

$$D(G_{f1}) = \sum_{\oplus 1} D_i(H_{f1})$$

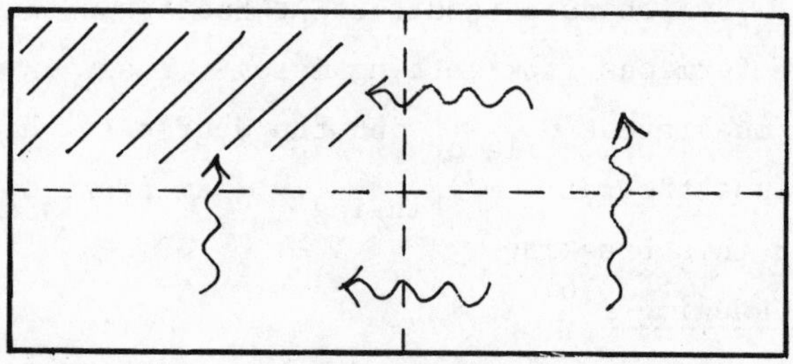

Fig. 1 The solid box represents the irreducible representation $D(G_{f1})$ and the small boxes are $D_i(H_{f1})$. The shaded area is the irreducible representation of H_{f1} with mass of order $\Lambda_{Binding}$.

by gauging a subgroup G_w of G_{fl} we can explicitly break G_{fl} to $G_w \otimes U(1)_x$ where $U(1)_x$ is a global chiral symmetry. $U(1)_x$ is then broken by weak instantons to the discrete symmetry $e^{i\frac{X2\pi}{N\nu}}$ where $N = \sum_i X_i$, X_i is the charge of the ith fermion emitted by an instanton with topological charge $\nu = 1$. The massless composites then obtain mass via weak exchanges in conjunction with a weak instanton. The mass is of order

$$m \sim \alpha(\Lambda_{Binding})\, e^{-4\pi/\alpha(\Lambda_{Binding})}\, \Lambda_{Binding} \quad . \tag{7}$$

3) Strong gauging[18]

We gauge a subgroup G_{TC} of G_{fl} such that at some scale $\Lambda_{TC} \ll \Lambda_{Binding}$ the forces associated with G_{TC} become strong. It is also assumed that at the scale Λ_{TC} those massless composite fermions carrying the strong force condense and obtain dynamical masses of order Λ_{TC}. We now argue that even those massless composite fermions which do not feel TC forces will obtain mass

$$m \sim \frac{\Lambda_{TC}^3}{\Lambda_{Binding}^2} \quad . \tag{8}$$

This is because at energies $E \ll \Lambda_{Binding}$ there probably exist

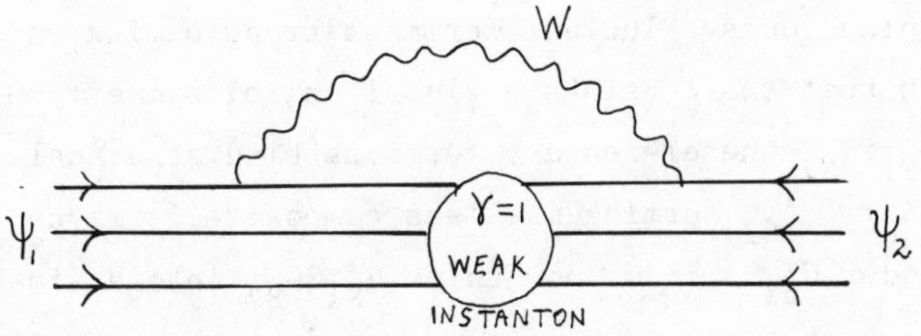

$X_1 + X_2 = N$

Fig. 2

residual nonrenormalizable interactions among the massless composities. For example, there can be four Fermi interactions of the form

$$\frac{1}{\Lambda^2_{\text{Binding}}}\,\psi\psi\psi\psi \tag{9}$$

with the global symmetry G_{fl}. Clearly, if two of the states carry G_{TC} forces and condense, we obtain a mass for the other two of the form

$$\frac{1}{\Lambda^2_{\text{Binding}}}\,\langle\psi_{TC}\psi_{TC}\rangle\psi\psi$$

or (10)

$$m \sim \frac{\Lambda^3_{TC}}{\Lambda^2_{\text{Binding}}} \quad .$$

4) <u>Unification</u>

This scenario includes the extended Technicolor ideas as a simple example (see Ref. 19 for further discussion). One just imagines that all the fermions in an ETC scenario are bound states of some strong interaction with a binding scale much larger than Λ_{ETC}.

In addition, we have the following possibility discussed recently by Abbott, et al.[20] Consider a strong group $G_{S1}\otimes G_{S2}$ where G_{Si} are simple groups. Include fermions transforming under $G_{S1}\otimes G_{S2}$ such that there exists a global chiral symmetry G_{fl}. As a result of $G_{S1}\otimes G_{S2}$ the elementary fermions bind at a scale $\Lambda_{\text{Binding}} \sim \Lambda_{S1} \sim \Lambda_{S2}$ forming massless composite fermions which transform under G_{fl}. If we now unify $G_{S1}\otimes G_{S2}$ into a simple group $G_S \supset G_{S1}\otimes G_{S2}$ at a scale $\Lambda_u \gg \Lambda_{\text{Binding}}$ and at the same time break G_{fl}, then some of the composite fermions will obtain mass. As an example, consider $G_S \equiv SU(5) \supset G_{S1}\otimes G_{S2} = SU(3)_S\otimes SU(2)_L$ where

$SU(2)_L$ can be thought of as a strong version of the standard weak $SU(2)_L$[21] and $SU(3)_S$ is a new strong interaction not to be confused with $SU(3)_{color}$. The breaking $SU(5) \to SU(3)_S \otimes SU(2)_L$ occurs at a scale $\Lambda_u >> \Lambda_{Binding} \sim \Lambda_S \sim \Lambda_{2L}$. Consider the following left-handed fermion states with their $SU(3)_S \otimes SU(2)_L$ quantum numbers:

$$\begin{array}{lll} T_{Si} & (3,2) & S = 1,2,3 \ \varepsilon \ SU(3)_S \\ & & i = 1,2 \quad \varepsilon \ SU(2)_L \\ \bar{A}^S & (\bar{3},1) & \\ \bar{B}^S & (\bar{3},1) & \\ \psi_i & (1,2) & \bar{e} \qquad (1,1) \end{array} \tag{11}$$

$\bar{e}$ is a spectator to the strong interactions. The flavor symmetry is

$$G_{f1} = SU(2)_{AB} \otimes U(1) \quad , \tag{12}$$

where the U(1) has no strong anomalies. 't Hooft's anomaly conditions can be solved and we find the massless composites

$$\bar{A}^{S*} T^*_{Si} \psi_i \qquad \bar{B}^{S*} T^*_{Si} \psi_i \tag{13}$$

transforming as a doublet under $SU(2)_{AB}$. Note that according to the scenario of a strong $SU(2)_L$,[21] these massless composites are to be associated with, for example, the ν e doublet and the standard weak interactions are associated with the residual 4 Fermi interactions which result from the binding forces $SU(3)_S \otimes SU(2)_L$. In this context, the singlet $\bar{e}$ is to be associated with the left-handed positron. We now show that the broken SU(5) interactions generate an electron mass. Under SU(5) the states of Eq. (11) transform as follows:

$$\begin{aligned} 10 &\supset \bar{A}\ T\ \bar{e} \\ \bar{5} &\supset \bar{B}\ \psi \quad . \end{aligned} \tag{14}$$

The broken SU(5) forces then induce the following mass term

$$\frac{1}{\Lambda_u^2}\ \bar{e}^*(T_{Si}\bar{B}^S\psi_i^*)$$

or (15)

$$m_e \sim \frac{\Lambda_{\text{Binding}}^3}{\Lambda_u^2} \quad .$$

To summarize, we have discussed four mechanisms to give a small mass $m \ll \Lambda_{\text{Binding}}$ to massless composite fermions. These are the tools for constructing realistic models. We note finally that they have been discussed separately for pedagogical purposes but clearly they can be used in tandem when building models.

3. GENERATION NUMBER

The origin of the different generations of quarks and leptons is a longstanding puzzle. In the context of composite models, the three generations of quarks and leptons must appear in the first approximation as three sets of massless composite fermions in identical representations of a continuous flavor symmetry G_{fl} with a conserved generation number distinguishing them. In addition, the generation symmetry is preferrably a discrete one in order to avoid massless Goldstone bosons that would occur if we would spontaneously break a continuous symmetry. Harari and Seiberg[22] have shown that in a class of rishon models there exist U(1) generation symmetries that are automatically broken to a discrete generation symmetry due to strong interaction instantons. Recently Eichten and Preskill[23] have found solutions to the anomaly conditions with this property. I would like to conclude this talk with one of their examples. We consider the strong interaction group $G_S = SU(N)$ with $N = 4(2m + 1)$, m a positive integer. The left-handed fermions are

$$\begin{array}{ll} \phi^{(ij)} & \\ \chi_{[ij]} & \\ \psi_{ia} & a = 1,\ldots,8 \end{array} \tag{16}$$

where $i,j = 1,\ldots,N \in SU(N)$. The symbols () ([]) mean (anti-) symmetrization. The flavor symmetry G_{fl} of the model is

$$G_{fl} = SU(8)\otimes U(1)\otimes U(1) \quad ,$$

which we assume is spontaneously broken to $H_{fl} = SU(8)\otimes U(1)_G$, where the generation number G is given by

$$G \equiv 8(m + 1)N_\phi - 8mN_\chi - 6(2m + 1)N_\psi \quad ,$$

and N_ϕ counts ϕ states, etc. The solutions to the anomaly conditions for H_{fl} are the massless composites

$$\begin{aligned} \phi^{ij}\psi_{ia}\psi_{jb} &= \xi^1_{[ab]} \\ \phi^{ij}(\chi\phi)_j{}^\ell\psi_{ia}\psi_{\ell b} &= \xi^2_{[ab]} \\ &\vdots \\ \phi^{ij}[(\chi\phi)^{2m}]_j{}^\ell\psi_{ia}\psi_{\ell b} &= \xi^{2m+1}_{[ab]} \quad . \end{aligned}$$

We thus find 2m + 1 generations in the 28 dimensional representation of SU(8) with generation numbers G = $-4(4m+1)$, $-4(4m-1)$,...,-4. If we now let m = 1 and N = 12, we find 3 generations. Preskill[23] has then shown one way for obtaining a nontrivial mass spectrum for these states by gauging the flavor group SU(8) in an extended Technicolor scenario. We note that when SU(8) is gauged, instantons of the strong Technicolor ($SU(8) \supset G_{TC} = SU(4)$) forces break $U(1)_G$ to the discrete symmetry

$$e^{2\pi i G/\Delta G} ,$$

where $\Delta G = 54$ and the 3 generations have multiplicative charges $e^{-2\pi i(20/54)}$, $e^{-2\pi i(12/54)}$, $e^{-2\pi i(4/54)}$. The discrete generation symmetry is then spontaneously broken by the strong Technicolor forces, enabling all the fermions to obtain mass.

To conclude, much progress has been made in understanding light composite fermions. Clearly, much more work is necessary before we shall obtain a realistic model of composite quarks and leptons.

REFERENCES

1. G. 't Hooft in Recent Developments in Gauge Theories, ed., 't Hooft, et al. (Plenum Press, New York, 1980).
2. J. Preskill and S. Weinberg, Phys. Rev. D24, 1059 (1981).
3. S. Raby, S. Dimopoulos, and L. Susskind, Nucl. Phys. B169, 373 (1980); M. E. Peskin (unpublished).
4. Y. Frishman, A. Schwimmer, T. Banks, and S. Yankielowicz, Nucl. Phys. B177, 157 (1981); S. Coleman and B. Grossman, to be published.
5. S. Coleman and E. Witten, Phys. Rev. Lett. 45, 100 (1980).
6. J.-M. Blairon, R. Brout, F. Englert, and J. Greensite, Brussels preprint 80-0724 (1980), to be published in Nuclear Phys. B.
7. H. Hamber and G. Parisi, Phys. Rev. Lett. 47, 1792 (1981); E. Marinari, G. Parisi, and C. Rebbi, Phys. Rev. Lett. 47, 1795 (1981).

8. D. Weingarten, Indiana University preprint (1981).
9. T. Banks and V. Kaplunovsky, Tel-Aviv University preprint TAUP 930-81 (1981).
10. J. Kogut, M. Stone, H. W. Wyld, J. Shigemitsu, S. H. Shenker, and D. K. Sinclair, University of Illinois preprint ILL-(TH)-82-5 (1982).
11. S. Dimopoulos and J. Preskill, Harvard preprint HUTP-81/A045 (1981).
12. I. Bars and S. Yankielowicz, Phys. Lett. 101B, 159 (1981).
13. S. Dimopoulos, S. Raby, and L. Susskind, Nucl. Phys. B173, 208 (1980).
14. E. Eichten and F. Feinberg, Fermilab Pub-81/62-THY (1981).
15. A. D'Adda, A. C. Davis, and P. Di Vecchia, Cern Preprint, Th. 3189-Cern (1981); A. D'Adda, A. C. Davis, P. Di Vecchia, and M. E. Peskin, Cern preprint (1981).
16. S. Weinberg, Phys. Rev. D13, 974 (1976); M. E. Peskin, talk at the Workshop on Gauge Theories and Their Phenomenological Implications, Chania, Crete (1980); S. Dimopoulos and L. Susskind, Stanford report ITP-681 (1980), Nucl. Phys. B, to be published; H. P. Nilles and S. Raby, Nucl. Phys. B189, 93 (1981).
17. S. Weinberg, Phys. Lett. 102B, 401 (1981).
18. P. Sikivie, Cern preprint TH. 3072-Cern (1981); University of Florida preprint UFTP-81-26 (1981).
19. S. Dimopoulos and L. Susskind, Nucl. Phys. B155, 237 (1979); E. Eichten and K. Lane, Phys. Lett. 90B, 125 (1980).
20. L. F. Abbott, E. Farhi, and A. Schwimmer, MIT preprint CTP-962 (1981).
21. L. F. Abbott and E. Farhi, Phys. Lett. 101B, 69 (1981); Nucl. Phys. B189, 547 (1981); H. Fritzsch and G. Mandelbaum, MPI-PAE/PTh 22/81 (1981); R. Barbieri, A. Masiero, and R. N. Mohapatra, CERN-Th-3089 (1981).
22. H. Harari and N. Seiberg, Phys. Lett. 102B, 263 (1981).

23. E. Eichten and J. Preskill, to be published; J. Preskill, HUTP-81/A051 (1981).

THE N-QUANTUM APPROXIMATION, CONCRETE COMPOSITE MODELS OF QUARKS AND LEPTONS, AND PROBLEMS WITH THE NORMALIZATION OF COMPOSITE MASSLESS BOUND STATES

O. W. Greenberg

University of Maryland

College Park, Maryland 20742

ABSTRACT

We discuss concrete composite models of quarks and leptons using the N-quantum approximation. The first section introduces the main ideas of this approximation, the second section describes the bound-state equations which follow when chiral symmetry is assumed to hold in the Wigner-Weyl mode so that the fermions of the theory, both elementary and composite, have zero mass, and the third section points out a problem with the normalization of composite zero-mass bound states, and offers a suggestion to resolve the problem.

We want to conclude this session on composite models of quarks and leptons by discussing concrete composite models. By "concrete" models, we mean models in which the bound states are described in terms of space-time or energy-momentum dependent amplitudes (or wave functions) for their constituents. One must go beyond the quantum number counting or algebraic approach to these models in order to make further progress. The formalism we use to describe the concrete models, the N-quantum approximation (NQA), although not new, is not well-known, so we describe this formalism in Sec. 1. In Sec. 2, we apply this formalism to derive bound-state equations for theories in which chiral symmetry is assumed to be exact, so that the composite fermions have zero mass. This work is incomplete, because of the appearance of a problem in the normalization of the bound-state amplitudes for zero mass particles. We describe this problem, and a suggestion to resolve it, in Sec. 3.

1. THE N-QUANTUM APPROXIMATION

The NQA uses the Haag expansion[1] of the Lagrangian fields of the theory in terms of asymptotic fields, including in or out fields for all stable bound states. Asymptotic fields for particles which are unstable, but are taken to be stable in the approximation being used also occur in the expansion; for example the pion might be considered stable in a study of the strong interactions, and then its asymptotic field would occur in the Haag expansion. Glaser, Lehmann, and Zimmermann[2] gave important properties of the Haag expansion, including the relation of the amplitudes which occur as coefficients of the normal-ordered asymptotic fields to vacuum-expectation values of multiple-retarded commutator functions with all but one leg on shell, and gave an infinite set of coupled equations which these amplitudes must obey, based only on general principles of quantum field theory. We proposed using the expansion to solve a specific theory,[3] and, in collaboration with Genolio, proposed treating bound states by introducing the corresponding asymptotic fields.[4] Similar approaches were proposed by Gross[5] and Johnson.[6] Pagnamenta[7] used the NQA to treat the higher sectors of the Lee model. Raychaudhuri[8] studied bound states in a scalar relativistic model, and showed that the relativistic corrections to the bound states of hydrogen which are usually derived using the Bethe-Salpeter equation can be simply derived using the NQA.[9] With Nussinov and Sucher we used the NQA to study the charge radii of the neutral kaon.[10] Bander, Chiu, and Shaw[11] used the NQA to study strongly-bound spinor-scalar states. The bound-state amplitudes in the NQA are on shell in all but one constituent, so that there are no relative times as in the Bethe-Salpeter amplitudes. The equations for the NQA amplitudes are covariant, but have the same three-dimensional character as the nonrelativistic Schrödinger equation. As we will see, the off-shell constituent corresponds to the Lagrangian field, and the on-shell constituents correspond to the asymptotic fields.

We assume asymptotic completeness, so that all states in the theory can be described either using the Lagrangian fields or using the asymptotic fields, and either set of fields is an irreducible set. (In theories with confinement this may not be true, and it may not be possible to expand, for example, the quark fields in asymptotic fields; however in such theories the gauge-invariant products of operators, which are local color singlets, such as

$$\bar{q}(x)e^{i\int_y^x A(w)\,dw}q(y) \quad ,$$

where the integral is ordered along a given path, can be expanded in normal-ordered asymptotic fields.) The Haag expansion arises when one asks how the Lagrangian field acts on a state expressed in the asymptotic field basis. Let

$$\Psi = \mathrm{poly}(A^\dagger_{in})|0\rangle \quad ,$$

then

$$A\Psi = \mathrm{poly}'(A^\dagger_{in})|0\rangle \quad ,$$

so that A has the form

$$A = \Sigma\int f\Pi A^\dagger_{in}\Pi A_{in} \quad .$$

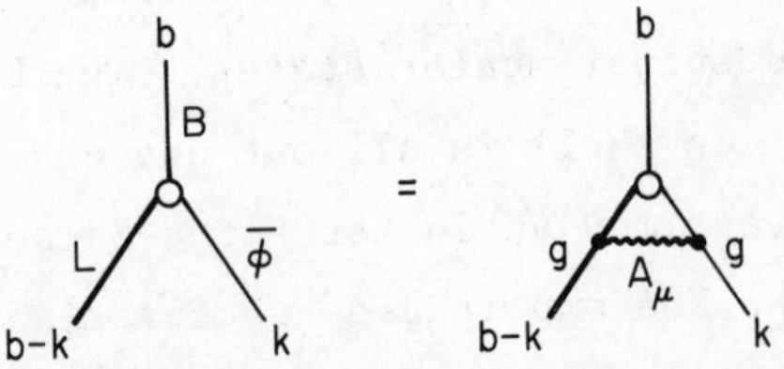

Fig. 1 Amplitude for a bound state B composed of an off-shell constituent L and an on-shell constituent $\bar{\phi}$. The equation shown corresponds to Eq. (9) in the text, and uses the singlet A_μ exchange approximation.

Here, "poly" or "poly'" stand for smeared polynomials, and the expressions are written very schematically.

It is convenient to represent the NQA amplitudes by graphs in which the on-shell lines, corresponding to the asymptotic fields, are thin, and the off-shell line, corresponding to the Lagrangian field, is thick. Fig. 1 shows the graph for a bound state B which is a composite of an off-shell constituent L and an on-shell constituent $\bar{\phi}$.

2. THE NQA APPLIED TO MASSLESS BOUND STATES IN A CHIRAL MODEL[12]

We consider a model with a (left-handed) chiral fermion and a massive scalar interacting with a gauge field, which in our approximations may be either abelian or nonabelian. For simplicity, we will write it as though it were abelian. The Lagrangian of the model is

$$= \bar{L}(i\partial\!\!\!/ - gA\!\!\!/)L + |(\partial_\mu + igA_\mu)\phi|^2 - 1/4\, G_{\mu\nu}G^{\mu\nu} - 1/2(\partial_\mu A^\mu)^2$$
$$- \mu^2\phi^\dagger\phi - (\lambda/4!)(\phi^\dagger\phi)^2 \quad , \tag{1}$$

where

$$L = 1/2\ (1+\gamma_5)\psi, \text{ and } G_{\mu\nu} = \partial_\mu A_\nu - \partial_\nu A_\mu \quad ,$$

and we have used Feynman gauge. The equations of motion are

$$i\not{\partial}L = g\not{A}L \ , \tag{2a}$$

$$(\Box+\mu^2)\phi = -ig(\partial_\mu A^\mu)\phi - 2igA_\mu\partial^\mu\phi + g^2A_\mu A^\mu\phi - (\lambda/12)\phi^\dagger\phi^2 \ , \tag{2b}$$

and

$$\Box A^\mu = g\bar{L}\gamma^\mu L + ig\phi^\dagger\overleftrightarrow{\partial}^\mu\phi - 2g^2\phi^\dagger A^\mu\phi \ . \tag{2c}$$

We have omitted renormalization counter terms, because they do not contribute in our approximation. We go to momentum space using, for example,

$$L(x) = \int L(p)e^{-ip\cdot x}d^4p \ . \tag{3}$$

The equations of motion, written in momentum space, are

$$\not{p}L(p) = g\int \not{A}(p-p')L(p')d^4p' \ , \tag{4a}$$

$$(\mu^2-k^2)\phi(k) = -g\int (k+k')^\mu A_\mu(k-k')\phi(k')d^4k' + \ldots, \tag{4b}$$

and

$$-q^2A^\mu(q) = g\int \bar{L}(p-q)\gamma^\mu L(p)d^4p +$$
$$g\int \phi^\dagger(k-q)(2k-q)^\mu\phi(k)d^4k + \ldots \ , \tag{4c}$$

where the dots indicate irrelevant terms, which we omitted. The chiral constraint on L in momentum space can be written

$$(1-\gamma_5)L(p) = 0 \ . \tag{5}$$

If we assume that there is a chiral composite state F made of $\bar{\phi}$

and L, then the Lagrangian field L must have a term

$$L(p) = \ldots + \frac{d^3k}{2\sqrt{\vec{k}^2+\mu^2}} \frac{d^3b}{2|\vec{b}|} \delta(b-p-k) f(k,b)$$

$$\bar{\phi}^{\dagger}_{in}(k) F_{in}(b) + \ldots \tag{6}$$

in its Haag expansion. The dots in Eq. (6) indicate the omitted terms in the Haag expansion, which include the in field of L, as well as infinitely many more terms which satisfy the constraints due to Poincaré covariance and conservation of the quantum numbers carried by the Lagrangian fields. The in fields in Eq. (6) satisfy the covariant three-dimensional commutation relations, such as

$$[F_{in}(b), \bar{F}_{in}(b')]_+ = 2|\vec{b}|\not{b}\delta\,(\vec{b}-\vec{b}') , \tag{7a}$$

and

$$[\bar{\phi}_{in}(k), \bar{\phi}^{\dagger}_{in}(k')] = 2\sqrt{\vec{k}^2+\mu^2}\,\delta(\vec{k}-\vec{k}') . \tag{7b}$$

The amplitude f is a 4×4 matrix, and transforms covariantly under the homogeneous Lorentz group. The relevant term in the Haag expansion for A_μ is

$$A_\mu(q) = \frac{g}{(2\pi)^3 q^2} \int d^4k\, d^4k'\ \delta(q-k+k')(k+k')_\mu\ \bar{\phi}^{\dagger}_{in}(k')\delta(k'^2-\mu^2)\ .$$

$$\cdot\ \bar{\phi}_{in}(k)\ \delta(k^2-\mu^2)\ . \tag{8}$$

Substitution of the relevant terms of the Haag expansion in Eq. (4) for L, normal ordering, and taking commutator with $\bar{\phi}_{in}$ and anticommutator with F_{in} gives the equation of motion for f. We emphasize that one cannot just "cross out" the in fields, but must take the commutators or anticommutators; otherwise one misses the $\not{b}$ factors in the equation, which are crucial for the zero-mass bound-state case. The result is

$$(\not{b}-\not{k})f(k,b)\not{b} = \frac{g^2}{8\pi^3}\int \frac{d^3k'}{2\sqrt{\vec{k}'^2-\mu^2}} \frac{\not{k}+\not{k}'}{(k-k')^2} f(k',b)\not{b} \quad . \tag{9}$$

The chiral constraints on the in fields translate into the constraints

$$(1-\gamma_5)\ f = 0 \quad , \quad \text{and } f(1-\gamma_5) = 0 \tag{10}$$

on f(k,b). Eq. (10) implies that f has the form

$$f = [A + iC_{\mu\nu}\ \sigma^{\mu\nu}]\ (1+\gamma_5) \quad , \tag{11}$$

where $C_{\mu\nu} = (k_\mu b_\nu - b_\mu k_\nu)C$ and A and C are invariant functions of k and b, i.e. functions of b·k. Inserting this form in Eq. (9) leads to the following equation for the single scalar amplitude a=A-2b·kC,

$$-b\cdot k\ a(b\cdot k) = \frac{g^2}{(2\pi)^3}\int \frac{d^3k'}{2\sqrt{k'^2+\mu^2}} \frac{b\cdot(k+k')\ a(b\cdot k')}{2\mu^2-2k\cdot k'-M_g^2} \quad . \tag{12}$$

Here we have arbitrarily added a mass M_g for the gluon. The transverse momentum integral on the right-hand side of Eq. (12) diverges. If we cut off this integral at k_T=K, we get

$$-xa(x) = \frac{g^2}{16\pi^2 x}\int_0^\infty dx'\ a(x')(x+x')\ln \frac{\mu^2(x-x')^2+M_g^2xx'}{K^2x^2} \quad . \tag{13}$$

In order to avoid this divergence, we square the gluon propagator, at the same time introducing a factor M^2 in the numerator to keep the dimensions the same; i.e. we replaced

$$(2\mu^2-2k\cdot k'-M_g^2)^{-1} \text{ by } M^2(2\mu^2-2k\cdot k'-M_g^2)^{-2} \quad .$$

The new equation is convergent, and with the replacement

$xa(x) = b(x)$, is

$$b(x) = -\lambda \int_0^\infty dx' \frac{(x+x')b(x')}{(x-x')^2+(M_g^2/\mu^2)xx'} \quad . \tag{14}$$

We noticed that the choice $M_g^2 = 4\mu^2$ gives the exactly soluble equation

$$b(x) = -\lambda \int_0^\infty dx' \frac{b(x')}{x+x'} \quad . \tag{15}$$

We solve this equation by letting the right-hand-side define a function b(z) which is analytic in the cut plane with a cut which we choose along the negative real axis. Taking boundary values across the negative real axis, we get

$$b(xe^{i\pi}) - b(xe^{-i\pi}) = 2\pi\, i\lambda\, b(x),\ x>0 \quad . \tag{16}$$

This is satisfied by z^s, provided $s=\pi^{-1}\arcsin\pi\lambda$. In order to pick out the solutions of any bound-state equation, one needs a normalization condition for the amplitude. We now leave the rather ad hoc discussion of this section, and consider the normalization of the amplitude for zero-mass bound states. We emphasize that our discussion of the normalization in Sec. 3 does not depend on the ad hoc approximations which we made in this section.

3. PROBLEMS WITH THE NORMALIZATION OF MASSLESS BOUND STATES

The chiral charge carried by the field L is

$$Q_5 = \int L^\dagger \gamma_5 L\, d^3x \quad . \tag{17}$$

This charge is carried by all states which contain quanta of the field L, whether these quanta are on or off shell. Since the elementary L particles and the composite F particles each contain one L quantum, they should each contribute one to the chiral charge. Using Eq. (17) and the Haag expansion, Eq. (6), this fact leads to the following normalization condition for $a(b\cdot k)$.

$$2(2\pi)^3 \int \frac{d^3k}{2(\vec{k}^2+\mu^2)^{1/2}} \, |a(b\cdot k)|^2 = 1 \quad . \tag{18}$$

Note that the same combination of A and C occurs in the normalization condition as in the equation of motion. Before discussing this condition for light-like b, i.e., for a massless bound state, we consider it for time-like b, i.e., for a massive bound state. In this case, the most convenient frame in which to evaluate the integral in Eq. (18) is the rest frame of b. Let $b^2=m^2$, and $b=(m,0,0,0)$. Then, the left-hand side of Eq. (18) becomes

$$\frac{2(2\pi)^4}{m^2} \int_{m\mu}^{\infty} dx \, [x^2-(m\ )^2]^{1/2} |a(x)|^2 \quad .$$

Clearly this will be finite if it decreases fast enough for large values of its argument. Geometrically, the integral at fixed $x=b\cdot k=m(k^2+m^2)^{1/2}$ in the rest frame of b is the intersection of the fixed-energy hyperplane with the mass shell of k, and is finite. For $b^2=0$, there is no rest frame; however we can choose $b=(1,0,0,1)$. Now the integral is

$$\int \frac{d^3k}{2(\vec{k}^2+\mu^2)^{1/2}} \, |a(b\cdot k)|^2 = \int_0^{\infty} \frac{dx}{2x} \, |a(x)|^2 \quad d^2k_T \quad . \tag{19}$$

The integral over the transverse momentum is unrestricted, and diverges. Geometrically, the integral at fixed $x=b\cdot k=(\vec{k}^2+\mu^2)^{1/2}-k^3$ corresponds to the intersection of a hyperplane parallel to the light cone with the mass shell of k, which is a noncompact set, so the integral diverges. In the massless case, the variable x is the x_- light-cone variable; the amplitude a depends on x_- alone, and is independent of k_T. A momentum on the k mass shell can be written

$$k = \left(\frac{\vec{k}_T^2+x^2+\mu^2}{2x}, \vec{k}_T, \frac{\vec{k}_T^2-x^2+\mu^2}{2x}\right), \qquad (20)$$

which makes explicit that there is a two-parameter set of k such that k is on-shell and $b \cdot k = x$. We can remove this redundancy by considering the quotient of the mass shell by two-dimensional transverse momentum space, which factors out the divergent integral in Eq. (19).[13] This will give a well-defined normalization for our amplitude; however, since, as we see below, for nonzero momentum transfers the matrix element of the current density does not diverge, we cannot remove this factor in that case, so this does not resolve our problem.

To write down the matrix element of the current density, we must decide whether the initial and final bound-state four momenta are linearly independent or are collinear. The former case, which is the generic one, corresponds to the assumption made by Weinberg and Witten[14] in their discussion of the nonexistence of high-spin massless particles; the latter case is the one considered by Sudarshan.[15] For the former case, $b+b'$ is timelike. In its rest frame, we can choose $b=(B,0,0,B)$, and $b'=(B,0,0,-B)$. The matrix element is

$$\langle b'|j^{\mu}(o)|b\rangle = \int \frac{d^3k}{2(\vec{k}^2+\mu^2)^{1/2}} \; a^*(k \cdot b')a(k \cdot b)$$

$$= \frac{\pi}{b' \cdot b} \int dx\, dx' \;\; a^*(x)\; a(x') \qquad , \qquad (21)$$

where the integral runs over $xx' > 1/2\, \mu^2\, b' \cdot b$. In the forward limit, which corresponds to the total charge, $b' \cdot b \to 0$, and the matrix element diverges. If the momenta b and b' are collinear, the matrix element diverges without going to the forward limit. It seems unlikely that this divergence can be removed by including higher terms in the Haag expansion, but we have not yet proved this.

It is instructive to see how this normalization problem arises and is resolved in other formalisms. We are studying the case of the Bethe-Salpeter amplitude in collaboration with J. Sucher. We describe this problem in the context of the Wightman framework using spin-zero objects since the difficulty does not seem to depend on spin. Assume that there is an amplitude F for which the composite mass-zero state can be constructed as a bound-state of quanta of two Lagrangian fields A_1 and A_2:

$$|B(b)\rangle = \int d^4p_1\, d^4p_2\, \delta(b-p_1-p_2)F(p_1,p_2)A_2^\dagger(p_2)A_1^\dagger(p_1)|0\rangle \ . \quad (22)$$

It is straightforward to calculate the inner product of two such states:

$$\langle B(b')|B(b)\rangle = \delta(b'-b)\int d^4p_1'\, d^4p_1\, F^*(p_1',b-p_1')G^{(4)}(p_1',b,p_1) \cdot F(p_1,b-p_1) \quad , \quad (23)$$

where, the Dirac δ is four-dimensional. (For those who are disconcerted by the four-dimensional δ, we mention that the usual three-dimensional normalization can be arranged using the asymptotic limit.) The distribution $G^{(4)}$ is the four-point vacuum expectation value in momentum space, and has support $p_1' \geqslant 0$, $b \geqslant 0$, and $p \geqslant 0$. The Lorentz invariance of G implies that it depends on the following scalar products: $p_1'^2$, $p_1'\cdot b$, $p_1'\cdot p_1$, $p_1\cdot b$, p_1^2, and b^2. The function F does not introduce any new scalar products. Thus the finiteness or divergence of the norm of the zero-mass bound state rests on whether or not the five scalar products above which depend on p_1 and p_1' can control the eight-dimensional integration in Eq. (23). The demonstration of the divergence is analogous to the demonstration of the divergence for the NQA amplitude, but more complicated. The essential point is that the scalar product $p_1'\cdot p_1$ must control both the $\vec{p}_{1T}$ and $\vec{p}_{1T}'$ integrations, but $p_1'\cdot p_1$ depends on only one linear combination of these vectors so that there is an unrestricted two-dimensional transverse momentum integral which necessarily diverges. This difficulty can be resolved using the construction of Haag,[16] Nishijima,[17] and Zimmermann.[18] Let the local field associated with the composite particle be defined by the limit

$$B(x) = \lim_{\vec{\varepsilon}\to o} \lim_{\varepsilon^o \to o} \frac{:A_2(x+\frac{1}{2}\varepsilon)A_1(x-\frac{1}{2}\varepsilon):}{\langle o|A_2(\frac{1}{2}\varepsilon)A_1(-\frac{1}{2}\varepsilon)|B(b)\rangle_{in}} , \tag{24}$$

where the double dots stand for the normal product, and B(x) is independent of b. The field $B^\dagger$ acting on the vacuum produces a superposition of states which includes a single B_{in} state with the standard normalization.

This work began in collaboration with S. Nussinov, and I have a great debt to him for the stimulation which started this project. I thank my colleagues at Maryland, including J. Bernstein, W. E. Caswell, A. J. Dragt, A. D. Kennedy, K. I. Macrae, M. Ogilvie, J. C. Pati, J. Sucher, and C. H. Woo for many helpful discussions. I also want to thank a number of colleagues at other institutions, including C. M. Bender, R. A. Brandt, S. J. Brodsky, A. De Rujula, R. Jackiw, G. P. Lepage, M. E. Peskin, and D. Zwanziger for their interest and help in this work. This work was supported in part by the National Science Foundation and by a Faculty Research Grant from the University of Maryland.

REFERENCES AND FOOTNOTES

1. R. Haag, Dan. Mat. Fys. Medd 29, no. 12 (1955).
2. V. Glaser, H. Lehmann, and W. Zimmermann, Nuovo Cimento 6, 1122 (1957).
3. O. W. Greenberg, Phys. Rev. 139, B1038 (1965).
4. O. W. Greenberg and R. J. Genolio, Phys. Rev. 150, 1070 (1966).
5. F. Gross, Phys. Rev. 186, 1448 (1969).

6. K. Johnson, Phys. Rev. D 4, 1101 (1972).
7. A. Pagnamenta, Ann. Phys. 39, 453 (1966).
8. A. Raychaudhuri, Phys. Rev. D 18, 4658 (1978).
9. A. Raychaudhuri, University of Maryland Technical Rep. No. TR-78-048 (unpublished), and University of Maryland Ph.D. thesis 1978 (unpublished).
10. O. W. Greenberg, S. Nussinov, and J. Sucher, Phys. Lett. 70B, 465, (1977).
11. M. Bander, Phys. Rev. Lett. 47, 549 (1981); erratum, op. cit. 47, 1419 (1981).
12. This work was done in collaboration with S. Nussinov.
13. The suggestion to go to the quotient space was made by J. Bernstein.
14. S. Weinberg and E. Witten, Phys. Lett. 96B, 59 (1980).
15. E. C. G. Sudarshan, Phys. Rev. D 24, 1591 (1981).
16. R. Haag, Phys. Rev. 112, 669 (1958).
17. K. Nishijima, Phys. Rev. 111, 995 (1958).
18. W. Zimmermann, Nuovo Cimento 10, 597 (1958).

MONTE CARLO EVALUATION OF HADRON MASSES

Don Weingarten

Indiana University

Bloomington, Indiana 47405

ABSTRACT

An improved Monte Carlo method is presented for lattice gauge theories with fermions. A variety of meson masses and decay constants are calculated on lattice up to 12^4 for gauge group $\bar{I}$, the best discrete approximation to SU(2). Arguments are given to show these predictions would be changed little by replacing $\bar{I}$ with SU(2) or SU(3).

1. INTRODUCTION

It appears rather probable that QCD governs the strong interactions. The old fashioned perturbation theoretic definition of QCD, however, is incomplete. It cannot be used to calculate most masses or scattering cross sections. The Euclidean lattice set up for QCD[1] potentially corrects this problem. The essence of lattice QCD is a mathematically well-defined path integral formula for vacuum expectation values with (1) t replaced by -it, (2) R^4 replaced by a lattice and (3) the lattice restricted to a finite box. The replacement of t by -it converts the oscillatory path integral of Minkowski QCD to an absolutely convergent, exponentially damped expression, and (2) and (3) cutoff UV and IR divergences, respectively.

Two main questions of lattice QCD are: can (1), (2) and (3) be reversed by a limiting process and analytic continuation to recover a sensible, continuum Minkowski theory and, if so, how can the Minkowski theory's consequences be determined? This talk assumes a Minkowski theory can be recovered from lattice QCD and is concerned only with calculating the Minkowski theory's predictions. For practical purposes, the task of obtaining predictions from the continuum Minkowski theory reduces to finding an approximation scheme for obtaining results from the Euclidean lattice theory which generates errors which are stable as the lattice spacing is made small and the box size made large. As I will mention later, many Minkowski predictions can be gotten from the Euclidean theory without continuing -it back to real t.

For a simplified version of QCD with quarks removed, some progress toward a method for calculating the Euclidean theory's consequences has been made over the past few years by adapting Monte Carlo methods, borrowed from statistical physics, to evaluate the Euclidean path integral. This strategy was first suggested by Wilson and first applied by Wilson and by Creutz, Jacobs and Rebbi. Ref. 2 is a highly incomplete list of work using this method. For pure gauge theories the Monte Carlo strategy has produced a number of striking results. It is natural to try to extend the method to theories including fermions. But there are difficulties which I will soon discuss. Possible solutions to these problems were suggested in late 1980 independently by Petcher and myself[3] and by Fucito, Marinari, Parisi and Rebbi.[4] Modifications of Fucito, Marinari, Parisi and Rebbi's method were suggested by Scalapino and Sugar and by Duncan and Furman.[5] More recently, a third class of strategies has been proposed by Hasenfratz and Hasenfratz and by Lang and Nicolai,[6] and Kuti[7] has suggested still another method. For reasons I will not really have time to discuss, I believe the methods of references 3,4,5,7 will require too much computer time to be used on large enough lattices to approximate the infinite volume continuum limit. The strategy of reference 6, on the other hand, depends on approximations which are probably unreliable unless the lattice spacing is fairly large.

What I want to describe today is a version of the method of reference 2 which has been speeded up by an extra approximation[8]. In effect virtual quarks are omitted and we include only valence quarks and all possible interactions of gluons. The result is a method which can be applied in reasonable amounts of time to large enough lattice to permit an estimate of the theory's behavior as the lattice spacing goes to zero and the lattice volume to infinity.

The remainder of this talk will consist of a brief review of Wilson's lattice path integral, followed by a little about how Monte Carlo is done for lattice gauge theories without fermions, then something on how Monte Carlo can be extended to include fermions, and finally a summary of the results of a meson mass calculation on lattices 4^4, 8^4, and 12^4. The gauge group I will use is $\bar{I}$, the best discrete approximation to SU(2). For reasons to be mentioned, the results for flavor nonsinglet meson masses using $\bar{I}$ should be nearly the same as we would have gotten using SU(3) as the gauge group. Work related to what will be described has also been done independently by Marinari, Parisi, and Rebbi and by Hamber and Parisi.[9]

2. PATH INTEGRAL

The theory lives on a periodic, hypercubic lattice with periodicity, say N, in each direction and lattice spacing a taken as the unit of length. On each nearest neighbor link (x,y) lives an element $U(x,y) = U(y,x)^{\dagger}$ of the gauge group G. For a continuous gauge group G, U(x,y) is related to the lattice version of the gauge potential A^a_μ by

$$U(x,y) = \exp[ig_o A^a_\mu (x-y)_\mu T^a] \quad ,$$

where g_o is the bare gauge coupling and the T^a are an orthonormal basis for the Lie algebra of G. At each lattice site x are a collection of anticommuting Grassmann variables $\Psi^{fa}_i(x)$ and $\bar{\Psi}^{fa}_i(x)$ with i=1,...4, a spin index, a=1,...N_c, a color index, and f=1,...N_f, a flavor index. The site variables are the lattice path integral's version of fermion fields.

From the U's and Ψ's, we construct an action

$$S = S_G + S_F \quad ,$$

$$S_G = \frac{\beta}{4} \sum_p \mathrm{Re}\, W(p) \quad ,$$

$$S_F = - \sum_x \bar{\Psi}(x)\Psi(x) + K \sum_{(x,y)} \bar{\Psi}(x)U(x,y)(1-\gamma_{xy})\Psi(y) \quad ,$$

where the sum in S_G is over all oriented nearest neighbor squares, p. For p = (w,x,y,z), W(p) is the trace of the ordered product of the link variables around the perimeter

$$W(p) = \frac{1}{N_c} \mathrm{Tr}\, [U(w,x)U(x,y)U(y,z)U(z,w)] \quad ,$$

beginning at some arbitrarily chosen starting point. The coefficient β is $4g_o^{-2}$. The sum in the first term of S_F is over all sites and in the second term is over all oriented nearest neighbor lattice links (x,y). For $x = y \pm \hat{\mu}$, γ_{xy} is $\pm\gamma_\mu$, where $\hat{\mu}$ is a unit lattice vector in the $+\mu$ direction and γ_1, γ_2, γ_3, and γ_4, are Hermitian Euclidean gamma matrices. The constant K is $(8+2m_o)^{-1}$ where m_o is the bare fermion mass.

If we restore a and take the formal continuum limit $a \to 0$, the nonvanishing terms in S_G and S_F become

$$S_G \to \text{constant} - \frac{1}{4}\int d^4x\, F^q_{\mu\nu} F^q_{\mu\nu} \quad ,$$

$$S_F \to 2 \int d^4x\, \bar{\Psi}(m + \not{\partial} + ig_o \not{A})\Psi \quad ,$$

which is the correct continuum action for Euclidean QCD after a renormalization of Ψ and $\bar{\Psi}$ by $\sqrt{2}$.

From S we define the vacuum expectation $\langle \mathcal{F} \rangle$ for any polynomial $\mathcal{F}$ in Ψ, $\bar{\Psi}$ and U as

$$\langle \mathcal{F} \rangle = Z^{-1} \int d\mu_G \int d\mu_F\, \mathcal{F} \exp(S_F + S_G) \quad , \tag{2.1}$$

where Z is defined by $\langle 1 \rangle = 1$, the gauge measure μ_G is a product of one copy of Haar measure on G for each independent U(x,y), and the fermion integral $\int d\mu_F$ is not really integration in the usual sense at all but instead represents a linear function from the Grassmann algebra of fermion fields to the complex numbers defined by the conditions

$$\int d\mu_F \prod_{a,f,i,x} \Psi_i^{fa}(x)\, \bar{\Psi}_i^{fa}(x) = 1 \quad ,$$

$$\int d\mu_F\, Q = 0 \quad ,$$

for any Q which does not contain each $\Psi^{fa}{}_i(x)$ and $\bar{\Psi}^{fa}{}_i(x)$ exactly once. If you have never seen fermion integration before, this may seem an odd definition but actually it turns out that it is really the natural choice. For a free fermion theory with all the link variables U(x,y) fixed at 1, it is not too hard to calculate the fermion propagator explicitly, take the continuum infinite volume limit and continue back to Minkowski space. The result will be the correct free fermion propagator for Minkowski space.

3. MONTE CARLO INTEGRATION WITHOUT FERMIONS

So that is the setup. I now want to back track and say a few sentences about how to evaluate the path integral without fermions by Monte Carlo. The pure gauge path integral is

$$<F> = Z^{-1} \int d\mu_G \, F \, \exp S_G \quad .$$

Now define a random field U(x,y), over all the links (x,y) of the lattice, with differential probability distribution

$$dP = Z^{-1} \exp S_G \, d\mu_G \quad .$$

Then the vacuum expectation <F> and the average of F(U) over an infinite ensemble of random U fields are the same

$$<F> = <F(u)>_u \quad .$$

The Monte Carlo method of evaluating $<F(U)>_U$ is simply to replace the infinite ensemble of U configurations with a finite ensemble. The finite ensemble is generated by a computer algorithm designed to provide U fields distributed according to dP. For sufficiently large, but finite, ensembles reliable values of $<F(U)>_U$ can be found just as reliable averages of any physical quantity can be gotten from a finite number of experimental events.

I do not really have time to tell you in detail how one goes about generating a sequence of random U. It has already been discussed in other talks. An essential feature of the methods that people have used, which I will need in the discussion later on, is that the computer time for an N^4 lattice grows as $O(N^4)$ as N becomes large. This growth is slow enough that lattice up to 16^4 or so have been done in reasonable amounts of machine time.

4. MONTE CARLO INTEGRATION WITH FERMIONS

How can the Monte Carlo method be applied with fermions present? It can not be used directly on the path integral of (2.1). The fermion integration in (2.1) is not integration in the usual sense and F is not necessarily a number, it could involve Grassmann variables. It is not possible to interpret $\langle F \rangle$ in (2.1) directly as the average of some random variable over a probability measure.

The first step in a way around this problem is simply to carry out the fermion integration in (2.1) explicitly. Recall that S_F is quadratic in the Ψ and $\bar{\Psi}$ fields. It can be written

$$S_F = - \Sigma \, \bar{\Psi}(x)\Psi(x) + K \, \Sigma \, \bar{\Psi}(x)B(x,y)\Psi(y)$$

for a certain coupling matrix B. Thus formally the fermion integration in (2.1) looks like a Gaussian integral. Such integrals for ordinary numbers can be done explicitly. It turns out the same is true for Grassmann variables. Rather than write down the general result, let me just give you an expression for the particular case we will need later to obtain hadron masses

$$\langle \bar{\Psi}(x_1)\Psi(y_1)\bar{\Psi}(x_2)\Psi(y_2) \rangle =$$

$$Z^{-1} \int d\mu_G \, F \, \det(1-KB) \exp S_G \quad , \tag{3.1}$$

$$F = - \, (1-KB)^{-1} \, (y_2,x_1) \; (1-KB)^{-1} \, (y_1,x_2) \quad ,$$

where the x's and y's in this expression are multiindices combining position, flavor, color and spin, and I have assumed for simplicity that the flavor of x_i differs from the flavor of y_i, $i = 1,2$. Without this restriction a second term would be required in F.

Now this expression could, in principle, be evaluated by Monte Carlo using $S_G + \ln \det(1-KB)$ as the action in place of S_G from before. Unfortunately this method would be hopelessly slow. It turns out a direct Monte Carlo evaluation of (3.1) would take $O(N^{16})$ steps, in place of $O(N^4)$ for pure gauge theories. Evaluating (3.1) for one value of K and one value of g_o on a 4^4 lattice by this method would take something like 60,000 hours on a CRAY-1. The Monte Carlo methods of references 3-7 all carry out further transformations of (3.1) to obtain, in effect, more efficient ways of handling det (1 - KB). The method Petcher and I developed requires $O(N^8)$ steps in place of $O(N^{16})$. Fucito, Marinari, Parisi, and Rebbi require $O(N^4(m_\pi/a)^4)$ steps, which becomes $O(N^8)$ if we take a to zero by setting a = L/N, for some fixed lattice periodicity L, and taking N to infinity. The methods of references 5,6 require $O(N^{12})$ steps. These methods are all probably too slow in their present forms for realistic calculations.

For some purposes, in particular evaluating flavor nonsinglet hadron masses, it may however be satisfactory to simply replace det(1-KB) by 1 in (3.1). Let me present a physical picture which supports this hypothesis. The effect of det(1-KB) in strong or weak coupling expansions is to contribute closed quark loops inside diagrams. With det(1-KB) replaced by 1, such loops are removed. With loops removed, the gauge field configurations between, say, the valence quark and antiquark in a meson is expected to be stringlike, characterized by some energy per unit length, the tension T. If we now restore det(1-KB) to the path integral, it is plausible that the field in a meson remains stringlike but at various points has holes where the string has been cut open by a quark loop. The string with breaks will have some new string tension T', with real part smaller than T, since the breaks tend to save string energy, and with an imaginary part, since for long strings some of the quarks and antiquarks occurring along the breaks might condense into real hadrons and contribute inelastic physical intermediate states to the string's propagator.

The hypothesis that the fields in mesons remain generally stringlike in the presence of quark loops is supported by the fact that meson Regge trajectories in the real world are nearly linear as predicted by classical or quantized forms of the Nambu-Goto string model. Omitting quark loops is also what is done to leading order in 1/N expansions.

If field configurations remain stringlike with loops present, then the theory without loops can be made to reproduce the theory with loops simply by shifting β in the loopless theory to get T to equal the real part of T'. Assuming the approximate validity of the string model, the required renormalization of β is automatically accomplished if β is chosen to yield the string tension which predicts the correct physically observed meson Regge slope α' according to the string model relation

$$\alpha' = (2\pi T)^{-1} \quad .$$

With this β, the loopless theory and the theory with loops should give approximately the same valence quark interaction and thus the same flavor nonsinglet mass spectrum. By some round about arguments which I will omit, it can be estimated that the accuracy of the loopless approximation to flavor nonsinglet hadron masses should be of the order of 10%. In any case, the approximation is testable. For example, masses could be calculated on small lattices with det(1-KB) included and compared with the results of the loopless approximation.

Using the loopless approximation, the vacuum expectation of (3.1) now becomes

$$<\bar{\Psi}(x_1)\ \Psi(y_1)\ \bar{\Psi}(x_2)\ \Psi(y_2)> =$$

$$- Z^{-1} \int d\mu_G\ (1-KB)^{-1}\ (y_2,x_1)\ (1-KB)^{-1}\ (y_1,x_2)\ \exp S_G \quad .$$

To evaluate this by Monte Carlo we still need an algorithm for obtaining $(1-KB)^{-1}$.

A fast method for finding $(1-KB)^{-1}$ was proposed in reference 3. It is generally called Gauss-Seidel iteration and goes as follows. For notational convenience let us now interpret 1-KB as a linear operator on a space of vectors with one vector component for each $\Psi(x)$ on the lattice. Let the vector h(z) in the space on which 1-KB acts have a single component of 1 for say z = x, and all other entries 0. Then

$$(1-KB)^{-1}(y,x) = [(1-KB)^{-1}h](y) \quad .$$

Define f to be $(1-KB)^{-1}h$. Then

$$(1-KB)f = h \quad ,$$

$$f = KBf + h \quad . \tag{3.2}$$

Gauss-Seidel iteration simply consists of solving (3.2) by iterating it site by site over the lattice, with some conveniently chosen initial f, for example f = h. If the resulting sequence of f converges, it must approach the required value $(1-KB)^{-1}h$.

5. MASSES

Finally let me show you the results of applying the loopless approximation to a calculation of flavor nonsinglet meson masses. As I mentioned earlier, in place of SU(3), I will use the gauge group $\bar{I}$. From a comparison of $\bar{I}$ with SU(2) by Petcher and myself[10] and by Bhanot and Rebbi,[11] it is clear that $\bar{I}$ results will be indistinguishable from SU(2). On the other hand, the argument I gave before for omitting quark loops can easily be adapted to show that SU(2) and SU(3) gauge groups would yield nearly the same meson spectrum.

To obtain masses we first need some field operators which create the particles in which we are interested out of the vacuum. For simplicity I will restrict the present discussion to pions. Similar calculations can be done for any other hadron. Define the π^+ and π^- fields to be

$$\pi^+(x) = \bar{\Psi}^u(x)\ \gamma^5\ \Psi^d(x) \quad ,$$

$$\pi^-(x) = \bar{\Psi}^d(x)\ \gamma^5\ \Psi^u(x) \quad .$$

Now choose some lattice direction, say x_4, and sum these fields over the hyperplanes orthogonal to x_4. Let the summed field at $x_4 = t$ be $\tilde{\pi}^+(t)$ and $\tilde{\pi}^-(t)$. Acting on the vacuum, these fields create states with zero momentum and pion quantum numbers. The lightest such state we define to be the pion. It can then be shown that if t and its periodic reflection N-t are both large we have

$$\langle\tilde{\pi}^-(t)\ \tilde{\pi}^+(0)\rangle \to Z_\pi\ [\exp(-m_\pi t) + \exp(-m_\pi(N-t))] \quad . \tag{5.1}$$

The second term in this expression is a result of the periodic boundary conditions which we have chosen. We can then calculate the correlation $\langle\tilde{\pi}^+(t)\ \tilde{\pi}^-(0)\rangle$ by the loopless approximation, fit the result to (5.1), and obtain a prediction for the pion mass.

Let me just mention here that there are two technical reasons for working with the hyperplane summed propagator rather than one between individual points. One reason is that the point-to-point correlation function at large distances looks like (5.1) but has as its first correction a term with m_π replaced by the energy of a pion state with minimal translational momentum. For large lattices this term can be close to the leading rest term and makes the leading term hard to pick out. The first correction to the hyperplane propagator has the π' mass, which lies well above the π and does not obscure the leading term. A second advantage of the summed correlation is that it collects more numbers from each individual gauge configuration U used to evaluate the path integral and, as a result, probably has smaller statistical fluctuations than the point-to-point correlation, evaluated over any fixed set of U. The Gauss-Seidel iteration automatically generates enough matrix elements of $(1-KB)^{-1}$ to evaluate the full hyperplane propagator. It takes no more work than a point-to-point correlation.

Now to the results. I measured m_π, m_ρ, m_A1, and f_π. I then chose β to give the correct physical Regge slope, and adjusted K to give m_π correctly. All the theory's parameters are then fixed and the remaining masses and decay constants are predictions. As I mentioned in the introduction, this has been done for lattice 4^4, 8^4, and 12^4. In each case the lattice periodicity was taken to be nearly $(90 \text{ MeV})^{-1}$, about five times larger than the electromagnetic radius of the π or ρ. The lattice spacing is therefore about $(360 \text{ MeV})^{-1}$ for 4^4, $(720 \text{ MeV})^{-1}$ for ^{4}, and $(1080 \text{ MeV})^{-1}$ for 12^4. Thus the 12^4 lattice has its high momentum cutoff well above m_ρ, and its size cutoff a good bit larger than the size of a pion. We would expect the predictions for π and ρ parameters from this lattice to be reasonably close to their infinite volume continuum limits. The final 12^4 predictions for a variety of quantities, and their observed values are:

	predicted	observed
m_ρ	670 ± 100 MeV	770 MeV
m_{A1}	1200 ± 400 MeV	1200 MeV
m_q	5.8 ± 2.0 MeV	6 MeV (?)
f_π	70 ± 50 MeV	90 MeV

The quark mass quoted here, m_q, incidentally, is an average of the up and down quark masses and defined by the relation

$$m_q = m_0 - \bar{m}_0 \quad ,$$

where m_0 is the bare quark mass giving the physical value for m_π and $\bar{m}_0$ is the bare quark mass giving an m_π of zero.

Fig. 1 shows a graph of m_ρ as a function of lattice spacing $a = (90 \times N \text{ MeV})^{-1}$. There is perhaps some sign of m_ρ leveling off

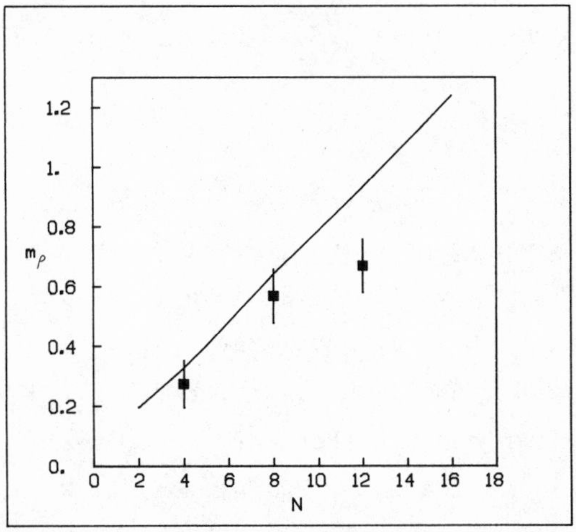

Fig. 1 The rho mass as a function of lattice spacing, $a = (90 \times N \text{ MeV})^{-1}$, with the pion mass and string tension tuned to their physical values.

as a goes to zero. The solid line in fig. 1 is the prediction for m_ρ from Wilson's leading order strong coupling expansion. The value of β at which m_ρ deviates from the strong coupling prediction is about 2.5, which is the same point at which the string tension found by Monte Carlo breaks off from the strong coupling prediction and weak coupling behavior seems to take over.[2]

In conclusion, I think the method I have described is a tractable procedure for obtaining a variety of physical predictions from realistic lattice gauge theories including fermions. Perhaps a convincing test of QCD for low energy phenomena is not too far away.

I would like to thank the high energy experimental physics group at Indiana University for time on their VAX 11/780, T. Sulanke for help using the VAX, and Fermi National Accelerator Laboratory for time on Fermilab's system of CDC 175's. I am grateful to the theory groups at Brookhaven and at Fermilab for their hospitality during the summer of 1981 while part of this work was being completed. For helpful conversations I would like to thank W. Bardeen, D. Brydges, W. Celmaster, M. Creutz, M. Feigenbaum,

B. Freedman, D. Lichtenberg, F. Paige, and D. Petcher. This work was supported in part by the U.S. Department of Energy.

REFERENCES

1. K.G. Wilson, Phys. Rev. D10, 2445 (1973).
2. K. Wilson, Cornell preprint (1979); M. Creutz, L. Jacobs, and C. Rebbi, Phys. Rev. Lett. 42, 1390 (1979); Phys. Rev. D20, 1915 (1979); M. Creutz, Phys. Rev. Lett. 43, 553 (1979); Phys. Rev. D21, 2308 (1980); Phys. Rev. Lett. 45, 313 (1980).
3. D.H. Weingarten and D.N. Petcher, Phys. Lett. 99B, 333 (1981).
4. F. Fucito, E. Marinari, G. Parisi, and C. Rebbi, Nucl. Phys. B180, 369 (1981).
5. D.J. Scalapino and R.L. Sugar, Phys. Rev. Lett. 46, 519 (1981); A. Duncan and M. Furman, Columbia University preprint (1981).
6. A. Hasenfratz and P. Hasenfratz, Phys. Lett. 104B, 489 (1981); C.B. Lang and H. Nicolai, CERN preprint (1981).
7. J. Kuti, Santa Barbara preprint (1981).
8. D. Weingarten, Phys. Lett. 109B, 57 (1982).
9. E. Marinari, G. Parisi, and C. Rebbi, Phys. Rev. Lett. 47, 1795 (1982); H. Hamber and G. Parisi, Phys. Rev. Lett. 47, 1792 (1982).
10. D. Petcher and D. Weingarten, Phys. Rev. D22, 2465 (1980).
11. G. Bhanot and C. Rebbi, Nucl. Phys. B180[FS2], 469 (1981).

REVIVAL OF THE OLD STRING MODEL

Yadin Y. Goldschmidt

Brown University

Providence, Rhode Island 02912

In this talk I would like to review the recent developments in the theory of strings induced by a novel approach due to Polyakov.[1,2] The "old" string model[3] of the early seventies had some severe problems that allowed a consistent treatment only when the space time dimensionality is 26. The important observation was that due to coordinate transformations invariance, also called "gauge" invariance, of the Nambu action, there are relations or constraints among the coordinates and momenta. In order to quantize the theory canonically, one can either eliminate redundant degrees of freedom and quantize noncovariantly or quantize covariantly and impose the constraints on the physical states. It turned out that in the first method, Lorentz covariance was recovered only in d=26, and in the second approach "ghost" states could be eliminated only at d = 26. In this space-time dimensionality both methods agree. But of course we are looking for a string theory which can be quantized in four dimensions.* Another problem of the "old" string model was the

* Brower (Phys. Rev. 06, 1655 (1972) and Goddard and Thorn (Phys. Lett. 40B, 235 (1972 actually showed that it is possible to eliminate ghosts for any d < 26. But there remains the fact that the Pomeron singularity becomes a pole rather than a cut at d = 26.

appearance of "tachyons" with imaginary mass.

Polyakov suggested a new way to quantize the string, with the goal of obtaining a sensible theory in four dimensions, hopefully without the difficulties of the previous method.

Let us review briefly the Nambu action and the classical equation of motion for the string. The string is described by a two dimensional surface $X^\mu(z_1,z_2)$ in a d dimensional space (μ=1, ..d). z_1, z_2 take their values in a two dimensional domain D. The Nambu action reads

$$A = \frac{1}{2\pi\alpha'} \int_D d^2z \sqrt{h_{ab}} \quad , \tag{1}$$

where

$$h_{ab} = \partial_a X^\mu \partial_b X^\mu \ ; \ h = \det h_{ab} \quad . \tag{2}$$

This action is proportional to the area of the surface. It is important to notice that this action is invariant under coordinate transformations

$$z_1 \to z_1^{\Omega}(z_1 z_2),\ z_2 \to z_2^{\Omega}(z_1,z_2),\ X^\mu(z_1,z_2) \to X^\mu(z_1^{\Omega-1}(z_1,z_2), z_2^{\Omega-1}(z_1,z_2)) \ . \tag{3}$$

The classical equations of motion are describing the surface of minimal area $\delta A = 0$. Using the relations

$$\delta h_{ab} = \partial_{(a} X_\mu \partial_{b)} \delta X_\mu \quad , \tag{4}$$

$$\delta\sqrt{h} = \frac{1}{2}\sqrt{h}\, h^{ab}\, \delta h_{ab} \ ; \ h^{ac} h_{cb} = \delta_{ab} \quad , \tag{5}$$

we obtain

$$\partial_a(\sqrt{h}\, h^{ab}\, \partial_b X_\mu) = 0 \quad . \tag{6}$$

Polyakov suggests to start with a different action which is manifestly invariant under coordinate transformation

$$W = \frac{1}{2} \int d^2Z \sqrt{g}\, g^{ab}\, \partial_a X_\mu \partial_b X \qquad (7)$$

(we have suppressed the factor $2\pi\alpha'$).

Here g_{ab} is a metric tensor on the domain D, and one considers X_μ and g_{ab} as independent degrees of freedom. The classical equations of motion become

$$\partial_a(\sqrt{g}\, g^{ab}\, \partial_b X_\mu) = 0 \text{ (by variation with respect to } \chi\text{),} \qquad (8)$$

$$\sqrt{g}\, \partial_a \chi_\mu \partial_b X - \frac{1}{2} \sqrt{g}\, g_{ab}\, g^{cd}\, \partial_c X_\mu\, \partial_d X_\mu = 0 \qquad (9)$$

(by variation with respect to g).

The second equation expresses the fact that the energy-momentum tensor vanishes identically. From Eq. (9) one obtains

$$g_{ab} = \partial_a X_\mu \partial_b X^\mu \qquad (10)$$

and hence Eq. (1) just coincides with Eq. (6) above. In order to quantize the theory Polyakov considers the functional integral

$$Z = \int [dg][dX] e^{-\frac{1}{2}\int d^2z\sqrt{g}\, X_\mu \Delta_g X_\mu \; - \; \mu_o^2 \int d^2z\sqrt{g}} , \qquad (11)$$

where

$$\Delta_g \equiv - \frac{1}{\sqrt{g}} (\sqrt{g}\, g^{ab}\, \partial_b) \qquad (12)$$

is the covariant Laplacian. The second term in the exponent of (11) is introduced since a similar counter term is generated by renormalization. We now fix the gauge (the appropriate Fadeev Popov determinant will be discussed later) and carry out the Gaussian X integration. The gauge we chose is the so called "conformal gauge":

$$g_{ab}(z) = \rho(z)\delta_{ab} \equiv e^{\phi(z)}\delta_{ab} \quad . \tag{13}$$

In this gauge the classical equations of motion become linear: $\partial^2 X^\mu = 0$. One notices that once Eq. (19) is substituted in the exponent of Eq. (11) it seems that the X integration completely decouples from the ϕ integration. This is not so because (11) is not defined without proper regularization which introduces ϕ through the back door. This phenomenon is the so called "trace anomaly" which was first noticed in the theory of gravity. The object we would like to calculate is

$$e^{-F} = \int DX_\mu \, e^{\frac{1}{2}\int_D d^2z\sqrt{g}X_\mu \Delta_g X_\mu - \mu_o{}^2 \int_D d^2z\sqrt{g}} \, , \tag{14}$$

which is the generating functional of a scalar field (μ is an internal symmetry index) in a background gravitational field. The energy momentum tensor is given by

$$T_{ab} = -\frac{2}{\sqrt{g}} \frac{\delta F}{\delta g^{ab}} \quad . \tag{15}$$

Hence

$$T^a{}_a = \frac{2}{\sqrt{g}} g^{ab} \frac{\delta F}{\delta g^{ab}} = -2\frac{\delta F}{\delta \rho} = d\cdot Y(z,z',t) \quad , \tag{16}$$

where $Y(z,z',t)$ is a solution of the "Heat equation"[4]

$$\frac{d}{dt} Y(z,z',t) = -\Delta_g Y(z,z',t) \quad , \tag{17}$$

$$Y(z,z',o) = \frac{1}{\sqrt{g}} \delta(z,z') \quad . \tag{18}$$

t is a "proper time" cutoff which has to be taken to zero at the end after divergent parts have been absorbed in renormalized coupling. The factor d on the r.h.s. of Eq. (16) comes from the fact that there are d scalar fields. One can solve Eq. 17 via a small t expansion to obtain

$$Y(z,z,t) = -\frac{i}{4\pi t} + \frac{1}{24\pi} R + O(t) \quad , \tag{19}$$

with

$$R = -\rho^{-1}\partial^2 \ell n\rho \quad . \tag{20}$$

This implies

$$T^a_{\ a} = \frac{d}{24\pi} R + \text{const} \quad ; \tag{21}$$

hence

$$-F = -\frac{d}{48\pi}\int d^2z \; [\tfrac{1}{2}(\partial_a \ell n\rho)^2 + \mu^2\rho] \quad . \tag{22}$$

It is possible to show that the Faddeev Popov determinant is given by

$$\Delta^{1/2}_{Fp}\;(\phi) = 1/2 \;\ell n \;\text{Det}\; L = \frac{26}{48\pi}\int d^2z \; [\tfrac{1}{2}(\partial_a\phi)^2 + \mu^2 e^\phi] \quad , \tag{23}$$

with

$$L_{ab} = \Delta_b\Delta_a - \Delta_a\Delta_b + g_{ab}\;\Delta_c\Delta^c \quad . \tag{24}$$

Consequently the integral on g becomes $\int [d\phi]\;\Delta^{1/2}_{Fp}\;(\phi)$ and this contribution combines with the previous determinant to yield

$$Z = \int [d\phi]\; e^{-\frac{26-d}{48\pi}\int d^2z[\frac{1}{2}(\partial_a\phi)^2 + \mu^2 e^\phi]} \quad . \tag{25}$$

Polyakov did not consider the contribution of boundary terms. The calculation of boundary terms for the case of Neuman boundary conditions

$$\partial_n X\ (z,z^*) = 0,\ z\varepsilon\partial D \tag{26}$$

(where ∂n denotes differentiation normal to the boundary in parameter space when conformal coordinates have been selected and $z = z_1 + iz_2$) has been performed by Durhuus, Nielsen, Olesen and Petersen.[5] The case of Neuman boundary conditions is important for the calculation of dual amplitudes. DNOP have found that one has to add to the classical action a fare coupling $\lambda_o \int_{\partial D} ds$ because a corresponding divergent counter-term arises in the calculation. Restricting themselves to the case where the domain D is the upper half plane with the real axis as its boundary DNOP found that effective action reads

$$S_{eff} = -\frac{d-26}{48\pi} \int_{\substack{\text{half}\\ \text{plane}}} d^2z\ [\tfrac{1}{2}(\partial_a\phi)^2 + \mu^2 e^{\phi}]$$

$$-\frac{d}{48\pi}\lambda \int_{\partial D} dz\ e^{(1/2)\phi} - \frac{d}{16\pi}\int_{\partial D} dz\partial_n\phi \quad . \tag{27}$$

The amplitude for the scattering of N particles is given by

$$G(P_1,\ldots,P_n) = g^N Z^{-1} < \prod_{i=1}^{N} \int d^2z_i\ \sqrt{g(z_i)}\ e^{ip_i X(z_i)} > \ , \tag{28}$$

where Z = <1>. Fourier transforming, this amplitude just reduces to the sum of all surfaces passing through N given points in X space. The average is with respect to the action (7). The X integration can now be performed to yield

$$G(p_1,\ldots p_N) = g^N\ Z^{-1}\int [D\phi]\ e^{-S_{eff}(\phi)} \int \prod_{i=1}^{N} d^2z_i\ e^{\phi(z_i)}\ e^{\frac{\pi\alpha}{\hbar}\sum_{i,j=1}^{N} p_i p_j\ K(z_i,z_j,\phi)} \ , \tag{29}$$

where $K(z,z'\phi)$ is the solution of Laplace equation on the upper half plane with Neuman boundary conditions

$$\partial_a(\sqrt{g}\, g^{ab}\, \partial_b K(z,z')) = \delta(z-z') \quad . \tag{30}$$

The solution of equation (30) is

$$\begin{aligned} K(z,z',\phi) &= \frac{1}{4\pi}\, \ell n|z'-z|^2 + \frac{1}{4\pi}\, \ell n|z'-z^*|^2, \quad z' \neq z \quad , \\ &= \frac{1}{4\pi}\ell n\varepsilon - \frac{\phi(z)}{4\pi} + \frac{1}{4\pi}\, \ell n|z-z^*|^2, \quad z'=z \ ; \end{aligned} \tag{31}$$

here ε is a cutoff. The important observation is that K depends on ϕ when z and z' coincides, again because of the need to properly regularize it. The last term on the right hand side of Eq. (31) is just the potential at z' due to the image charge to the charge at the point z. Since for the time being it is not known how to solve the quantum Liouville theory it was proposed to substitute in Eq. (29) a classical solution of the Liouville equation satisfying the boundary conditions. Fluctuations around the saddle point vanish in the limit $d \to -\infty$.

The Liouville equation with the boundary terms reads (here $y = z_2$ is the direction perpendicular to the boundary)

$$\partial^2\phi = \mu^2 e^{\phi} + \delta(y)\, [\tfrac{1}{2}\lambda e^{\frac{1}{2}\phi} - \frac{\partial\phi}{\partial y}] + 3\delta'(y) \quad . \tag{32}$$

In the interior of the domain Eq. (32) just reduces to the usual Liouville equation. This equation has the conformal invariance

$$z \to W(z) \quad ,$$

$$\phi(z) \to \phi(W(z)) + \ell n\, \left|\frac{\partial W}{\partial z}\right|^2 \quad . \tag{33}$$

Therefore, once a particular solution

$$\phi(f) = -2\ell n(\frac{\mu^2}{2} - ff^*) \tag{34}$$

is found, the general solution follows from (3)

$$\phi(z,z^*) = n \frac{|f'(z)|^2}{[(1/2)\mu^2 - |f(z)|^2]^2} \quad . \tag{35}$$

For $\mu^2 > 0$ this solution is always singular on a curve in the complex plane.

The singularity can be arranged on the boundary in such a way that Eq. (32) with the boundary terms will be satisfied. The appropriate solution is

$$\phi(x,y) = \ell n \frac{2}{\mu^2 y^2} \quad , \tag{36}$$

provided

$$\lambda = \sqrt{2}\,\mu \quad . \tag{37}$$

Substituting (36) in (29) one obtains

$$G(P_1,\ldots,P_n) = \bar{g}^N \int_D \prod_{j=1}^{N} d^2 z_j \prod_{i<j} |z_i - z_i|^{\alpha' P_i P_j} |z_i - z_j^*|^{\alpha' P_i P_j}$$

$$\times \prod_{\ell=1}^{N} |z_\ell - z_\ell^*|^{-2 + \alpha' P_\ell^2} \quad . \tag{38}$$

We observe that there are poles in external legs occurring as $z_\ell \to z_\ell^*$ and the external momenta obey the relation

$$\alpha' P_i^2 = 1,0,-1,-2,\ldots \quad . \tag{39}$$

The lowest state is a tachyon. The hope is that this problem occurs only in the saddle point approximation and will be eliminated in the exact solution of the Liouville action. After poles in the external lines have been identified it is possible to write down the S matrix

$$S(P_1,\ldots,P_n) = g^{-N} \int_{-\infty}^{\infty} \prod_{j=1}^{N} dz_j \prod_{i<j} |z_i - z_j|^{2\alpha' P_i P_j} \;;(40)$$

the integration is now along the real axis which is the domain boundary, and Eq. (40) is just the well known Koba-Nielson amplitude. The singularities of this amplitude lie on linear Regge trajectories which describe infinitely narrow states.

A saddle point analysis of the Liouville action has been carried out by C. I. Tan and author,[6] in order to probe the Pomeron singularity for the case of the string with a cylinder topology.

Polyakov also suggested a method to quantize the supersymmetric string[2] which was first introduced by Neveu and Schwarz.[7] Following Brink and Schwarz,[8] he introduced four fields: $X^\mu(z)$ and $\psi^\mu_\alpha(z)$, bosonic and fermionic fields, respectively, describing the string, together with a metric tensor $g_{ab}^{(z)}$ and a spin 3/2 gravitino field χ_a^α. The action reads

$$S = \frac{1}{2}\int d^2a \sqrt{g}\,\{g^{ab}\partial_a x \partial_b + \bar{\psi}\, i\gamma^a\, \partial_a\psi + \bar{\chi}_a \gamma^b\gamma^a \psi\partial_b x + \bar{\chi}_a \gamma^b\gamma^a \chi_b \bar{\psi}\psi\}. \tag{41}$$

If we choose the conformal gauge

$$g_{ab}(z) = e^2(z)\delta_{ab} ,$$

$$\chi_a(z) = \gamma_a \chi , \tag{42}$$

then using the identity $\gamma_a\gamma^b\gamma^a = 0$, we observe that the action does not depend on e and χ. Again, one has to be careful about the anomalies. Integrating out x and ψ one defines

$$e^{-W(e,x)} = \int dx d\psi\, e^{-S} . \tag{43}$$

Calculating the anomalies Polyakov shows that

$$w(e,x) = A \int d^2x \left[\frac{1}{2}(\partial_\mu \phi)^2 + \frac{1}{2}\mu^2 e^{2\phi} + \frac{1}{2}\bar{\chi} i \partial\!\!\!/ \chi + \frac{1}{2}\mu\bar{\chi}\chi e^{\phi} \right], \quad (44)$$

where $\phi = 2\ln e$ and $A = \frac{10 - d}{8\pi}$ after including the appropriate Faddeev Popov determinant. Again we observe the appearance of ten dimensions where the theory becomes free. Eq. (44) is the supersymmetric Liouville action. Polyakov conjectured[2] that the 3D Ising model can be expressed as a supersymmetric string and therefore its critical behavior can be obtained from the renormalization properties of the supersymmetric Liouville action and its implication for the 3D Ising model is given in ref. (9).

To summarize, Polyakov introduced a new method to quantize the string using a proper treatment of conformal anomalies. The usefulness of this approach will depend on our ability to solve or treat systematically the quantum Liouville theory, and to find out if the resulting string theory at d = 4 is a "healthy" theory without tachyons, branch cuts, singularities, and ghosts. In that case it can be the first step towards an analytic solution of QCD as the first term in some systematic approximation.

REFERENCES

1. A. M. Polyakov, Phys. Lett. 103B, 207 (1981).
2. A. M. Polyakov, Phys. Lett. 103B, 211 (1981).
3. For a review of the "old" string theory see S. Mandelstam, Phys. Rep. 13C, 261 (1974); C. Rebbi, Phys. Rep. 12C, 1 (1974); and "Dual Theory", edited by M. Jacob, North Holland, and references therein.
4. For a review of the trace anomaly and the proper time method see S. W. Hawking, Comm. Math. Phys. 55, 133 (1977) and V. N. Romanov and A. S. Schwarz, Theo. Mat. Fiz. 41, 190 (1979).
5. B. Durhuus, H. B. Nielsen, P. Olesen and J. L. Petersen, Niels Bohr Institute preprint NBI-ME-81-40.

6. Y. Y. Goldschmidt and C. I. Tan, Physics Lett. B. - to be published. (Brown University preprint HET 473).
7. A. Neveu and J. Schwarz, Nucl. Phys. B31, 86 (1971).
8. L. Brink and J. Schwarz, Nucl. Phys. B121, 285 (1977).
9. Y. Y. Goldschmidt, Physics Lett. B. - to be published. (Brown University preprint HET 463).

QCD ON A RANDOM LATTICE

N.H. Christ

Columbia University

New York, New York 10027

The usual lattice gauge theory developed by Wilson[1] and discussed by many of the previous speakers has two important advantages over the continuum theory: i) The number of degrees of freedom per-unit-volume is finite, so that the theory is regularized and numerical calculation is possible. ii) Both the weak- and strong-coupling regimes can be evaluated analytically. The long distance behavior of the weak-coupling limit agrees with continuum perturbation theory while the strong-coupling limit shows confinement.

However, as conventionally formulated, this theory is based on a hypercubic lattice; the theory is Lorentz invariant only in the $g \to 0$, continuum limit. The physically attractive strong-coupling limit is very rotationally asymmetric - the confining force between two external color charges varies by factors of two as the line between the charges varies relative to the lattice. Furthermore, the regularity of the hypercubic lattice may have additional, more subtle, effects such as the "roughening transition"[2,3] which separates the tractable, strong-coupling region from the physical, $g \to 0$ limit.

This research was supported in part by the U.S. Department of Energy.

The lattice gauge theory which is the subject of this talk is based on a random lattice. Such a theory is intrinsically rotationally symmetric and quite irregular, avoiding the difficulties just described. The work which I will describe was done in collaboration with Richard Friedberg and T.D. Lee.[4] It will first discuss the definition and properties of our random lattice and then consider a pure gauge theory defined on such a lattice, examining in particular its strong-coupling limit.

Let us begin with a volume Ω in D-dimensional space-time which contains N points distributed at random. These points are our lattice sites. For convenience we will impose periodic boundary conditions on Ω. In order to define a gauge theory on such a lattice its necessary to introduce links connecting "neighboring" sites. As an illustration of the many possible definitions, examine the links shown in Figure 1 which were drawn so that each lattice

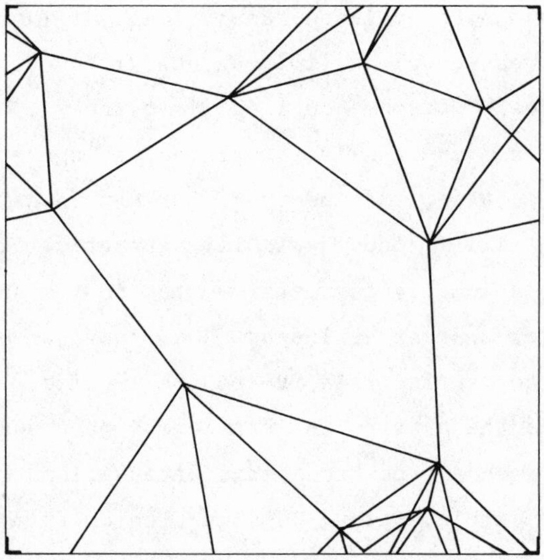

Fig. 1 A lattice of ten random sites defined with periodic boundary conditions. The links are drawn so that each site is connected to each of its five nearest neighbors.

site is connected to (at least) its five nearest neighbors. Although such a lattice may indeed be satisfactory as the basis of a gauge theory, the center of the figure contains a rather large region with five sites on its boundary but traversed by no links. In addition the complex of overlapping lines in the lower right provides no natural choice of plaquettes to be used in defining a gauge theory action.

A more satisfactory choice of links is based on aggregates of D+1 sites which we call "clusters". A set of D+1 sites forms a cluster if the unique D-1 dimensional sphere which passes through the D+1 sites contains no other lattice sites in its interior. Each pair of sites in a cluster is then connected by a link. In two dimensions, these clusters contain three sites and the resulting links form a triangle. For D=3 the linked cluster is a tetrahedron and for D=4 a four-simplex. If the collection of sites shown in Figure 1 is linked according to this prescription, the lattice of Figure 2 results. As is evident, the result is an irregular triangular lattice made up of nonoverlapping triangles filling the entire volume. The circle appearing in that figure contains no sites, hence the three sites on its circumference form a cluster and are linked to form one of the triangles in the lattice. It is easy to prove that this is the general result of our procedures: in D dimensions the links so defined make up an array of nonoverlapping D-simplexes filling the entire volume. The lattice dual to the one just defined is represented by the dashed lines in Figure 2. This dual lattice is made up of an array of convex polyhedra and is discussed in the literature of disordered media,[5] where it is referred to as the lattice of "Voronoi polyhedra".

Given this particular construction of a random lattice, we can explicitly compute some of its average properties. For example the total number, n_o, of D-simplexes present in the volume Ω is given by

$$n_o = \frac{1}{(D+1)!} \int e^{-\rho V(\vec{r}_1, \ldots, \vec{r}_{D+1})} \prod_{i=1}^{D+1} d^D r_i \quad . \qquad (1)$$

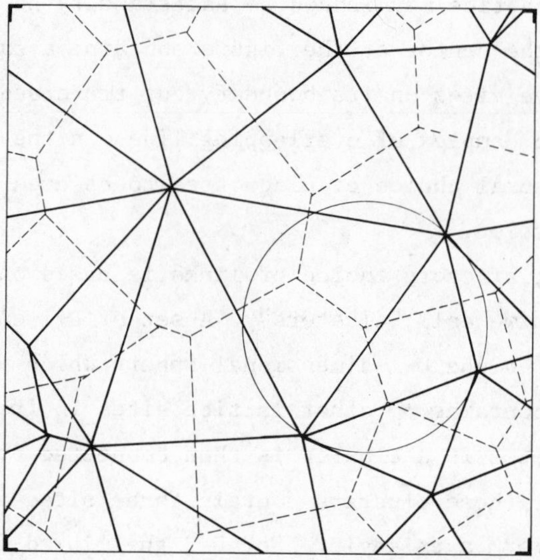

Fig. 2 The lattice shown in Figure 1 with links redrawn so that each group of points which form a cluster is linked together. The dashed lines are the links of the dual lattice.

Here ρ is the density of lattice sites and $V(r_1,\ldots,r_{D+1})$ is the volume of the interior of the (D-1)-dimensional sphere passing through the D+1 points $r_1,\ldots,r_{D+1}$. The exponential of $(-\rho V)$ is the probability that the volume V contains no lattice sites - hence the probability that those D+1 sites form a D-simplex. The integral over all possible positions of the vertices of this D-simplex then gives (D+1)! times the average number of such simplices and eq. (1) results. Define $N_{d/o}$ as the average number of d-simplices connected to a single site in our lattice. Thus $N_{d/o} = (d+1)\, n_d/N$. If the integral in eq. (1) is explicitly evaluated for D=4 and Euler's theorem (relating the number of points, edges, faces, etc. appearing in the division of a surface into simplexes) is repeatedly applied we obtain the following results:

$$N_{4/0} = \frac{1430}{9}, \quad N_{2/0} = \frac{590}{3},$$
$$N_{3/0} = \frac{2860}{9}, \quad N_{1/0} = \frac{340}{9}. \tag{2}$$

These numbers are large compared to the corresponding quantities for a rectangular lattice. For example, the number of links connected to a single site in a rectangular lattice is 8, less than one quarter of the result 340/9 = 37.7 obtained above. In a similar fashion we can introduce additional factors into the integral in eq. (1) and compute the average length of a link $\bar{\ell}$, area of a triangle $\bar{\Delta}$, volume of a tetrahedron $\bar{\tau}$, or total 4-volume $\bar{V}$ found in an average 4-simplex:

$$\bar{\ell} = \left(\frac{2}{\pi^2}\right)^{1/4} \frac{27}{77\pi} \frac{16!!}{15!!} \Gamma\left(\frac{1}{4}\right) = 1.3825,$$
$$\bar{\Delta} = (2)^{1/2} \frac{2}{13} \left(\frac{8}{11}\right)^2 \frac{17!!}{16!!} \Gamma\left(\frac{1}{2}\right) = .6809,$$
$$\bar{\tau} = \left(\frac{2}{\pi^2}\right)^{3/4} \frac{32}{385} \left(\frac{14}{13}\right)^2 \frac{18!!}{17!!} \Gamma\left(\frac{3}{4}\right) = .1923, \tag{3}$$
$$\bar{V} = \frac{9}{286} = .0314.$$

It is straightforward to define a gauge theory on such a random lattice. Following the usual procedure, we assign a group element U(i,j) to a link going from the site j to the site i with U(j,i) = U(i,j). For QCD the variable U(i,j) would be an SU(3) matrix. Next, to each triangle Δ_{ijk} made of the vertices i,j,k belonging to a cluster, we associate the product

$$U_{ijk} = U(i,j)\, U(j,k)\, U(k,i). \tag{4}$$

Each product U_{ijk} appears once in the action

$$A = \frac{1}{g^2} \sum_{\Delta ijk} \kappa_{ijk} \{tr(I-U_{ijk}) + c.c.\} \quad , \tag{5}$$

where the sum ranges over all such triangles, and g is the coupling constant. The weights κ_{ijk} are chosen so that the action (5) reduces to the continuum Yang-Mills action if U(i,j) is replaced by the path-ordered integral

$$U(i,j) \rightarrow P \exp(-ig \int dx^\mu A^a_\mu \lambda^a) \quad , \tag{6}$$

where the vector potential $A^a_\mu(x)$ is a slowly varying function of the position x. Here the matrices λ^a are the generators of the group.

A quantum lattice gauge theory can be defined for a particular random lattice by incorporating the action (5) in the usual path-integral formula for a quantum mechanical amplitude. We expect that such a theory defines a regularized version of the standard continuum gauge theory. Just as for a regular lattice, the long-wave-length properties of this random lattice gauge theory should approach those of the continuum theory in the limit of vanishing lattice coupling g. (Of course, we must require that the particular random lattice which we are using satisfies some conditions of randomness in order to rule out pathological possibilities.) The weak-coupling limit of this random lattice gauge theory appears difficult to analyze beyond the "tree" level and I will not discuss it here. Instead, let us examine the strong-coupling limit.

Consider first the Wilson loop expectation value corresponding to a large planar loop C:

$$W(C) = \langle tr \prod_{(i,j)\varepsilon C} U(i,j)\rangle \quad . \tag{7}$$

The matrices U(i,j), (i,j) ε C correspond to a sequence of links forming a curve in the lattice which approximates the original loop C; the matrices are ordered in the same sequence as the corresponding links. In a theory where color charges in the fundamental representation are confined by a linear potential, this expectation value should behave for a large loop C as

$$W(C) = \exp(-K\ A) \quad , \tag{8}$$

where A is the area of the loop and K the string tension.

Since the integral of a single matrix variable U(i,j) vanishes, each link appearing in the product in (7) must also be present in a factor U_{ijk}/g^2 coming from the expansion of the exponentiated action. Likewise, a link (i,k) present in U_{ijk}, but not lying on the Wilson loop, must also appear in a second factor $U_{ik\ell}/g^2$ coming from the action. Thus the leading strong-coupling contribution to (7) must come from a surface of triangles bounded by the original sequence of links. With appropriate normalization, the leading contribution to (7) in the strong-coupling approximation is simply

$$W(C) \sim (1/g^2)^{N'} \quad , \tag{9}$$

where N' is the number of triangles in this minimal surface.

We can significantly simplify the determination of this minimal surface if we generalize somewhat our random lattice theory. Let us add to the Feynman path integral which defines the quantum gauge theory an average over the entire ensemble of random lattices with fixed site density ρ. That is, in addition to the integration over the link variables U(i,j), we will also integrate over the positions $\vec{r}_i$ of the sites. Since almost every lattice in the ensemble is presumed to have the same weak-coupling, long-wavelength, behavior as the continuum theory, the ensemble average should also. Thus we consider a lattice approximation to the continuum theory in which the lattice sites themselves appear as dynamical variables.

With this modification, the strong-coupling limit of the Wilson loop expectation value, is simply the average of (9) over the ensemble of all random lattices. For large g, we can evaluate that average using the saddle-point approximation. The largest contribution to the average comes from lattices containing a nearly planar array of triangles bordered by the Wilson loop. If $\vec{r}_1,\ldots,\vec{r}_m$ are the positions of the vertices of these triangles, then to leading order, the probability of finding a lattice in the ensemble containing these triangles is

$$\exp(-V(\vec{r}_1,\ldots,\vec{r}_m)) \quad , \tag{10}$$

where $V(\vec{r}_1,\ldots,\vec{r}_m)$ is the minimum volume that must be free of other sites in the lattice if the sites $\vec{r}_1,\ldots,\vec{r}_m$ are to be linked according to our prescription (for simplicity we have adopted units in which $\rho = 1$). For each triangle in the array, let S_{ijk} be the D-1 dimensional sphere of minimum volume on whose surface the vertices of the triangle lie; $V(\vec{r}_1,\ldots\vec{r}_m)$ is the volume of the union of all of these spheres. The saddle-point is determined by that planar configuration of triangles, specified by $r_1,\ldots,r_m$ for which the product of the probability (10) and the factor $(1/g^2)^{N'}$ is a maximum. For a fixed number of triangles the minimum empty volume V is achieved when the triangles form a regular equilateral array. We can uniquely associate with each triangle in the array that portion of the sphere S_{ijk} which lies directly above or below the triangle (see figure 3), that is, the volume of the sphere S_{ijk} minus the three caps. This volume can be readily calculated in terms of the area of the associated triangle. For four dimensions the result is

$$v = \frac{A^2}{N'^2}\,\frac{\pi}{\sqrt{3}} \quad , \tag{11}$$

where A is the area of the entire array. Thus we must choose the number of triangles N' to minimize the product

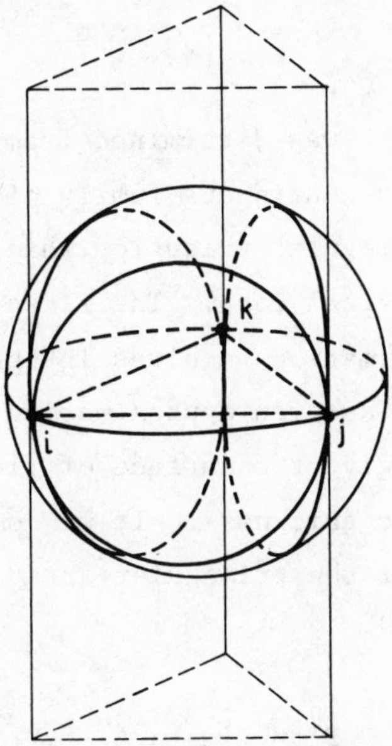

Fig. 3 The sphere S_{ijk} of minimum volume that must be empty if the triangle Δ_{ijk} is to occur in our lattice. That portion of the sphere lying inside the vertical, right triangular prism contains the volume v, uniquely associated with that triangle.

$$e^{-N' \operatorname{Ln} g^2}\left[e^{-\frac{A}{N'^2}\frac{\pi}{\sqrt{3}}}\right]^{N'} = e^{-A(n \operatorname{Ln} g^2 + \frac{1}{n}\frac{\pi}{\sqrt{3}})} , \qquad (12)$$

where $n = N'/A$. This expression is a minimum when

$$n \cong \sqrt{\frac{\pi}{\sqrt{3}}\frac{1}{\operatorname{Ln} g^2}} , \qquad (13)$$

so the saddle point approximation to the expectation value (7) is

$$W(C) = \exp -A\, 2 \sqrt{\frac{\pi}{\sqrt{3}} \ln g^2} . \qquad (14)$$

The area law behavior in (14) implies the confinement of color charges in the standard way with a string tension K given by

$$K = 2\sqrt{\frac{\pi}{\sqrt{3}}\ln g^2} \quad . \tag{15}$$

Although this result was determined from a particular surface of triangles, there is a continuous family of lattices containing surfaces of slightly irregular triangles which do not lie precisely in the plane of the original loop. The first order correction to our saddle-point approximation requires integration over these nearby configurations in a Gaussian approximation. These fluctuations about the saddle-point give the surface of flux bounded by the Wilson loop an effective thickness. If we compute the rms displacement, T of the center of the triangular array out of the plane of the Wilson loop, we find

$$T^2 = \frac{1}{2\pi K}\ln A \quad . \tag{16}$$

This relation between the string tension and the string thickness is the same as that found for the relativistic string.[6] The logarithmic dependence of the result (16) on the area of the Wilson loop follows from the variation of $V(\vec{r}_1,\ldots,\vec{r}_m)$ coming from long wavelength fluctuations of the vertices $\vec{r}_1,\ldots,\vec{r}_m$. For long wave length, the resulting fluctuation in V is simply K times the variation in the area of the fluctuating surface. Thus, in the long wavelength approximation, the dynamics of our fluctuating surface is the same as that of a relativistic string with the action exactly proportional to the space-time area swept out by the string.

Using a similar style of argument, we can also compute the glue-ball mass in the strong-coupling approximation. Consider the correlation function

$$< \mathrm{tr}\, U_{ijk}\ \mathrm{tr}\, U_{\lambda mn} > \quad , \tag{17}$$

where Δ_{ijk} and $\Delta_{\ell mn}$ are two triangles in our lattice separated by a distance τ. For large τ this expectation value should behave as

$c_1 + c_2 \exp(-\mu\tau)$ where μ is the mass of the lowest lying excitation in this pure gauge theory. Again we can compute this expectation value in the strong-coupling approximation, if we evaluate the average over an ensemble of random lattices using the saddle-point method. Thus, to find the connected term $c_2 \exp(-\mu\tau)$ we need to find that chain of four-simplices joining the two triangles Δ_{ijk} and $\Delta_{\ell mn}$ appearing in (17) which makes the largest contribution to that expectation value. Presumably, a prism of four-simplices similar to the three-dimensional structure of figure 4 is this most favorable configuration. (Actually, a small twist of approximately 4° per tetrahedron makes the configuration more probable.) Just as before, we can find that prism of length τ containing N four-simplices which is most probable to occur in the lattices of our ensemble. The volume v, uniquely associated with each four-simplex, that must be empty of other lattice sites if the links required by such a prism (of fixed length) is to occur according to our rules is given by $19.83\ (\tau/N)^4$. Hence, the saddle-point contribution to the connected part of the correlation function (17) coming from a prism containing N simplices is

$$\exp\{-2N\ \ell n\ g^2 - (\tfrac{\tau}{N})^2\ 19.38\ N\} \quad , \tag{18}$$

which can be rewritten in terms of the number of four-simplices per unit length $n = N/\tau$ as

$$\exp - \tau(2n\ \ell n\ g^2 + 19.83/n^3) \quad . \tag{19}$$

This is a maximum for $n = 2.34\ (\ell n\ g^2)^{-\frac{1}{4}}$. If this value of n is substituted into (19), that expression becomes the saddle-point approximation to the connected part of the correlation function in question. We conclude that the glue-ball masss is

$$\mu = 6.19\ \ell n^{\frac{3}{4}}\ (g^2) \tag{20}$$

in the strong-coupling approximation. From the dimensionless ratio μ^2/K we can obtain the coupling constant g:

$$\ell n\, g^2 = \frac{(\mu/6.19)^2}{K\sqrt{\sqrt{3}/\pi^2}} . \tag{21}$$

Substituting, for example, the Montecarlo result of Berg, Billoire, and Rebbi[7] for the ratio $\mu/\sqrt{K} = 2.4\pm0.6$ into eq. (21) implies $\ell n\, g^2 = .41\ (+.23/-.18)$, a value not entirely incompatible with a strong-coupling approximation.

As a final application of these techniques (which makes significant use of the rotational symmetry of our random lattice), let us discuss the masses of higher angular momentum glue-ball states.

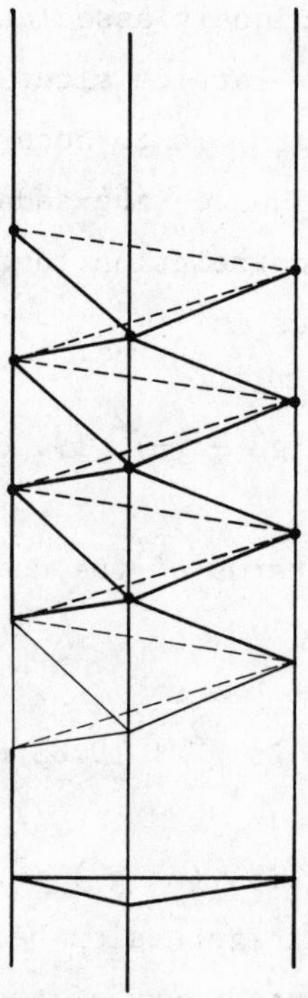

Fig. 4 The prism of tetrahedra which links the triangles Δ_{ijk} and $\Delta_{\ell mn}$. In the strong coupling approximation to the expectation value in (17), only those terms in the action corresponding to triangles lying on the three faces of this prism contribute.

This can be done by computing an angular momentum projection of the correlation function (17). For simplicity, consider a system with one time and two space dimensions so that spatial rotation are one parameter rotations about the time axis. In this case, the mass m(1) of the lowest-lying glue-ball state with angular momentum ℓ can be extracted from the large τ behavior of the following integral:

$$\int_0^{2\pi} d\Phi\, e^{-i\ell\Phi} \langle R_\Phi\{trU_{ijk}\}\, trU_{\ell mn}\rangle \quad . \tag{22}$$

Again we can identify that configuration of tetrahedra which links the rotated triangle Δ_{ijk} with the second triangle and which occurs with the greatest probability in our ensemble of random lattices. Presumably a prism of N tetrahedra like that in Figure 4, but with a net longitudinal twist through an angle Φ is selected out. If $V(\Phi/N)(\tau/N)^3$ is the minimum volume per tetrahedron that must be free of other lattice sites if the links are drawn as required, then in the strong-coupling limit the expectation value (22) becomes

$$\sum_N \int_{-\infty}^{\infty} d\Phi\, e^{-i\ell\Phi}\, e^{-2N\ell n\, g^2}\, e^{-(\tau/N)^3 V(\Phi/N)N} \quad . \tag{23}$$

By allowing the integral over Φ to extend from $-\infty$ to $+\infty$, we add together the contribution of configurations which are twisted through $\Phi + 2\pi q$ for a positive or negative integer q. Change variables in (23), letting $n = N/\tau$, the number of tetrahedra per unit length of the chain, and $\phi = \Phi/N$, the net twist per tetrahedron; then (23) becomes

$$\int_{-\infty}^{\infty} d\phi \int_0^{\infty} dn\; e^{-\tau[+i\ell n\phi + 2n\, \ell n(g^2) + V(\phi)/n^2]} \quad . \tag{24}$$

If $\ell << \ell n(g^2)$, then for two spatial dimensions, $n \sim \ell n^{-\frac{1}{3}}(g^2)$ and ϕ is forced to be near 0, the minimum of $V(\phi)$. The lowest order effect of ℓ can be obtained by Taylor expanding $V(\phi)$ through second order

$$V = V_o + \frac{1}{2} V_2 \phi^2 \quad . \tag{25}$$

With this approximation and the integral over n performed, the integral over ϕ becomes

$$\int_{-\infty}^{\infty} d\phi \; e^{-\tau[in\ell\phi+\mu+\frac{1}{2}V_2 \frac{\phi^2}{n^2}]} = e^{-\tau[\mu + \frac{1}{2}\ell^2 n^4/V_2]} \quad , \tag{26}$$

where n is to be replaced by its saddle-point value $(2.34\cdot(\ell n\, g^2)^{-\frac{1}{4}}$ if this were three spatial dimensions). Thus the glue-ball spectrum looks like that of a two-dimensional rigid rotor with moment of inertia

$$I = V_2/n^4 \quad . \tag{27}$$

We have also carried out the above calculation for three spatial dimensions with analogous results. For $\ell \ll \ell n(g^2)$ the spectrum is that of a rigid rotor with two different moments of inertia. In addition, the case of large ℓ can be analyzed. For $\ell \gg \ell n(g^2)$ the minimum cancellation in the ϕ integral occurs if $V(\phi)$ is a rapidly varying function of ϕ. The optimum ϕ dependence results if the prism of Figure 4 is stretched into a double sheet of triangles of width $2\sqrt{1/\pi K}$ and length τ. The resulting mass is given by

$$m(1) = 8\tau\, \ell \sqrt{\frac{\pi}{\sqrt{3}} \log(g^2)} \quad . \tag{28}$$

Hence for large ℓ, the glue-ball spectrum shows a linear Regge behavior. The Regge slope is simply $4\pi K$ if expressed in terms of the string tension K, just as is the case of a relativistic string.

In conclusion, I have outlined a generalization of the usual hypercubic lattice gauge theory based on a lattice constructed from a random array of sites. As we have seen, the strong-coupling limit of this random-lattice gauge theory is particularly atractive, giving a string thickness and high angular momentum glue-ball mass identical to that found for a relativistic string. Such a construction has at least two applications: i) As a basis for Montecarlo calculation using a fixed random lattice that is changed infrequently and ii) As a framework for further strong-coupling computations where, as above, an average over lattice sites is added to the Feynman path integral. It is our hope that the greater continuity and rotational symmetry of this formulation increase the range of couplings for which this lattice theory is a good approximation to the continuum.

REFERENCES

1. K. Wilson, Phys. Rev. D10, 2455 (1974).
2. Anna Hasenfratz, Etelka Hasenfratz and Peter Hasenfratz, Nucl. Phys. B180, 353 (1981).
3. C. Itzykson, Michael E. Peskin and J.B. Zuber, Phys. Lett. 95B, 259 (1980).
4. N.H. Christ, R. Friedberg and T.D. Lee, Columbia University preprint CU-TP-205, to appear in Nucl. Phys.
5. J.M. Ziman, Models of Disorder (New York, Cambridge University Press, 1979).
6. M. Luscher, Nucl. Phys. B180, 52 (1981).
7. B. Berg, A. Billoire, and C. Rebbi, Brookhaven preprint, BNL 30826. (This paper contains references to other glue-ball mass calculations).

COLLECTIVE FIELD THEORY*

B. Sakita

City College of the City University of New York

New York, New York 10031

ABSTRACT

The ideas of collective field theory are reviewed by using Kogut-Susskind Hamiltonian of U(N) gauge theory. It is shown that the large N limit is the classical limit of the collective Hamiltonian. A solution is obtained for the strong coupling case. As an exactly solvable example we discuss the one plaquette Kogut-Susskind model in detail. In the final section the possibility of weak coupling expansion is briefly discussed.

I. INTRODUCTION

The collective field theory[1,2] is an extension of the Bohm-Pines theory of plasma oscillations[3] to various quantum systems, including Yang-Mills gauge field theory. The theory is especially suitable for the investigation of the large N limit of various models. In the Bohm-Pines theory this limit corresponds to the high density limit of plasma.

*Supported in part by NSF-PHY-78-12399 grant and Faculty Research Award, PSC-BHE.

Although the original intention was to explore a new nonperturbative method for non-Abelian gauge theory, the most successful applications of the theory are so far confined to the models whose large N limit can be obtained by other means, such as the matrix model,[4] nonlinear σ model,[5] one plaquette Kogut-Susskind model,[6] and QCD in two dimensions.[7] The starred references are the works which do not use the collective field method.

Originally the collective field theory was formulated in the Hamiltonian canonical formalism. However, the same idea can be used for the path integral formulation of Euclidean field theories based on the Lagrangian.[8] It has been shown in reference 8 that an application of the collective field idea leads to the same large N classical equation derived from the Dyson-Schwinger equation such as the Migdal- Makeenko equation[9] in the large N limit of U(N) gauge theory.

The other related works include the approach of the so-called "large N classical dynamics" first developed by Jevicki, Papanicolaou and Levine[10] and later greatly advanced by Bardakci, Halpern and Schwarz.[11] This approach uses the master field from which the solutions of large N collective field equations can be constructed. The theory involves the equation of the master field (Euler equation) and the constraints.

It is impossible to review all these works in a single talk. In what follows, therefore, I concentrate on the basic ideas of the collective field theory using the Kogut-Susskind model Hamiltonian of U(N) gauge theory.[12] The present article is based on the lecture notes published in Japanese in Soken 62, 214-259 (1981).

II. THE COLLECTIVE FIELD THEORY OF U(N) GAUGE FIELDS

The Kogut-Susskind lattice gauge theory is a canonical formalism for non-Abelian gauge theories formulated algebraically in terms of operators assigned on each link of the square lattice in space. In this theory the dynamical variables are N × N unitary matrices (one matrix on each link). We denote the matrix on link ℓ by $u(\ell)$ and the corresponding U(N) momenta by $E_\alpha(\ell)$ ($\alpha = 0,1,\ldots, N^2-1$). They obey the following commutation relations:

$$[E_\alpha(\ell),\ E_\beta(\ell')] = i\ f_{\alpha\beta\gamma}\ E_\gamma(\ell')\ \delta_{\ell\ell'} \quad , \tag{2.1}$$

$$[E_\alpha(\ell),\ u(\ell')] = t_\alpha\ u(\ell)\ \delta_{\ell\ell'} \quad , \tag{2.2}$$

where the $f_{\alpha\beta\gamma}$'s are structure constant of a U(N) Lie algebra and the t_α's are the N × N Hermite matrix representation (fundamental) of U(N) generators, which are a generalization of the Pauli matrices ($\sigma_0,\sigma_1,\sigma_2,\sigma_3$) to U(N). Matrices t_α have the following orthogonality and Fierz recombination relations:

$$\begin{aligned} \mathrm{tr}\ (t_\alpha t_\beta) &= \delta_{\alpha\beta} \quad , \\ (t_\alpha)_{ij}\ (t_\alpha)_{k\ell} &= \delta_{i\ell}\ \delta_{jk} \quad . \end{aligned} \tag{2.3}$$

The Kogut-Susskind Hamiltonian is defined in terms of these operators as

$$H = \frac{g^2}{a}\ [\frac{1}{2} \sum_\ell E_\alpha^2(\ell) - \frac{1}{2g^4} \sum_\gamma \{\mathrm{tr}\ (u(\ell_1)\ u(\ell_2)\ u(\ell_3)\ u(\ell_4)) + \mathrm{h.c.}\}] \quad , \tag{2.4}$$

where g is the coupling constant and a is the lattice distance. We denote the smallest plaquettes (fundamental plaquettes) by γ, which consists of four links $\ell_1\ \ell_2\ \ell_3\ \ell_4$.

The gauge transformations are defined by

$$u(\ell_{pq}) \rightarrow U(p)\ u(\ell_{pq})\ U^\dagger(q) \quad , \tag{2.5}$$

where ℓ_{pq} is the link between sites p and q. A gauge transformation is a direct product of a U(N) transformation on each site. The K-S Hamiltonian (2.4) is invariant under gauge transformation.

The Schrödinger equation,

$$H|\psi > = E|\psi > \quad , \tag{2.6}$$

defines the dynamics of the theory. In gauge theories the physical states are defined to be singlets under the gauge transformations. In the coordinate representation, ($u(\ell)$ diagonal representation) this last statement is expressed by

$$\psi [u] = \psi [U u U^{\dagger}] \quad , \tag{2.7}$$

where the inside of [ ] is an abbreviation of (2.5). In operator form we can express it as

$$| \psi > = \Phi [W] |0 > \quad , \tag{2.8}$$

where

$$E_{\alpha} (\ell) |0 > = 0 \quad , \tag{2.9}$$

and $\Phi[W]$ is a function of gauge invariants W.

The most general gauge invariants constructed from $u(\ell)$'s are Wilson loops. We define a loop Γ by a set of successively connected links (see Fig. 1):

$$\Gamma \equiv \ell_1 \ell_2 \ldots\ldots \ell_m \quad . \tag{2.10}$$

The corresponding Wilson loop operator is given by

$$W(\Gamma) = \frac{1}{N} \operatorname{tr}(u(\ell_1) u(\ell_2)\ldots\ldots u(\ell_m)) \quad . \tag{2.11}$$

The conjugate loop $\tilde{\Gamma}$ is defined by the loop in opposite

$$\frac{1}{2}\sum_{\ell} E_\alpha^2(\ell)|\psi> = \frac{1}{2}\sum_{\ell}[E_\alpha(\ell), [E_\alpha(\ell), \Phi[W]]]\ |0>$$

$$= \frac{1}{2}\sum_{\Gamma}\sum_{\ell}[E_\alpha(\ell), [E_\alpha(\ell), W(\Gamma)]]\ \frac{\delta\Phi[W]}{\delta W(\Gamma)}\ |0>$$

$$+ \frac{1}{2}\sum_{\Gamma_1\Gamma_2}\sum_{\ell}[E_\alpha(\ell), W(\Gamma_1)]\ [E_\alpha(\ell), W(\Gamma_2)]\ \frac{\delta^2\Phi[W]}{\delta W(\Gamma_1)\delta W(\Gamma_2)}\ |0>$$

$$= (\frac{1}{2}\sum_{\Gamma}\omega(\Gamma;W)\ \pi(\Gamma) + \frac{1}{2}\sum_{\Gamma_1\Gamma_2}\Omega(\Gamma_1,\Gamma_2;W)\ \pi(\Gamma_1)\ \pi(\Gamma_2)\ )\ |\ \psi>\ , \qquad (2.16)$$

where

$$\pi(\Gamma) = -i\ \frac{\partial}{\partial W(\Gamma)}\ , \qquad (2.17)$$

and ω and Ω are defined by

$$\omega(\Gamma;\ W) = \sum_{\ell}[E_\alpha(\ell), [E_\alpha(\ell), W(\Gamma)]]\ , \qquad (2.18)$$

$$\Omega(\Gamma_1,\ \Gamma_2\ W) = -\sum_{\ell}[E_\alpha(\ell), W(\Gamma_1)]\ [E_\alpha(\ell), W(\Gamma_2)]\ . \qquad (2.19)$$

A straightforward calculation using (2.2) and (2.3) shows that ω and Ω are expressed again in terms of Wilson loops, and their N dependence is explicitly exhibited:

$$\omega(\Gamma;\ W) = N\ \tilde{\omega}\ (\Gamma;\ W)$$

$$= N\ [\ell(\Gamma)\ W(\Gamma) + \sum_{\Gamma_1\Gamma_2}\Delta(\Gamma;\ \Gamma_1,\Gamma_2)W(\Gamma_1)\ W(\Gamma_2)]\ , \qquad (2.20)$$

$$\Omega(\Gamma_1,\Gamma_2;\ W) = \frac{1}{N}\ \tilde{\Omega}(\Gamma_1,\Gamma_2;\ W)$$

$$= \frac{1}{N}\ [\tilde{\ell}\ (\Gamma_1)\ \delta(\Gamma_1,\tilde{\Gamma}_2) + \sum_{\Gamma_3}\tilde{\Delta}(\Gamma_3;\ \Gamma_1,\Gamma_2)\ W(\Gamma_3)]. \qquad (2.21)$$

$\ell(\Gamma)$ is the length of Γ, namely

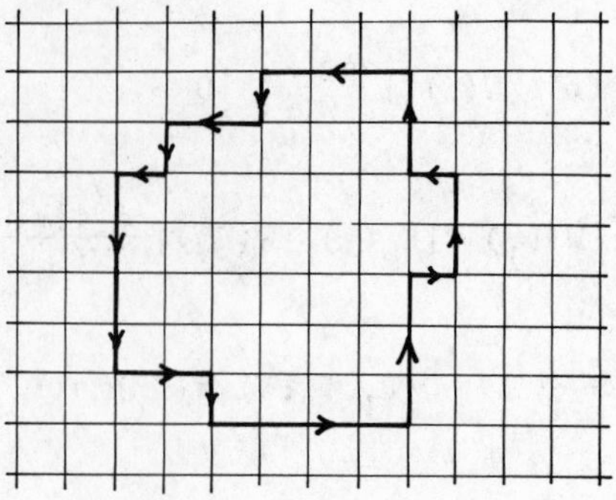

Fig. 1 Wilson loop.

direction, i.e.

$$\tilde{\Gamma} \equiv \tilde{\ell}_m \tilde{\ell}_{m-1} \cdots\cdots \tilde{\ell}_2 \tilde{\ell}_1 \quad .$$

If we define

$$u(\tilde{\ell}) = u^\dagger(\ell) \quad ,$$

we obtain

$$W(\tilde{\Gamma}) = W^\dagger(\Gamma) \quad .$$

We note that the fundamental plaquettes appearing in the po
term (the second term) in the K-S Hamiltonian are expressed
Wilson loops:

$$\mathrm{tr}(u(\ell_1)\, u(\ell_2)\, u(\ell_3)\, u(\ell_4)) = NW(\gamma) \quad .$$

The essence of collective field theory is to express t
Hamiltonian in terms of invariant variables. We have alrea
pressed the potential term. The kinetic energy term (the fi
term in (2.4)) can be expressed as follows:

$$\ell(\Gamma) = m \quad \text{for} \quad \Gamma \equiv (\ell_1 \ell_2 \ldots \ell_m) \quad , \tag{2.22}$$

while $\tilde{\ell}(\Gamma) = m$ if Γ is a simple uncrossed loop

$$= 2^2 m \text{ if } \Gamma \equiv (\ell_1 \ell_2 \ldots \ell_m \ell_1 \ell_2 \ldots \ell_m) \quad , \tag{2.23}$$

etc.

Δ and $\tilde{\Delta}$ are loop splitting and recombining Δ-functions. For example, when Γ is given by Fig. 2a, $\Delta(\Gamma; \Gamma_1 \Gamma_2)$ takes the value of the length of the overlap with a negative sign and Γ_1 and Γ_2 are given by Fig. 2b. When Γ is given by Fig. 3a, $\Delta(\Gamma; \Gamma_1, \Gamma_2)$ is the value of overlap with a positive sign and Γ_1 and Γ_2 are given by Fig. 3b. On the other hand, the recombining Δ-function $\tilde{\Delta}$ is illustrated by Fig. 4 and Fig. 5 for two typical cases. Note that Fig. 5 is the reverse process of Fig. 3, and indeed in this case $\Delta(\Gamma_3; \Gamma_1, \Gamma_2) = - \tilde{\Delta}(\Gamma_3; \Gamma_1, \Gamma_2)$.

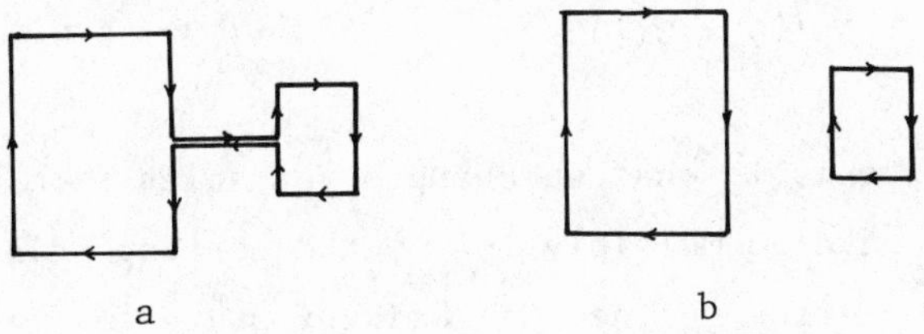

Fig. 2 Loop splitting - example 1.

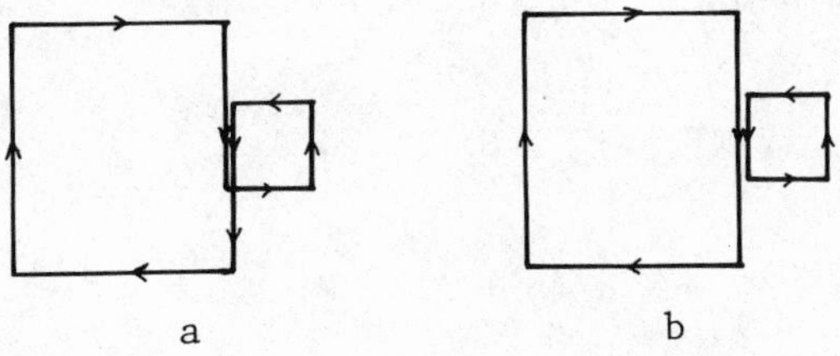

Fig. 3 Loop splitting - example 2.

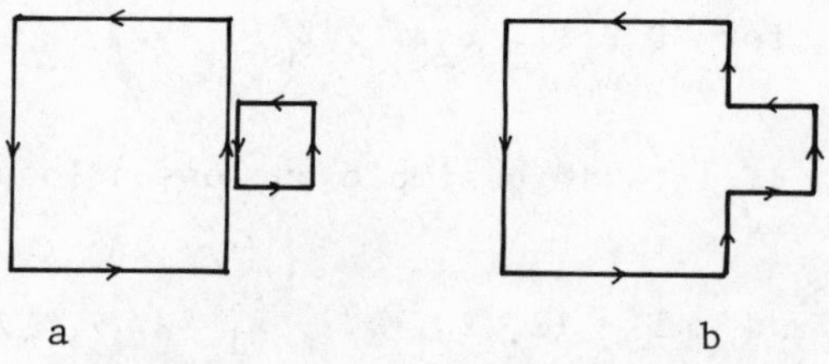

Fig. 4 Combining of two loops into one loop.

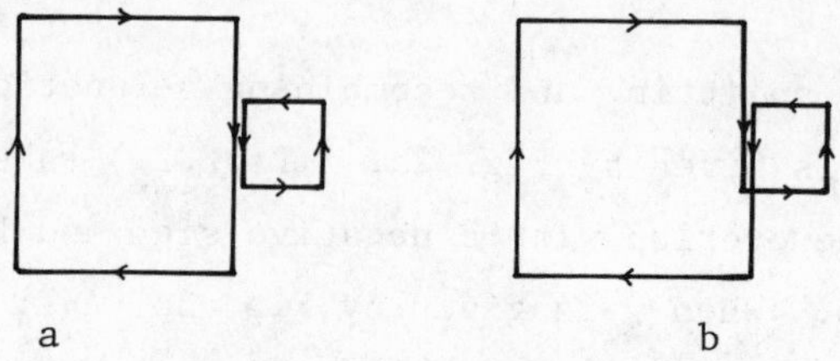

Fig. 5 Combining of two loops into one loop: Inverse process of Fig. 3

The kinetic energy term obtained in (2.16) is not Hermite under the operation

$$W^{\dagger}(\Gamma) = W(\tilde{\Gamma}) \quad , \quad \pi^{\dagger}(\tilde{\Gamma}) = \pi(\tilde{\Gamma}) \quad . \tag{2.24}$$

The reason for this is that we changed variables from $u(\ell)$'s to $W(\Gamma)$'s so we have to multiply $|\psi>$ by the corresponding Jacobian factor $J^{1/2}$ to define the new Schrödinger state $|\psi_{new}>$,

$$|\psi_{new} > = J^{1/2}|\psi > \quad . \tag{2.25}$$

Accordingly, the Hamiltonian should be similarly transformed.

$$H_{eff} [\pi, W] = J^{1/2} H [\pi, W] J^{-1/2}$$

$$= H (\pi + i C, W] \tag{2.26}$$

where

$$C(\Gamma) = \frac{1}{2} \frac{\delta \ln J}{\delta W(\Gamma)} \quad . \tag{2.27}$$

Since the Jacobian is Hermite,

$$C^{\dagger}(\Gamma) = C(\tilde{\Gamma}) \quad . \tag{2.28}$$

In practice it is difficult to calculate J, but it is easy to obtain an equation that $C(\Gamma)$ should satisfy by requiring the Hermiticity of H_{eff};

$$2 \sum_{\Gamma'} \Omega\,(\Gamma,\Gamma';\, W)\, C(\Gamma') + \omega(\Gamma;\, W) = 0 \quad . \tag{2.29}$$

Using (2.4), (2.15), (2.16), (2.26) and (2.29) we obtain

$$H_{eff} = K + V_{eff} \quad , \tag{2.30}$$

$$K = \frac{g^2}{a} \frac{1}{2} \,(\sum_{\Gamma,\Gamma'} \pi(\Gamma)\, \Omega(\Gamma,\Gamma';\, W)\, \pi(\Gamma') \quad , \tag{2.31}$$

$$V_{eff} = \frac{g^2}{a} [- \frac{1}{2} \,(\sum_{\Gamma\Gamma'} C(\Gamma)\, \Omega(\Gamma,\Gamma';\, W)\, C(\Gamma') + \sum_{\Gamma} \omega(\Gamma,W)\, C(\Gamma))$$

$$- \frac{N}{2g^4} \sum_{\gamma} (W\,(\gamma) + W(\gamma))] \quad . \tag{2.32}$$

The Wilson loops are not necessarily independent. However, in the process of rewriting the kinetic energy term of the Hamiltonian we used the chain rule for differentiation for which we must treat the W's as if they were independent. The interdependence of W's should be imposed after all the differentiations are carried out in the Schrödinger equation. Therefore, it is natural to define the collective field theory by replacing $W(\Gamma)$ in the Hamiltonian (2.30) by $\phi(\Gamma)$, which is defined to be independent,

$$[\pi(\Gamma),\ \ \phi(\Gamma')] = - i\delta\,(\Gamma,\Gamma') \quad . \tag{2.33}$$

Once the Schrödinger equation of the collective field theory is solved, by imposing the interrelations among $\phi(\Gamma)$'s, one can obtain the solutions of the original theory. In practice, however, it is difficult to solve the collective field theory with boundary conditions, which are consistent with the boundary conditions of the original theory. We therefore consider only the limiting case in which all the Wilson loops are independent, namely the large N limit.

III. LARGE N LIMIT

The large N limit of U(N) gauge theory is a limit of $N \to \infty$ and $Ng^2 \equiv \lambda$ being kept finite.

We note first that $\omega = N\tilde{\omega}$, $\Omega = \frac{1}{N}\tilde{\Omega}$ and $\tilde{\omega}$ and $\tilde{\Omega}$ are 0(1) operators. In order to eliminate the N dependence from the Hermiticity condition (2.29) we scale $C(\Gamma)$ as

$$\tilde{C}(\Gamma) = N^{-2}\, C(\Gamma) \quad . \tag{3.1}$$

The Hermiticity condition is then

$$2\,\tilde{\Omega}\,\tilde{C} + \tilde{\omega} = 0 \quad . \tag{3.2}$$

The effective potential (2.32) is then given by

$$V_{eff} = N^2\,\frac{Ng^2}{a}\,[-\frac{1}{2}\,(\tilde{C}\,\tilde{\Omega}\,\tilde{C} + \tilde{\omega}\,\tilde{C}) - \frac{1}{2(Ng^2)^2}\,\sum_{\gamma}\,(\phi(\gamma) + \phi(\tilde{\gamma}))] \quad . \tag{3.3}$$

It is of order N^2.

In order to make the kinetic energy term the same order in N as V_{eff}, we define $\tilde{\pi}(\Gamma)$ by

$$\pi(\Gamma) = N^2\,\tilde{\pi}(\Gamma) \quad . \tag{3.4}$$

We then obtain

$$K = N^2 \frac{Ng^2}{a} \frac{1}{2} \tilde{\pi} \tilde{\Omega} \tilde{\pi} \quad . \tag{3.5}$$

Substituting (3.4) we obtain

$$[\tilde{\pi}(\Gamma), \quad \phi(\Gamma')] = - \frac{i}{N^2} \delta(\Gamma,\Gamma') \quad . \tag{3.6}$$

Since H_{eff} is expressed as N^2 times a function of Ng^2, $\tilde{\pi}(\Gamma)$'s and $\phi(\Gamma)$'s, and since the commutation relation between $\tilde{\pi}$ and ϕ is proportional $1/N^2$, the large N limit is given by the classical limit of H_{eff} Hamiltonian system. The classical solution $\phi_0(\Gamma)$ is interpreted as usual as the vacuum expectation value of Wilson loop operator in the large N limit:

$$\phi_0(\Gamma) = \langle W(\Gamma) \rangle_0 \quad . \tag{3.7}$$

If we restrict ourselves to time independent solutions, the large N classical equation is given by

$$\frac{\delta V_{eff}}{\delta \phi(\Gamma)} = 0 \quad , \tag{3.8}$$

namely

$$\sum_{\Gamma_1\Gamma_2} \tilde{C}(\Gamma_1) \tilde{\Delta}(\Gamma; \Gamma_1 \Gamma_2)\tilde{C}(\Gamma_2) + \sum_{\Gamma'} \frac{\delta\tilde{\omega}(\Gamma')}{\delta \phi(\Gamma)} \tilde{C}(\Gamma')$$

$$+ \frac{1}{(Ng^2)^2} \sum_{\gamma} (\delta(\Gamma,\gamma) + \delta(\Gamma,\tilde{\gamma})) = 0 \quad . \tag{3.9}$$

We cannot solve this equation for an arbitrary $\lambda = Ng^2$ because of the complication of the loop sum involved in the equation. However, the solution is expected to have an exponential fall-off consistent with confinement as shown in Fig. 6. Since this curve indicates a denser condensation of small loops than large loops in

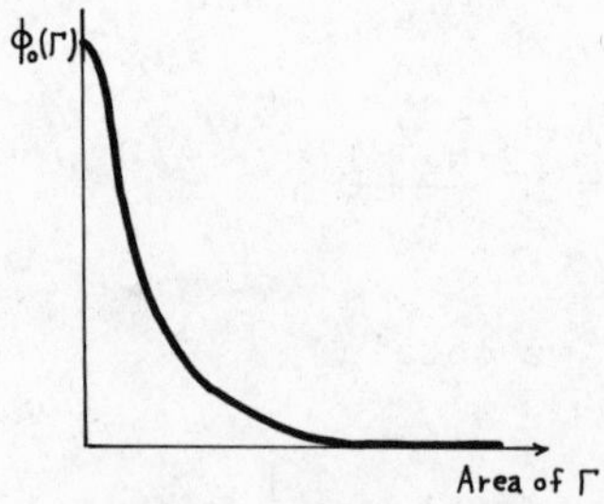

Fig. 6 Vacuum expectation value of Wilson loop which is consistent with the quark confinement.

the vacuum, the simplest confinement picture would be the condensation of small Wilson loops, namely the gluon field condensation.

In order to illustrate how the collective field theory works, let us consider the strong coupling case:

$$N \gg Ng^2 \equiv \lambda \gg 1 \quad . \tag{3.10}$$

In this case one can solve equation (3.2) and (3.9) by perturbation. The solution is

$$\phi_0(\Gamma) \sim \frac{1}{2\lambda^2} \sum_{\gamma} [\delta(\Gamma,\gamma) + \delta(\Gamma,\gamma)] + O(\frac{1}{\gamma^4}) \quad . \tag{3.11}$$

Thus in the strong coupling case, predominantly the fundamental plaquette Wilson loops condense in the vacuum.

We can see the collective excitations by looking at the small fluctuations about the classical solution (3.11). We set

$$N(\frac{\lambda}{a}\,\tilde{\ell}\,(\Gamma))^{1/2}\,\tilde{\pi}(\Gamma) = p(\Gamma) \quad ,$$

$$\phi(\Gamma) = \phi_0(\Gamma) + \frac{1}{N}\,(\frac{\lambda}{a}\,\tilde{\ell}(\Gamma))^{1/2}\,q(\Gamma) \quad . \tag{3.12}$$

Using the commutation relation (3.6) we obtain

$$[p(\Gamma),\, q(\Gamma')] = -i\,\delta(\Gamma,\Gamma') \quad . \tag{3.13}$$

Inserting (3.11) and (3.12) into (3.3) and (3.5), and using (3.10) we obtain the quadratic part of the Hamiltonian given by

$$H_{coll} = \sum_{\Gamma} \frac{1}{2} (p(\Gamma)\ p(\tilde{\Gamma}) + E^2(\Gamma)\ q(\Gamma)\ q(\tilde{\Gamma})) \ , \qquad (3.14)$$

where

$$E(\Gamma) = \frac{1}{2} \frac{\lambda}{a} \ell(\Gamma) \quad . \qquad (3.15)$$

Thus, in the strong coupling case of this theory there exists a set of glue balls which are specified by the loop Γ and its energy is proportional to the length of the loop.

IV. ONE PLAQUETTE MODEL

One of the exactly solvable models in the large N limit is the one plaquette model which assumes that the lattice consists of only one plaquette. In this model the Wilson loop operators are classified by the winding number n. Accordingly, we denote the corresponding collective field by ϕ_n, where n takes values of positive and negative integers. n=0 is a special case that corresponds to the trace of the zero-th power of $u(\ell_1)\ u(\ell_2)\ u(\ell_3)\ u(\ell_4)$. Thus,

$$\phi_0 = 1 \quad . \qquad (4.1)$$

The calculation of $\tilde{\omega}$ and $\tilde{\Omega}$ can be done in the same way as in the previous section. We obtain

$$\tilde{\omega}(n) = n \sum_{n_1, n_2} \frac{1}{2} (\varepsilon(n_1) + \varepsilon(n_2))\ \delta_{n, n_1+n_2}\ \phi_{n_1}\ \phi_{n_2} \quad , \qquad (4.2)$$

$$\tilde{\Omega}\ (n_1,\ n_2) = -\ n_1 n_2 \phi_{n_1+n_2} \quad . \qquad (4.3)$$

The Hermiticity condition (3.2) can be solved to find

$$n\,\tilde{C}_n = -\frac{1}{2}\,\varepsilon(n)\,\phi_{-n} \quad . \tag{4.4}$$

Inserting this result into (3.3) we obtain the following effective potential:

$$V_{eff} = N^2 \frac{\lambda}{a}\left[\frac{1}{24}\sum_{n_1 n_2 n_3} \delta_{0,n_1+n_2+n_3}\,\phi_{n_1}\phi_{n_2}\phi_{n_3} - \frac{1}{2\lambda^2}(\phi_1 + \phi_{-1})\right] . \tag{4.5}$$

In order to solve the large N classical solution it is convenient to introduce the Fourier transform of the collective field

$$\phi(\sigma) = \frac{1}{2\pi}\sum_n e^{-in\sigma}\,\phi_n \tag{4.6}$$

and express V_{eff} in terms of it:

$$V_{eff} = N^2 \frac{\lambda}{a}\int_{-\pi}^{\pi} d\sigma\,\left\{\frac{\pi^2}{6}\,\phi^3(\sigma) - \frac{1}{\lambda^2}\,\phi(\sigma)\cos\sigma\right\} \quad . \tag{4.7}$$

We minimize (4.7) with the conditions

$$\int_{-\pi}^{\pi} d\sigma\,\phi(\sigma) = 1 \tag{4.8}$$

and

$$\phi(\sigma) \geq 0 \quad . \tag{4.9}$$

Condition (4.8) is equivalent to (4.1) while (4.9) is due to the definition of Wilson loop. Let $e^{i\sigma_i}$ $(i=1,\ldots,N)$ be eigenvalues of the unitarity matrix $u \equiv u(\ell_1)\,u(\ell_2)\,u(\ell_3)\,u(\ell_4)$. Then,

$$\phi(\sigma) = \frac{1}{2\pi}\sum_n e^{-in\sigma}\,\frac{1}{N}\,\mathrm{tr}(u^n) = \frac{1}{2\pi N}\sum_i\sum_n e^{-in(\sigma-\sigma_i)}$$

$$= \frac{1}{N}\sum_i \delta(\sigma-\sigma_i) \geq 0 \quad . \tag{4.10}$$

The solution of the large N classical equation is given by the following expression:

For $\lambda \geq \lambda_c = \frac{8}{\pi}$ (strong coupling)

$$\phi(\sigma) = \frac{2}{\pi\lambda}\frac{1}{\sin\alpha}\left(1 - \sin^2\alpha\,\sin^2\frac{\sigma}{2}\right)^{1/2} , \tag{4.11}$$

where α is a solution of

$$\frac{8}{\pi\lambda}\;\frac{1}{\sin\alpha}\int_0^{\pi/2} d\psi\,\left(1 - \sin^2\alpha\,\sin^2\psi\right)^{1/2} = 1 . \tag{4.12}$$

For $\lambda \leq \lambda_c = \frac{8}{\pi}$ (weak coupling)

$$\phi(\sigma) = \frac{2}{\pi\lambda}\left(\sin^2\alpha - \sin^2\frac{\sigma}{2}\right)^{1/2} , \tag{4.13}$$

where α is given by

$$\frac{8}{\pi\lambda}\int_0^{\alpha} d\psi\,\left(\sin^2\alpha - \sin^2\psi\right)^{1/2} = 1 . \tag{4.14}$$

The solutions are illustrated in Fig. 7. From this figure one sees that something discontinuous occurs by decreasing the value of λ at $\lambda = \lambda_c$. Indeed it has been shown by S. Wadia[6] that the second

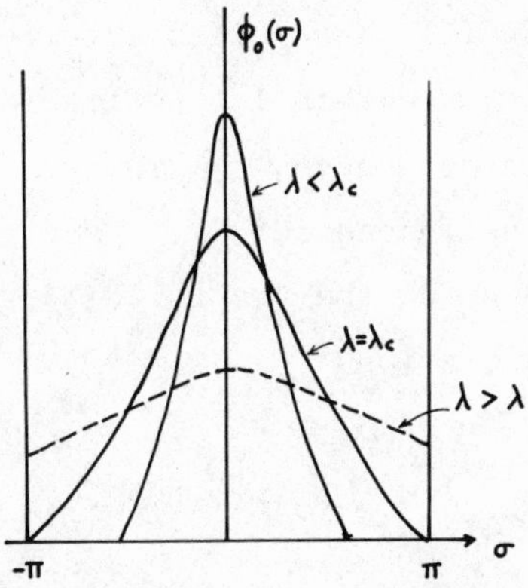

Fig. 7 Solution for the large N collective field equation of the one plaquette Kogut-Susskind model.

derivative of the ground state energy with respect to λ is discontinuous precisely at this point (3rd order phase transition).

V. REMARKS

As explained in detail in the previous sections, the essence of the collective field theory is to use a (over) complete set of invariants as collective variables and transform the Hamiltonian in a form suitable for the large N limit. The large N limit then becomes a classical limit of the effective Hamiltonian. The information on the ground states are obtained from the solution of this large N classical equation for the collective field. In this theory, the excited states are obtained from the collective excitations which can be calculated as the small quantum fluctuations about the classical solution.

In this approach the basic equation to solve is the equation (3.2) and (3.9), which we solved in III for the strong coupling case. It would be very interesting to solve them in the weak coupling case also. However, at first sight the equation (3.2) and (3.9) are incompatible with the weak coupling expansion, because one would expect $\phi(\Gamma) = 1 + O(\lambda^2)$, so from (3.9) one obtains $\tilde{C}(\Gamma) \sim \frac{1}{\lambda}$ which seems to be incompatible with (3.2).

One resolution[13] to this dilemma is that det $\tilde{\Omega} = 0$ as $\lambda \to 0$ and the λ^{-1} coefficient of $\tilde{C}(\Gamma)$ is an eigenvector of $\tilde{\Omega}\big|_{\lambda=0}$ with zero eigenvalue. In order to see this let us first look at the one plaquette model. If we set $\phi_{n_1+n_2} = 1$ in (4.3), we obtain $\tilde{\Omega}(n_1,n_2)\big|_{\lambda=0} = -n_1 n_2$, so indeed det $\tilde{\Omega}\big|_{\lambda=0} = 0$. It is not difficult to see that this happens for the general U(N) gauge theory if we realize the following relations

$$\begin{aligned} \tilde{\Delta}(\Gamma_3; \Gamma_1, \Gamma_2) &= -\tilde{\Delta}(\Gamma_3'; \Gamma_1, \tilde{\Gamma}_2) , \\ \tilde{\Delta}(0; \Gamma_1, \tilde{\Gamma}_1) \equiv \tilde{\ell}(\Gamma_1) &= -\tilde{\Delta}(\Gamma_3; \Gamma_1, \Gamma_1) . \end{aligned} \tag{5.1}$$

Thus, in principle the weak coupling expansion can be done for the collective field of U(N) gauge theory.

Finally, it should be mentioned that some efforts[14] are being spent to solve numerically the large N collective field equations.

ACKNOWLEDGEMENT

I acknowledge the help that Dr. A. Jevicki has given me in preparing this talk.

REFERENCES

1. B. Sakita, Phys. Rev. D21, 1067 (1980).
2. A. Jevicki and B. Sakita, Nucl. Phys. B165, 511 (1980).
3. D. Bohm and D. Pines, Phys. Rev. 92, 609 (1953).
   N.N. Bogoliubov and D.N. Zuberev, JETP 1, 83 (1955).
4. *E. Brezin, C. Itzykson, G. Parisi and J.B. Zuber, Comm. Math. Phys. 59, 35 (1978).
   A. Jevicki and B. Sakita, Nucl. Phys. B165, 511 (1980).
   M. Mondello and E. Onofri, Phys. Lett. B98, 277 (1981).
   J. Shapiro, Nucl. Phys.
   R. Jackiw and A. Strominger, Phys. Lett. B99, 133 (1981).
   I. Affleck, Nucl. Phys. B185, 346 (1981).
5. *T.H. Berlin and M. Kac, Phys. Rev. 86, 821 (1952).
   *H.E. Stanley, Phys. Rev. 176, 718 (1968).
   *S. Coleman, R. Jackiw and H.D. Politzer, Phys. Rev. D10, 2491 (1974).
   A. Jevicki and N. Papanicolaou, Nucl. Phys. B171, 362 (1980).
   A. Guha and B. Sakita, Phys. Lett. B100, 489 (1981).
6. *S. Wadia, Phys. Lett. 93B, 403 (1980).
   A. Jevicki and B. Sakita, Phys. Rev. D22, 467 (1980).
   I. Affleck, Proc. Brown Workshop on Non-perturbative Studies in QCD, Brown HET 457 (1981).
   O.A. Karim, NSF-ITP-81-109.

7. *G. 't Hooft, Nucl. Phys. B75, 461 (1974).
*C.G. Callan, N. Coote, and D.J. Gross, Phys. Rev. D13, 1649 (1976).
W. Gutierrez, Nucl. Phys. B176, 192 (1980).
K. Kikkawa, Ann. Phys. to be published.
Xi-te, Proc. Brown Workshop on Non-perturbative Studies in QCD, Brown HET 457 (1981).
S.B. Libby and Chung-I Tan, Proc. Brown Workshop on Non-perturbative Studies in QCD, Brown HET 457 (1981).
8. A. Jevicki and B. Sakita, Nucl. Phys. B185, 89 (1981).
9. Yu. M. Makeenko and A.A. Migdal, Phys. Lett. 88B, 135 (1979).
10. A. Jevicki and N. Papanicolaou, Nucl. Phys. B171, 362 (1980).
A. Jevicki and H. Levine, Phys. Rev. Lett. 44, 1443 (1980).
11. K. Bardakci, Nucl. Phys. B178, 263 (1981).
M.B. Halpern, Nucl. Phys. to appear.
M.B. Halpern and C. Schwartz, UCB-PTH-81/5.
12. J.B. Kogut and L. Susskind, Phys. Rev. D11, 395 (1975).
13. I owe this point to the discussion with A. Jevicki.
14. Brown group (Private communication with A. Jevicki and G. Guralnik).
L. Yaffe (Private communication with E. Witten).

PROGRAM

ORBIS SCIENTIAE 1982

MONDAY, January 18, 1982

Opening Remarks and Welcome

SESSION I: OVERVIEW OF CURRENT GRAVITATIONAL AND PARTICLE THEORY

Moderator: Abraham Pais, The Rockefeller University

Dissertators: P. A. M. Dirac, Florida State University
"THE PRESENT STATE OF GRAVITATIONAL THEORY"

David Gross, Princeton University
"QUANTUM CHROMODYNAMICS - PROSPECTS AND PROBLEMS"

Howard Georgi, Harvard University
"GRAND UNIFICATION: A STATUS REPORT"

SESSION II: WEAK INTERACTIONS, ETC. - THEORY AND EXPERIMENT

Moderator: Gianni Conforto, University of Florence and University of Michigan

Dissertators: Boris Kayser, National Science Foundation
"MAJORANA NEUTRINOS - THEIR ELECTROMAGNETIC PROPERTIES AND NEUTRAL CURRENT WEAK INTERACTIONS"

William Marciano, Brookhaven National Laboratory
"PROTON DECAY - 1982"

Gianni Conforto, University of Florence and University of Michigan
"PROMPT NEUTRINOS: PRESENT ISSUES AND FUTURE PROSPECTS"

Ronald Diamond, Florida State University
"RECENT RESULTS IN CHARMED PARTICLES LIFETIMES AND PRODUCTION"

E. A. Paschos, University of Dortmund
"THE WEAK MIXING ANGLE AND ITS RELATION TO THE MASSES M_Z AND M_W"

	Zohreh Parsa, New Jersey Institute of Technology "INTERMEDIATE VECTOR BOSONS AND NEUTRINO COSMOLOGY"

January 19, 1982

SESSION III:	RECENT PROGRESS IN ASTROPHYSICS AND COSMOLOGY
Moderator:	Remo Ruffini, University of Rome
Dissertators:	Francis Everitt, Stanford University "RECENT PROGRESS IN TESTING OF GENERAL RELATIVITY"
	F. Ciatti, University of Padova "PRECESSIONAL PERIOD AND MODULATION OF RELATIVISTIC RADIAL VELOCITIES IN SS 433"
	George W. Collins II, Ohio State University "PERIODIC AND SECULAR CHANGES IN SS 433"
	Jonathan Grindlay, Harvard University "X-RAY OBSERVATIONS OF SS 433"
	Remo Ruffini, University of Rome "MASSIVE NEUTRINOS AND GAMOW COSMOLOGY"
SESSION IV:	SUPERSYMMETRY
Moderator:	John Schwarz, California Institute of Technology
Dissertators:	John Schwarz, California Institute of Technology "SUPERSTRINGS"
	Savas Dimopoulos, Harvard University "MODELS WITH DYNAMICAL SYMMETRY BREAKING"
	Peter van Niewenhuizen, State University of New York, Stony Brook "A GEOMETRICAL APPROACH TO QUANTUM FIELD THEORY"

January 20, 1982

SESSION V:	GLUEBALLS AND QCD
Moderator:	Sydney Meshkov, National Bureau of Standards and California Institute of Technology
Dissertators:	Daniel L. Scharre, Stanford Linear Accelerator Center "GLUEBALLS - A STATUS REPORT"
	Carl Carlson, William and Mary University "GLUEBALLS - SOME SELECTED THEORETICAL TOPICS"
	Claudio Rebbi, Brookhaven National Laboratory "MONTE CARLO COMPUTATIONS OF THE HADRONIC MASS SPECTRUM"

Richard Brower, University of California, Santa Cruz
"LEARNING FROM MASS CALCULATIONS IN QUARKLESS QCD"

SESSION VI: QCD: PROS AND CONS

Moderator: Giuliano Preparata, University of Bari

Dissertators: Claudio Rebbi, Brookhaven National Laboratory
"FOR QCD"

Giuliano Preparata, University of Bari
"BEYOND QCD: WHY AND HOW"

Asim Barut, University of Colorado
"LARGE MAGNETIC MOMENT EFFECTS AT HIGH ENERGIES AND COMPOSITE PARTICLE MODELS"

January 21, 1982

SESSION VII: THE NEXT LAYER OF THE ONION? COMPOSITE QUARKS AND LEPTONS

Moderator: O. W. Greenberg, University of Maryland

Dissertators: O. W. Greenberg, University of Maryland
"THE NEXT LAYER OF THE ONION? COMPOSITE MODELS OF QUARKS AND LEPTONS"

Gordon Shaw, University of California, Irvine
"COMPOSITE FERMIONS - CONSTRAINTS AND TESTS"

Stuart Raby, Los Alamos National Laboratory
"LIGHT COMPOSITE FERMIONS: AN OVERVIEW"

E. Eichten, Fermi National Accelerator Laboratory
"RULES FOR COMPOSITE FERMIONS"

O. W. Greenberg, University of Maryland
"THE N-QUANTUM APPROXIMATION, CONCRETE COMPOSITE MODELS OF QUARKS AND LEPTONS, AND PROBLEMS WITH THE NORMALIZATION OF COMPOSITE MASSLESS BOUND STATES"

SESSION VII: QUANTUM FIELD THEORY

Moderator: Gerald Guralnik, Brown University

Dissertators: Donald Weingarten, University of Indiana
"MONTE CARLO EVALUATION OF HADRON MASSES"

Yadin Goldschmidt, Brown University
"REVIVAL OF THE OLD STRING MODEL"

Norman Christ, Columbia University
"QCD ON RANDOM LATTICE"

Bunji Sakita, City College of the CUNY
"COLLECTIVE FIELD THEORY"

PARTICIPANTS

Hadi H. Aly
Southern Illinois University

Robert C. Ball
University of Michigan

Asim Barut
University of Colorado

Carl M. Bender
Washington University

David L. Bintinger
Fermi National Accelerator Laboratory

Richard C. Brower
University of California Santa Cruz

Arthur A. Broyles
University of Florida

Richard H. Capps
Purdue University

Carl Carlson
College of William and Mary

Peter Carruthers
Los Alamos National Laboratory

Norman Christ
Columbia University

F. Ciatti
University of Padova, Italy

George W. Collins II
Ohio State University

Gianni Conforto
Fermi National Accelerator Laboratory

A. P. Contogouris
McGill University, Canada

Fred Cooper
Los Alamos National Laboratory

John M. Cornwall
University of California Los Angeles

Thomas Curtright
University of Florida

Richard H. Dalitz
Oxford University, England

Stanley R. Deans
University of South Florida

Ronald N. Diamond
Florida State University

Savas Dimopoulos
Stanford University

P. A. M. Dirac
Florida State University

Gabor Domokos
Johns Hopkins University

Susan Kovesi-Domokos
Johns Hopkins University

Bernice Durand
University of Wisconsin, Madison

Loyal Durand
University of Wisconsin, Madison

E. Eichten
Fermi National Accelerator Laboratory

C. W. Francis Everitt
Stanford University

W. F. Fry
University of Wisconsin
Madison

Henning Genz
University of Miami and
University of Karlsruhe, Germany

Howard Georgi
Harvard University

Maurice Goldhaber
Brookhaven National Laboratory

Yadin Goldschmidt
Brown University

O. W. Greenberg
University of Maryland

Jonathan E. Grindlay
Harvard University

David Gross
Princeton University

Gerald S. Guralnik
Brown University

C. R. Hagen
University of Rochester

Vasken Hagopian
Florida State University

M. Y. Han
Duke University

J. Kandaswamy
Cornell University

Gabriel Karl
University of Guelph, Canada

Boris Kayser
National Science Foundation

Anthony D. Kennedy
University of Maryland

Nicola N. Khuri
The Rockefeller University

Behram N. Kursunoglu
University of Miami

Thaddeus Francis Kycia
Brookhaven National Laboratory

K. MacRae
University of Maryland

K. T. Mahanthappa
University of Colorado

William J. Marciano
Brookhaven National Laboratory

R. E. Marshak
Virginia Polytechnic Institute

Jay Marx
Lawrence Berkeley Laboratory

Sydney Meshkov
California Institute of
Technology

Kimball A. Milton
Oklahoma State University

Stephan L. Mintz
Florida International University

Gerald H. Newsom
Ohio State University

John N. Ng
University of British Columbia

Martin Olsson
University of Wisconsin, Madison

Heinz R. Pagels
The Rockefeller University

A. Pais
The Rockefeller University

Zohreh Parsa
New Jersey Institute of
Technology

E. A. Paschos
University of Dortmund, Germany

J. Patera
Universite de Montreal, Canada

Arnold Perlmutter
University of Miami

Stephen S. Pinsky
Ohio State University

C. Piron
University of Massachusetts

Giuliano Preparata
Universita di Bari, Italy

Stuart Raby
Los Alamos National Laboratory

Claudio Rebbi
Brookhaven National Laboratory

Remo Ruffini
Universita di Roma, Italy

Bunji Sakita
City College of New York

Daniel L. Scharre
Stanford Linear Accelerator Center

John H. Schwarz
California Institute of Technology

Gordon Shaw
University of California Irvine

L. M. Simmons, Jr.
Los Alamos National Laboratory

George A. Snow
University of Maryland

Charles M. Sommerfield
Yale University

Katsumi Tanaka
Ohio State University

Kunihiko Terasaki
University of Maryland

Robert Thews
Department of Energy

Y. Tosa
Virginia Polytechnic Institute

Peter van Nieuwenhuizen
State University of New York

K. C. Wali
Syracuse University

Don Weingarten
Indiana University

Louis Witten
University of Cincinnati

G. B. Yodh
University of Maryland

F. Zachariasen
Los Alamos National Laboratory

Arnulfo Zepeda
Centro de Investigacion y de Estudios Avanzados, Mexico

INDEX

Anisotropic Chromo Dynamics (ACD)	305-321
Anisotropic Space-Time (AST)	313ff
Anomalous magnetic moments	350,351-352
Appelquist-Carrazone theorem	367
Asymptotic Freedom (AF)	11ff,309-312
Atomic time	5,6,7,8,9
Axion	32-33
Bardakci, K.	436
Bars, I.	43,367
Barut, A.O.	323-332
Baryon octet	333-335
Basdevant, J.L.	318
BCS superconductivity	319
BEBC collaboration	99ff,132ff
Becchi-Rouet-Stora-Tyutin (BRST) transformations	213ff
Bender, C.M.	390
Bernstein, J.	390
Bethe-Salpeter amplitudes	381,389
Bhanot, G.	301,402
Bjorken, J.	340
Bjorken scaling	310
B-L symmetry and neutrino masses	33-34
Bohm, D.	435
Boson masses	76-77,123-137
Brandt, R.A.	390
Breit-interaction	324-326
Brodsky, S.J.	390
Brower, R.C.	293-303
Carlson, C.E.	263-275
Cartan, E.	214,224
Cartan-Maurer equations	223
CDHS collaboration	99ff,129ff
Centauro events	24
CESR results	308
CHARM collaboration	99ff,131ff

Charmed baryons 106,117ff
Charmed particle decays 96ff
Charmed particle lifetimes and production 105-121
Chiral 19
Chiral symmetry 333ff
Chiral symmetry breaking 18-20
Christ, N.H. 619-433
Ciatti, F. 151-161
Coleman, S. 366
Collective Field Theory 435-453
Collins II,G.W. 151ff,163-191
Color-Flavor Model 342
Composite particle models and large magnetic moment effects at high energies 323-332
Composite quarks and leptons 42-43,333-347,379-391
Composite fermions 349-363,365-378
Confinement 14-18
Conforto, G. 95-104
Coyne, J. 273
CP violation and QCD 32-33
Creutz, M. 16ff,294,301-394
Crystal ball detector 243ff,307

D'Adda, A. 369
Dashen, R. 301
DeRujula, A. 390
DeWitt, B.S. 217
Diamond, R.N. 105-121
Dimopoulos, S. 340,367-368
Dirac neutrinos 49ff
Dirac, P.A.M. 1-10
Dirac particle 325
D-mesons 96ff,105-121
Dover, C. 91
Dragt, A.J. 390
Drell-Yan Process 12,13
Dynamics of Relativistic Spin Interactions 324-327

E(6) 37ff
E(1420) 246,248
Effective Field Theories 34-37,194
Effective technicolor interaction(eTC) 43-44
Eichten, E. 365,369,375
Einstein, A. 27-28
Einstein-de sitter model 4-5
Einstein Theory of Gravitation 1-10
Einstein time 7,8,9

Electromagnetic Properties of Dirac Neutrinos 49ff
Electroweak effects 125ff
Exotic decays of $W^{\pm}$ and Z^{o} 142-144
Exotic hadrons 23
Extended technicolor interaction(ETC) 41-42,372ff

f(1270) 252,255
f'(1515) 249,253
Fadeev-Popov ghosts 214,215,219,220
Feinberg, F. 369
Feinberg, G. 346
Fermions and Chiral Symmetry Breaking 18-20
Fierz identities 228
Fine structure constant 74-75,324
Fishbane, P. 273
F-mesons 102,107,117
FMOWW experiment 100ff
Form Factors 55ff,352-353,358ff
Friedberg, R. 420
Friedmann model 4
Fritzsch, H. 341

Gauss-Seidel iteration 403
Gell-Mann, M. 39
Gell-Mann Low equation 294
Gell-Mann-Nishijima formula 343
Generation mixing 84
Geometrical approach to Quantrum Field Theory 213-237
Georgi, H. 27,31-47,71
Georgi-Glashow SU(5) model 71ff
Glaser, V. 380
Glashow, S. 33,64,71
Global symmetries 39
Glue jets 308-309
Glueballs 17,278,282,284,296,307,432
  A status report 239-261
  Some selected theoretical topics 263-275
Gluonium-Quarkonium mixing 271-272
Gluons 12,16,17,23,26,28,239ff
296,307-309

Goldhaber, M. 91
Goldschmidt, Y. 407-417
Grand Unification: a status report 31-47
Grand Unified Mass 338
Grand Unified Theories(GUTS) 73,78,334
Grassmann field 233
Gravitation 214

Gravitational theory, present state of 1-10
Green, M. 194ff
Greenberg, O.W. 333-347,379-391
Green's Functions 127,213,278ff
Gross, D.J. 11-29
Gross, F. 273,380

Haag, R. 380ff
Hadronic mass spectrums, Monte Carlo computations of 277-292
Hadronic Masses, Monte Carlo evaluation of 393-406
Halliday, I. 297
Hall, L. 35
Han-Nambu Quarks 329
Hansson, H. 273
Harari, H. 345-346,374
Hasenfratz, A and P. 394
"Hermaphrodite" states of gluons 307-309
Higgs bosons 78ff,125,337,341
Higgs mechanism 17,25,38-39,41,43
Higgs potential 336,340
Holdom, B. 44
Homestake gold mine collabration 72

Intermediate Vector Bosons 76-77,123-137,139-149
Iota(1440) 242-248,265-270

Jackiw, R. 390
Jacobi identities 214,221,223,224,227
Jevicki, A. 435ff
Johnson, K. 380
Josephson Effect 74
Judek effects 24

Kaluza-Klein theory 29
Kayser, B. 49-69
Kennedy, A.D. 390
Kessler, D. 303
Killing form 224
Kim, J. 132-133
Kinoshita, T. 74
Kogut-Susskind lattice gauge theory 435ff
Kolar gold field experiment 72

Landau, L. 26
Langacker, P. 91
Large Numbers Hypothesis 5-10
Large N limit of U(N) gauge theory 21,444-447
Lattice approximation to QCD 15-18,277,292,293-303
393ff,419ff
Lee, T.D. 420
Lehmann, H. 380
Lemaitre model 4
Lepage, G.P. 390
Leptons, composite 42-43,333-347
Levine, H. 303
Lifetime of the proton 71-93
Lifetimes of charmed particles 105-121
Light composite fermions 365-378

Mack, G. 297
MacRae, K.I. 390
Magnetic moment effects at high energies and composite particle models 323-332
Magnetic resonances and structure of hadrons and leptons 327-329
Majorana neutrinos 46-69
Majorana representation 196
Mammano, A. 151-161-186
Marciano, W. 35,71-93,133
Marinari, E. 394-395
Mark II detector 241ff
Mass generation mechanisms 369-374
Mass of W- and Z-bosons 76-77,123-137,338
Massless bound states 382ff
Meissner effect (electric) 14
Mercury, motion of perihelion of 1-2
Meshkov, S. 273
Microwave black-body radiation 8
Migdal, A.A. 21-22,436
Minkowski space field 230,393ff
Mixing angle, weak 123-137
Monte Carlo simulations of QCD 15-18
Monte Carlo Computations of the Hadronic Mass Spectrum 277-292
Monte Carlo Evaluation of Hadron Masses 393-406
Moon, angular acceleration of 9
Mueller, A. 13

Nambu, Y. 19
Nambu-Goldstone boson 320
Nambu-Goto action 195,400,407ff

Ne'eman, Y. 215
Neutral Currents Weak Interactions of Majorana Neutrinos 49ff
Neutrino cosmology and intermediate vector bosons 139-149
Neutrino masses 33-34, 145-148
Neutrino species 144-145
Neutrinos, Dirac 49-69
Neutrinos, Majorana 49-69
Neutrinos, prompt 95-104
Newsom, G.H. 151ff,163-191
Nielson-Kallosh ghosts 215-216,219
Nishijima, K. 390
Normalization of composite massless bound states 379-391
N-Quantum Approximation 379-391
Nussinov, S. 347

Oddballs 264
Ogilvie, M. 390
O(N) gauge group 366
Onion, next layer of the 333-347
Ore, Jr.,F.R. 213-237
OZI rule 21
OZI suppression 240,241,242,255

Pagnamenta, A. 380
Parisi, G. 394,395
Parsa, Z. 139-149
Paschos, E.A. 123-137
Pati, J.C. 340,341,342,346,347,390
Peccei, R. 32
Petcher, D. 394,402
Periodic and Secular Changes in SS 433 163-191
Perturbative QCD 12-14,311
Peskin, M. 390
Peterson, C. 273
PETRA 308,350,358,359
Phase transitions of quark matter 23
Photon-gluon fusion model 119
Phi Phi glueballs 255-257
Pines, D. 435
Planck scales 28,31,194,338
Plaquette 279ff,297,439ff
PLUTO 359
Polyakov, A.M. 22,407ff
Pound, R.V. 3

Precessional Period and Modulation of Relativistic Radial Velocities in SS 433 151-161
Preparata, G. 305-321
Preskill, J. 43,365,367,375
Prompt Neutrinos 95-104
Proton Decay 71-93,123,135
Psi decays 254-255,265ff

Quantized Hall Effect 74
Quantum Chromodynamics (QCD)
  Analytical Solutions of 20-22
  Beyond 305-321
  CP violation and 32-33
  lattice approximation to 15-18,21
  Mass calculations in quarkless 293-303
  Monte Carlo computations and 277,292,393-406
  On a Random Lattice 419-433
  Perturbative 12-14,311
  Problems and Prospects of 11-29
Quantum Field Theory (QFT) 24-25
  A Geometrical Approach to 213-237
Quark fusion Model 119
Quarkless QCD, mass calculations in 293-303
Quarks, composite 42-43, 333-347
Quarks, Han-Nambu 329
Quarks, phase transitions of 23
Quinn, H. 32

Raby, S. 340,365-378
Ramond, P. 39
Random Lattice, QCD on a 419-433
Raychaudhuri, A. 380
Rebbi, C. 16,19,277-292,394,402
Rebka, G.A. 3
Recent Results on Charmed Particle Lifetimes and Production 105-121
Regge behavior 22,193,195,400ff,432
Realistic Spin Interactions 324-327
Renormalization 25-26
Retardation of electromagnetic waves 3
Riemann space 3
Rishon model 345-346,374
Ruffini, R. 186

Sakita, B. 435-453
Sakurai, J.J. 340
Salam, A. 340,341,342

Scalapino, D.J. 394
Scharre, D.L. 239-261
Scherk, J. 194ff
Schwarz, J.H. 193-212
Semileptonic decays of charmed particles 95ff
Senjanovic, G. 91
Shapiro, I.I. 2,3,10
Shaw, G. 338,347,349-363
Silverman, D. 349-363
Sirlin, A. 90,133
Slansky, R. 39
Spin effects in particle physics 323ff
Størmer's problem 327
Strathdee, J. 341
String Model, Revival of the Old 407,417
Strong gauging 371-372
SU(2) 16,17,278ff,293ff
SU(2|1) 214
SU(2) × U(1) symmetry breaking 32,33,41
$SU(2)_L$ × U(1) electroweak model 75-77
SU(3) 16,240,278ff
SU(3) × U(1) 36
SU(3) × SU(3) 19,369
SU(3) × SU(2) × U(1) 35,37,43
$SU(3)_c \times SU(2)_L \times U(1)$ couplings 74ff,139ff
SU(5) 16,32,33,35,38ff,71ff,135
$SU(5) \to SU(3)_s \otimes SU(2)_L$ 373
$SU(6) \to SU(3)^s \otimes SU(3)^L \otimes U(1)$ 368
SU(8) 376
SS 433 151-161,163-191
Sucher, J. 347,389,390
Sudarshan, G. 388
Sugar, R. 394
Supergravity 193ff,214ff
Superspace 225-235
Superstrings 193-212
Supersymmetry (SS) 40-41,87-89,193-212
Susskind, L. 19,255,287,340,367-368,436ff

TASSO 359
Tau-neutrino 102,358ff
Technicolor interaction 40,41-42
Technicolor interaction, extended (ETC) 41-42,372ff
Technicolor interaction, effective (ETC) 43-44
Theta (1640) 249-254,270-271
't Hooft, G. 32,43,340,353,365ff
Trilling, G. 106
Turolla, R. 151-161

U(1) axial transformation 32
U(1) problem 18,19
U(1) forces 369

Van Flandern, T.C. 9
van Nieuwenhuizen, P. 213-237
Vector-meson dominance model 119
Veneziano dual string theory 195,197,201

Wadia, S. 449
Ward identities 213,223
W-boson 337
W-boson masses 76-77,123-137,139-149,338
Weak gauging 370-377
Weak mixing angle 123-137
(see also Weinberg Angle)
Weinberg angle 27,75ff,123-137
Weinberg, S. 32,36,43,365,388
Weinberg-Witten theorem 344-345
Weingarten, D. 19,393-406
Wilczek, F. 32,36
Wilson, K.G. 16,19,21,34,280,285,290,294 298,299,300,301,302,394ff 419ff,438ff
Wirkel, M. 133
Witten, E. 39,366,388
Woo, C.H. 390

X-boson 337

Y-boson 337
Yang-Mills gauge theory 28,193,200ff,214,219,221 231,424,435

Z_2-Higgs Model 297,299
Z-boson mass 76-77,123-137,139-149
Z-boson 337
Zee, A. 36
Zimmerman, W. 380,390
Zwanziger, D. 390